Testing of Body Armor Materials
Phase III

Committee on Testing of Body Armor Materials for Use by the U.S. Army—Phase III

Board on Army Science and Technology
Division on Engineering and Physical Sciences

and

Committee on National Statistics
Division of Behavioral and Social Sciences and Education

NATIONAL RESEARCH COUNCIL
OF THE NATIONAL ACADEMIES

THE NATIONAL ACADEMIES PRESS
Washington, D.C.
www.nap.edu

THE NATIONAL ACADEMIES PRESS 500 Fifth Street, NW Washington, DC 20001

NOTICE: The project that is the subject of this report was approved by the Governing Board of the National Research Council, whose members are drawn from the councils of the National Academy of Sciences, the National Academy of Engineering, and the Institute of Medicine. The members of the committee responsible for the report were chosen for their special competences and with regard for appropriate balance.

This material is based on work supported by the National Science Foundation under Grant No. SES-0453930, Amendment #012. Any opinions, findings, and conclusions or recommendations expressed in this material are those of the author(s) and do not necessarily reflect the views of the National Science Foundation.

International Standard Book Number ISBN-13: 978-309-25599-8
International Standard Book Number ISBN-10: 0-309-25599-6

Limited copies of this report are available from:

Board on Army Science and Technology
National Research Council
500 Fifth Street, NW, Room 940
Washington, DC 20001
(202) 334-3118

Additional copies are available from:

The National Academies Press
500 Fifth Street, NW
Keck 360
Washington, DC 20001
(800) 624-6242 or (202) 334-3313
http://www.nap.edu

Printed in the United States of America

THE NATIONAL ACADEMIES

Advisers to the Nation on Science, Engineering, and Medicine

The **National Academy of Sciences** is a private, nonprofit, self-perpetuating society of distinguished scholars engaged in scientific and engineering research, dedicated to the furtherance of science and technology and to their use for the general welfare. Upon the authority of the charter granted to it by the Congress in 1863, the Academy has a mandate that requires it to advise the federal government on scientific and technical matters. Dr. Ralph J. Cicerone is president of the National Academy of Sciences.

The **National Academy of Engineering** was established in 1964, under the charter of the National Academy of Sciences, as a parallel organization of outstanding engineers. It is autonomous in its administration and in the selection of its members, sharing with the National Academy of Sciences the responsibility for advising the federal government. The National Academy of Engineering also sponsors engineering programs aimed at meeting national needs, encourages education and research, and recognizes the superior achievements of engineers. Dr. Charles M. Vest is president of the National Academy of Engineering.

The **Institute of Medicine** was established in 1970 by the National Academy of Sciences to secure the services of eminent members of appropriate professions in the examination of policy matters pertaining to the health of the public. The Institute acts under the responsibility given to the National Academy of Sciences by its congressional charter to be an adviser to the federal government and, upon its own initiative, to identify issues of medical care, research, and education. Dr. Harvey V. Fineberg is president of the Institute of Medicine.

The **National Research Council** was organized by the National Academy of Sciences in 1916 to associate the broad community of science and technology with the Academy's purposes of furthering knowledge and advising the federal government. Functioning in accordance with general policies determined by the Academy, the Council has become the principal operating agency of both the National Academy of Sciences and the National Academy of Engineering in providing services to the government, the public, and the scientific and engineering communities. The Council is administered jointly by both Academies and the Institute of Medicine. Dr. Ralph J. Cicerone and Dr. Charles M. Vest are chair and vice chair, respectively, of the National Research Council.

www.national-academies.org

COMMITTEE ON THE TESTING OF BODY ARMOR FOR THE U.S. ARMY – PHASE III

LARRY G. LEHOWICZ, MG, U.S. Army (retired), *Chair*, Quantum Research International, Arlington, Virginia
CAMERON R. BASS, Duke University, Durham, North Carolina
THOMAS F. BUDINGER, E.O., NAE/IOM,[1] Lawrence Berkeley National Laboratory, Berkeley, California
MORTON M. DENN, NAE, City College of the City University of New York
WILLIAM G. FAHRENHOLTZ, Missouri University of Science and Technology, Rolla
RONALD D. FRICKER, JR., Naval Postgraduate School, Monterey, California
YOGENDRA M. GUPTA, Washington State University, Pullman
DENNIS K. KILLINGER, University of South Florida, Tampa
VLADIMIR B. MARKOV, Advanced Systems and Technologies, Inc., Irvine, California
JAMES D. McGUFFIN-CAWLEY, Case Western Reserve University, Cleveland, Ohio
RUSSELL N. PRATHER, Survice Engineering Company, Bel Air, Maryland
SHELDON WIEDERHORN, NAE, National Institute of Standards and Technology, Gaithersburg, Maryland
ALYSON GABBARD WILSON, Institute for Defense Analyses, Alexandria, Virginia

Staff
BRUCE A. BRAUN, Director, Board on Army Science and Technology
ROBERT LOVE, Study Director
HARRISON T. PANNELLA, Senior Program Officer
NIA D. JOHNSON, Senior Research Associate, Board on Army Science and Technology
JAMES MYSKA, Senior Research Associate, Board on Army Science and Technology
DEANNA P. SPARGER, Program Administrative Coordinator, Board on Army Science and Technology
ANN LARROW, Research Assistant
JOSEPH PALMER, Senior Program Assistant
ALICE WILLIAMS, Senior Program Assistant (until September 10, 2010)
CONSTANCE CITRO, Director, Committee on National Statistics
DENNIS CHAMOT, Acting Director, National Materials Advisory Board
JAMES P. McGEE, Director, Laboratory Assessments Board

[1]NAE/IOM, National Academy of Engineering/Institute of Medicine.

COMMITTEE ON NATIONAL STATISTICS

LAWRENCE D. BROWN, *Chair*, Department of Statistics, Wharton School, University of Pennsylvania
JOHN M. ABOWD, School of Industrial and Labor Relations, Cornell University
DAVID CARD, Department of Economics, University of California, Berkeley
ALICIA CARRIQUIRY, Department of Statistics, Iowa State University
CONSTANTINE GATSONIS, Center for Statistical Sciences, Brown University
JAMES S. HOUSE, Survey Research Center, Institute for Social Research, University of Michigan
MICHAEL HOUT, Survey Research Center, University of California, Berkeley
SALLIE ANN KELLER, University of Waterloo, Ontario, Canada
LISA LYNCH, Heller School for Social Policy and Management, Brandeis University
SALLY C. MORTON, Department of Biostatistics, Graduate School of Public Health, University of Pittsburgh
RUTH D. PETERSON, Criminal Justice Research Center, Ohio State University
EDWARD H. SHORTLIFFE, Columbia University and Arizona State University
HAL STERN, Donald Bren School of Information and Computer Sciences, University of California, Irvine
JOHN H. THOMPSON, National Opinion Research Center, University of Chicago
ROGER TOURANGEAU, Westat, Rockville, Maryland

Staff
CONSTANCE CITRO, Senior Board Director
JULIA KISA SHAKEER, Financial Associate
JACQUI SOVDE, Program Associate

Preface

This report is the final volume of a three-phase study commissioned by the Director of Operational Test and Evaluation (DOT&E) of the Department of Defense (DoD) to assist in addressing shortcomings that had been reported by the Government Accountability Office (GAO) and the DoD Inspector General in DoD's body armor testing process. Independent committees were empanelled for the three study phases. Each committee produced an independent report, although this final Phase III report builds on the results of the letter reports delivered in Phases I and II, both of which provided findings and recommendations on key issues that required near-term resolution by DOT&E. The study was conducted under the auspices of the National Research Council (NRC) Board on Army Science and Technology (BAST) and Committee on National Statistics.

The Phase I letter report, released in January 2010, addressed the adequacy of laser instrumentation for evaluating ballistics tests in clay material. The Phase II report, released in May 2010, focused on the behavior of ballistics clay used as a recording medium during live-fire testing. The Phase III committee had more time for meetings and data gathering than the two previous committees and was able to use the substantial amount of data collected throughout the entire study. As a result the committee was able to delve more deeply into all available data than had been possible in the earlier phases of the effort.

This Phase III report provides a wide range of recommendations designed to help enable the entire body armor community to utilize an effective testing process leading to fielding the best equipment possible that meets performance specifications while reducing the weight burden placed on soldiers in training or combat.

The Phase III committee deserves special thanks for its hard work. Several committee members went well beyond the norm in interviewing numerous experts, assessing the pertinent issues, and developing recommendations to address the many demands of the committee's statement of task. In particular, committee member Thomas Budinger deserves special credit for leading the Phase III ad hoc instrumentation committee subgroup that produced a thoughtful review of the data and information related to instrumentation. The committee is also grateful to the many DoD, Army, Marine Corps, industry, and contractor personnel engaged in body armor testing for the useful information they provided.

Finally, the committee also greatly appreciates the support and assistance of the NRC staff members who assisted the committee in its fact-finding activities and in the production of the three separate committee reports. In particular, thanks are due to the BAST staff, principally Bruce Braun, Margaret Novack, and Robert Love, who ably facilitated the committee's work.

Larry Lehowicz, *Chair*
Committee on Testing of Body Armor
Materials for Use by the U.S. Army—Phase
III

Acknowledgment of Reviewers

This report has been reviewed in draft form by individuals chosen for their diverse perspectives and technical expertise, in accordance with procedures approved by the National Research Council's Report Review Committee. The purpose of this independent review is to provide candid and critical comments that will assist the institution in making its published report as sound as possible and to ensure that the report meets institutional standards for objectivity, evidence, and responsiveness to the study charge. The review comments and draft manuscript remain confidential to protect the integrity of the deliberative process. We wish to thank the following individuals for their review of this report:

Morris E. Fine (NAE), Northwestern University
John S. Foster, Jr. (NAE), GKN Aerospace Transparency Systems
David Higdon, Los Alamos National Laboratory
Peter Matic, Naval Research Laboratory
Erik Novak, Veeco Instruments,
Henry Smith (NAE), Massachusetts Institute of Technology
Leslie J. Struble, University of Illinois
Stephen F. Vatner, New Jersey Medical School
Emmanuel Yashchin, IBM Watson Research Center
Laurence R. Young (NAE/IOM), Massachusetts Institute of Technology

Although the reviewers listed above have provided many constructive comments and suggestions, they were not asked to endorse the conclusions or recommendations, nor did they see the final draft of the report before its release. The review of this report was overseen by Lawrence D. Brown, NAS, Wharton School, University of Pennsylvania, and Arthur H. Heuer, NAE, Case Western Reserve University. Appointed by the National Research Council, they were responsible for making certain that an independent examination of this report was carried out in accordance with institutional procedures and that all review comments were carefully considered. Responsibility for the final content of this report rests entirely with the authoring committee and the institution.

Contents

SUMMARY 1

1 INTRODUCTION 20
Background, 20
Study Tasks, 20
Study Context, 22
Study Implementation, 22
Report Organization, 23
References, 24

2 OVERVIEW OF BODY ARMOR 25
Background, 25
Ceramic Plates in Body Armor, 26
Fiber and Resin Composites in Helmets, 26
Survivability vs. Mobility, 27
Medical Study Basis for Testing Body Armor, 27
Body Armor Testing Process, 28
Body Armor Testing Range, 31
Government Accountability Office Report, 32
References, 33

3 HISTORICAL BASIS FOR CURRENT BODY ARMOR TESTING 34
Background, 34
Evolution of Clay Usage, 38
High-Energy Threats, 41
Rifle Threats for Hard Body Armors, 42
Work Performed after the Prather Study, 42
Current Standard, 44
References, 44

4 CLAY AND BACKING MATERIALS 46
Use of Backing Material as a Recording Medium, 46
Characteristics and Properties of RP #1, 51
Behavior in Testing, 51
Influence of Structure on Properties of Oil-Base Modeling Clay, 63
Short-Term Development of an Interim Standard Clay Formulation for Ballistic Testing, 69
Conditioning and Handling of Clay, 71
Calibration Drop Test, 74
Alternative Backing Materials and Systems, 76

Medium-Term and Long-Term Replacements for Modeling Clay, 76
Electronic Recording Systems, 78
Roadmap for Improving the Testing Process, 79
Near-Term Actions, 81
Medium-Term Research Needs, 85
Long-Term Goal, 88
References, 88

5 INSTRUMENTATION AND PROCEDURES FOR MEASURING AN INDENT IN A CLAY BACKING MATERIAL 92
Conceptual Steps Toward Improvements in the Measurement of BFD, 92
Step One, 92
Step Two, 93
Step Three, 93
Step Four, 93
Steps Five and Six, 94
Instrumentation Performance Based on Statistical Analysis, 94
Overview of Current Instrumentation and Measurements, 96
Coordinate Measuring Machine, 96
Digital Caliper, 96
Faro System, 97
BFD Measuring Procedures, 99
Human Operator Considerations, 99
Compensating for Offset between the Point of Aim and the Deepest Indent, 101
Variability (Noise) in the Overall Testing Process, 101
Need for a Stand-Alone BFD Artifact or Standard Model for InterorganizationVerification, 103
Characteristics of a "Best Utility" Measuring Instrument, 104
References, 106

6 STATISTICAL CONSIDERATIONS IN BODY ARMOR TESTING 107
Introduction, 107
Statistically Principled Testing, 107
Uncertainty and Variation Drive Overdesign, 108
Key Test Protocol Design Requirements, 109
Background, 111
Historical First Article Testing Protocols, 111
DOT&E Protocol for Body Armor FAT, 113
Test Protocol Assumptions, 119
Lot Acceptance Testing, 123
Protocol Design Trade-offs and Comparisons, 128
Recommendations, 132
References, 135

7 HELMET TESTING 137

Ballistic Helmet Test Methodolgies, 137
  Injury Modes, 139
Existing Human Injury Criteria, 140
  Head Injury from Ballistic Impact, 142
  Neck Injury, 143
Helmet Design and Suspension Systems, 145
  Existing Helmet Test Methodologies, 146
  NIJ Standard 0106.01, 148
Army Clay Head Form, 150
  Testing Standards, 151
  Test Procedures, 152
  H.P. White Laboratory Test Procedure, 157
  Peepsite Head Form, 160
  UVA Head Form, 163
  BLS Head Form, 164
References, 165

8 MEDICAL BASIS FOR FUTURE BODY ARMOR TESTING 169

Thoracic Ballistic Test Methodologies, 169
  Introduction to Behind Armor Blunt Trauma, 169
  Ergonomics of Body Armor, 173
  Injury Biomechanics, 173
  Injury Criteria and Experimentation, 177
Blast Injury Criteria and Blastlike Mechanisms, 183
  Low-Rate Blunt Trauma Mechanisms, 185
  Human Epidemiology for Battlefield BABT, 185
Injury Scales, 186
  Current Epidemiology for Battlefield/Law Enforcement BABT, 191
  Large Animal Experiments for Behind-Armor Blunt Trauma, 191
Potential Adverse Effects of Body Armor in Blast Exposures, 206
  Cadaveric Experiments for Behind-Armor Blunt Trauma, 209
  Rationale for Large-Animal, Live-Fire Experiments, 212
Instrumented Determination of Backface Deformation—Research Directions, 213
  Bulk Tissue Simulants, 215
  Instrumented Response Elements, 215
  Instrumented Detailed Anatomical Surrogates, 221
  Developmental Testing Requirements, 225
Medical Research Needs, 229
  Near-Term Actions, 229
  Medium-Term Actions, 230
References, 231

9 FUTURE IMPROVEMENTS IN TESTING METHODOLOGY 240
Building on the Prather Study, 240
Synopsis of Near-Term Improvements, 241
Linking Medical Research Data to Product Testing Criteria, 243
Dynamics and Measurement of Behind-Armor Forces, 244
Synchronizing the Stakeholders, 248
Military and Law Enforcement Personnel, 249
Technologists, 250
Medical Researchers, 250
Testers—Developmental and Operational, 250
Aligning Stakeholders, 251
Establishing National Standards, 252
References, 254

APPENDIXES

A Biographical Sketches of Committee Members 257
B Committee Meetings 264
C Additional Phase III Tasks 268
D Report Sections Cross-Referenced to the Statement of Task 269
E Ballistic Body Armor Insert Composition and Defeat Mechanisms 272
F Committee Responses to the Government Accountability Office Report 274
G Determining the Necessary Level of Precision for Body Armor Testing 280
H Statistical Tolerance Bounds 299
I Analytical Approaches for Comparing Test Protocols 302
J Contemporary Methods for Assessing Behind-Armor Blunt Trauma in Live Animals 305
K Phase I Findings 316
L Phase II Recommendations 317
M Estimating the Accuracy and Precision of the Digital Caliper and Faro Laser 323

Tables, Figures, and Boxes

TABLES

3-1 Strengths and Weaknesses of the Prather Methodology, 45

4-1 Elastic Recovery in Modified Charpy Testing of Oil-Based Modeling Clay, 49

6-1 60-Plate Protocol, 115
6-2 Proposed FAT Standards, 117
6-3 Proposed LAT Standards, 124
6-4 Risk Comparisons for Probability of Complete Penetration, 129

7-1 Characteristics of Test Rounds from NIJ Standard-0106.01, 149
7-2 H.P. White Laboratory Test Procedure, 158
7-3 Helmet Test Matrix, 161

8-1 Muzzle Parameters for Various Types of Rounds, 170
8-2 Energy/Momentum for Various Typical Thoracic Trauma Situations, 171
8-3 Description of Levels of Thoracic Trauma, 186
8-4 Combat Effectiveness vs. Levels of Thoracic Trauma, 187
8-5 Bullet Specifications and Injury Outcome, 194

9-1 Strengths and Weaknesses of the Prather Methodology, 241

FIGURES

S-1 Road map showing suggested near-term actions, medium-term research needs, and a long-term goal to develop a more consistent backing material and a more reliable process for evaluating hard armor, 8
S-2 Road map showing suggested near-term and medium-term research needs, and a long-term goal to provide the fundamental medical basis for injury risk assessment behind helmets and hard body armor, 17

2-1 The clay appliqué applied to the clay box, 29
2-2 Surface of the BFD as measured by a laser scanning system, 30
2-3 The body armor test range at ATC, 32

3-1 Overview of development of Prather clay methodology, 35

3-2 Blunt deformation profiles into gelatin using seven-ply K29 armor samples mounted on gelatin and tested with the .38-cal LRN bullet at 213 m/sec (800 ft/sec), 37
3-3 Deformation depth vs. time of candidate materials in a goat thorax using a blunt impactor at 55 m/sec, 39
3-4 Logistic regression model of death vs. deformation for blunt impact into goat chests, 39
3-5 Logistic regression model of death vs. deformation for blunt impact into clay using deformation response into goat chests and clay, 41
3-6 Clay deformation behind hard armor with rifle round threats, 42
3-7 Variation of clay penetration depth with velocity for behind-body armor deformation (7.62-mm NATO round, UHMWPE body armor), 43
3-8 Variation of clay penetration area with velocity for behind-body armor deformation (7.62-mm NATO round, UHMWPE body armor), 43

4-1 A schematic illustration of the stress-strain curves for two idealized solids, 47
4-2 A contour gage in use, 48
4-3 Column drop test as performed at ATC, 53
4-4 The results of drop tests on clay boxes allowed to naturally cool from 40°C in a room at normal room temperature (roughly 23°C), 55
4-5 Drop test results using the standard Army right-circular cylinder with a solid hemispherical cap (44.5 mm [1.75 in.] in diameter with a mass of 1 kg [2.2 lb]), a similar non-standard double-length cylinder of the same diameter with the same type of hemispherical cap, and sphere with the diameter specified in the National Institute of Justice Standard (NIJ), 63.5 mm (2.5 in.) in diameter, 56
4-6 Spatial pattern used in a series of experiments to determine the effect of position on the size of the cavity produced during a drop test, 57
4-7 Illustrative results from a study on the effect of radial position on depth of penetration during a drop calibration test, 58
4-8 A clay box used for .32-cal rubber sting ball testing, 61
4-9 Results of sting ball experiments with projectiles ranging from 55 m/sec (180 ft/sec) to 168 m/sec (550 ft/sec), 62
4-10 A schematic illustration of the “thixotropic cycle” of a two-phase system, 65
4-11 Optical micrographs of a three-dimensional network of spherical latex particles, 66
4-12 Road map showing suggested near-term actions, medium-term research needs, and a long-term goal to develop a more consistent backing material and a more reliable process for evaluating hard armor, 81

5-1 Digital calipers used in armor testing, 97
5-2 Faro Quantum laser scan arm, 98

6-1 USSOCOM FAT shot pattern, 113

6-2 Plot of the actual coverage level achieved by a lower confidence bound calculated according to the Clopper-Pearson method for $n = 60$ and various Pr(*nP*), 122
6-3 Probability a lot passes LAT first shot requirements for Pr(*nP*) for the S-4 and S-3 inspection levels for various lot sizes and an AQL of 4 percent, 126
6-4 Probability a lot passes LAT second shot requirements for Pr(*nP*) for the S-4 and S-3 inspection levels for various lot sizes and an AQL of 4 percent, 127
6-5 Plot of the manufacturer's risk for various Pr(*nP*) under the DOT&E protocol, 130
6-6 Risk comparisons for BFD assuming in the left plot that the manufacturer's true mean BFD is 38 mm and in the right plot is 40 mm; the associated fraction of variation is shown on the *x*-axis, 131

7-1 Effective momentum of high-rate ballistic impacts at muzzle velocity and low-rate football impact, 138
7-2 Ballistic impact injury timescales, 138
7-3 Likeness of a deformed personnel armor system for ground troops helmet, 139
7-4 Cadaver instrumentation overview, 142
7-5 Neck injury assessment value for 9-mm FMJ test round at various velocities into helmeted human cadavers, 144
7-6 Residual head/neck velocity from momentum transfer to the helmet/head system, 145
7-7 Impact energy for helmet standards, 147
7-8 Ballistic (high-rate) skull fracture data vs. impact injury criteria for typical blunt injury, 147
7-9 NIJ sagittal penetration head form, 148
7-10 Army clay head form, 150
7-11 ATC head form with clay, 153
7-12 Head form clay conditioning by analogy, 154
7-13 Test impact locations, 156
7-14 Test frame, 156
7-15 H.P. White head form, 159
7-16 Peepsite head forms: different head forms for different shot directions, 160
7-17 Left, UVA head form; right, risk assessment, 163
7-18 BLS head form, 164
7-19 Arrangement and dimensions of load cells in the BLS head form, 165

8-1 Initial energy and momentum for ballistics and other blunt impacts, 171
8-2 Superimposed high speed X-rays of the initial shock wave and deformation of the thorax during a 7.62 mm projectile live-fire test in a pig protected by hard body armor, 172
8-3 Development of surrogate injury model, 175
8-4 Volunteer experimentation, 177

8-5 Human upper torso, 179
8-6 Threshold pressures and exposure times needed to damage drosophila larvae using various loading devices, 180
8-7 Threshold damage for various tissues, 181
8-8 ASII values versus peak inward chest wall velocity, 183
8-9 Overpressure/duration blast injury criteria, 184
8-10 Kinetic energy vs. injury severity, 188
8-11 Time of delivery of wounded to the CMH (average 1983-1984), 189
8-12 Severity of wounds for patients delivered to the CMH (average 1983-1984), 190
8-13 Lateral dog thorax impacted by nonpenetrating missiles, 192
8-14 Impact energy (scaled to a 75 kg man) vs. increased lung mass, 193
8-15 Body armor for Oksbøl trials, 196
8-16 Average first and second peak pressure, Oksbøl trials, 197
8-17 Average postmortem lung mass, Oksbøl trials, 198
8-18 Oksbøl first peak on Bowen curve, 199
8-19 Animal fatalities during monitoring period, 201
8-20 BABT flash X-ray, 202
8-21 Relationship between area of lung surface contusion and maximum back-face deformation of body armor, 202
8-22 Relationship between area of lung surface contusion and pressure 6 cm from point of impact, 203
8-23 Examples of BABT assessment devices and methodologies, 214
8-24 DERA BABT simulator displacement sensor system, 216
8-25 DERA tissue viscoelastic stimulant concept as described by Mirzeabassov et al., 2000, 218
8-26 Hybrid III 50th percentile male dummy, 219
8-27 ATM with mounted body armor; ATM instrumented response element with padding; oblique view within torso, 220
8-28 Human CT scan; finite-element model, 221
8-29 AUSMAN upper torso, 222
8-30 AUSMAN thorax with body armor in place, prior to testing, 223
8-31 Road map showing suggested near-term and medium-term research needs, and a long-term goal to provide the fundamental medical basis for injury risk assessment behind helmets and hard body armor, 228

9-1 Road map showing suggested near-term actions, medium-term research needs, and a long-term goal to develop a more consistent backing material and a more reliable process for evaluating hard armor, 242
9-2 Flow chart showing suggested near-term and mid-term research needs, and a long-term goal to provide the fundamental medical basis for injury risk assessment behind helmets and hard body armor, 243
9-3 Schematic of conceptual approach used by both testers and researchers showing a projectile impacting normally onto hard body armor (A), soft body armor (B), and a recording medium surrogate for a human body (C), 244

9-4 Schematic of the dynamic measurement method, 246
9-5 Schematic of stakeholder relationships, 249

G-1 Plot of normally distributed BFDs from a design that just meets the 90 % upper tolerance limit requirement with μ_{pop} = 42.7 mm and σ_{pop} = 1 mm, 281
G-2 The consequence of measurement error on the apparent depths of BFDs, 282
G-3 The relationship between measurement error and the overall variance in armor testing, 286
G-4 How improving the performance of armor relates to the probability of passing FAT assuming a lot size of 60 plates, 289
G-5 Plot of the difference between the two FAT failure curves, 291
G-6 Photograph, laser scan, and cross section of cavity in RP #1 produced by armor testing, 293
G-7 Digital calipers used in armor testing, 294
G-8 Two images of typical BFD cavities in RP #1 produced by the Faro laser scanner, 295
G-9 The probability a manufacturer will pass the first article, first shot BFD test (solid line) for various population mean BFD levels (μ) versus the probability that a plate will have a BFD greater than 50 mm from the same population (dotted line), 297

H-1 Ninety-fifth quintile distribution, 300

J-1 Comprehensive protocol for live-animal live-fire tests, 306

M-1 Plot of the paired BFD measurements made by ATC, 324
M-2 Plot of the paired BFD measurements made by Chesapeake Testing, 324
M-3 Absolute value of offsets for caliper measurements from Realistic Clay III, 330

BOXES

S-1 Statement of Task, 2

1-1 Statement of Task, 21

L-1 Phase II Recommendations to Improve Body Armor Testing, 321

Acronyms and Abbreviations

AIS	abbreviated injury scale
AP	armor piercing
AQL	acceptance quality level
ARDS	adult respiratory distress syndrome
ARL	U.S. Army Research Laboratory
ASTM	American Society of Testing and Materials
ATC	U.S. Army Aberdeen Test Center
ATD	anthropometric test device
ATM	anthropomorphic test module
BABT	behind-armor blunt trauma
BFD	backface deformation
BLS	ballistic load sensing
CMH	central military hospital
CMM	co-ordinate measuring machine
DAI	diffuse axonal injury
DERA	Defense Evaluation and Research Agency
DGA	Délégation Générale pour L'Armement
DoD	Department of Defense
DOT&E	Office of the Director, Operational Test and Evaluation
DREV	Defense Research Establishment Valcartier
ECG	electrocardiogram
ESAPI	enhanced small arms protective insert
FAT	first article testing
FMJ	full metal jacket
GAO	Government Accountability Office
HIC	head injury criteria
IG	Inspector General
ISS	injury severity score
kPa	kilopascal

LAT	lot acceptance testing
LRN	lead round nose
MPa	megapascal
NATO	North Atlantic Treaty Organization
NIJ	National Institute of Justice
NIST	National Institute of Standards and Technology
NRC	National Research Council
PEO-S	U.S. Army Program Executive Office Soldier
RCC	right circular cylinder
RP #1	Roma Plastilina #1
TAB	trauma-attenuating backing
TBI	traumatic brain injury
TOP	test operating procedure
UHMWPE	ultra-high molecular weight polyethylene
USSOCOM	United States Special Operations Command
UVA	University of Virginia
XSAPI	X Small Arms Protective Inserts

Summary

In 2009, the Government Accountability Office (GAO) released the report *Warfighter Support: Independent Expert Assessment of Army Body Armor Test Results and Procedures Needed Before Fielding*, which commented on the conduct of the test procedures governing acceptance of body armor vest-plate inserts worn by military service members (GAO, 2009). This GAO report, as well as other observations—for example, the Army Audit Agency report to the Program Executive Officer Soldier on Body Armor Testing (AAA, 2009)—led the Department of Defense (DoD) Director, Operational Test & Evaluation (DOT&E) to request that the National Research Council (NRC) Division on Engineering and Physical Sciences (DEPS) conduct an ad hoc study to investigate issues related to the testing of body armor materials for use by the U.S. Army and other military departments. Box S-1 contains the statement of task for the three-phase study. Phases I and II resulted in two NRC letter reports: one in 2009 and one in 2010.[1] This is the Phase III report.

[1]Findings and recommendations from the Phase I and Phase II reports are in Appendixes K and L respectively.

Box S-1 Statement of Task

The National Research Council will convene specialists in committee form to consider the technical issues relating to the testing of body armor. To do this the National Research Council shall conduct a 3-phase study:

In Phase I a committee will comment on the validity of using laser-profilometry/ laser-interferometry techniques to determine the contours of an indent made by a ballistic test in a non-transparent clay material at the level of precision established in the Army's procedures for testing personal body armor. If laser-profilometry / laser-interferometry is not a valid method, the committee will consider whether a digital caliper can be used instead to collect valid data. The Committee will also provide interim observations regarding the column drop performance test described by the Army for assessing the part to part consistency of a clay body used in testing body armor. The committee will prepare a letter report documenting the findings from its Phase I considerations. This is a six week effort beginning November 1 2009 and ending mid December 2009.

In Phase II a committee will consider in greater detail the validity of using the column drop performance test described by the Army for assessing the part-to-part consistency of a clay body within the level of precision that is identified by the Army test procedures. The committee will prepare a letter report documenting the findings from its Phase II considerations. This is a three months effort beginning November 1 2009 and ending early February 2010.

In Phase III a committee will consider test materials, protocols and standards that should be used for future testing of personal armor by the Army. The committee will also consider any other issues associated with body armor testing that the committee considers relevant, including issues raised in the Government Accountability Office Report---Warfighter Support, Independent Expert Assessment of Body Armor Test Results and Procedures Needed Before Fielding (GAO-10-119).The committee will prepare a final report. This is a 14-months effort beginning November 1 2009 and ending January 2011.

The final report will document the committee's findings pertaining to the following issues that are of particular immediate concern to DOT&E including the following:

- The best methods for obtaining consistency of the clay, and of conditioning and calibrating the clay backing used currently to test armor.
- The best instrumentation (e.g., laser scanning system, digital caliper, etc.) and procedures to use to measure the back face deformation (BFD) in the clay.
- The appropriate use of statistical techniques (e.g., rounding numbers, choosing sample sizes, or test designs) in gathering the data.
- The appropriate criteria to apply to determine whether body armor plates can provide needed protection to soldiers; this includes the proper prescription for determining whether a test results in a partial or complete penetration of body armor, including, as appropriate, the soft armor underlying hard armor.

The final report will also document the committee's findings regarding any other issues regarding body armor testing that the committee found relevant. The study team will have access to all data with respect to body armor testing that the team needs for the conduct of the study.

The last task for Phase III of the study was to document in its final report any other issues regarding body armor testing that the committee found relevant. In response, this report also addresses the following tasks:

- Provide a road map to reduce the variability of clay processes and show how to migrate from clay to future solutions.
- Consider the use of statistics to permit a more scientific determination of sample sizes to be used in body armor testing.
- Develop ideas for revising or replacing the Prather study methodology.
- Review and comment on methodologies and technical approaches to military helmet testing.
- Consider the possibility of combining various national body armor testing standards.

The preponderance of body armor testing is conducted by the U.S. Army Aberdeen Test Center (ATC) in support of the body armor acquisition authority, which is the U.S. Army Program Executive Office Soldier (PEO Soldier). In developing its report, the Phase III Committee on Testing of Body Armor Materials for Use by the U.S. Army (the Phase III committee) built on the work of the Phase I and Phase II committees, conducting data-gathering sessions at the ATC in Maryland and visiting testing facilities of the Army and commercial testers. Appendix B provides a list of committee briefings and activities.

The broad purposes of the study were to verify and validate current test procedures for body armor plates, to investigate long-standing issues related to the testing process, and to recommend approaches that will improve testing methodologies and procedures in the future. Committee responses to specific issues raised in the GAO Report are contained in Appendix F. This summary includes the numbered recommendations from each chapter of the report with principal findings of the study highlighted in italic typeface.

OVERVIEW OF BODY ARMOR TESTING

Ceramic materials have been used successfully in personal armor systems to defeat small-arms threats in both Iraq and Afghanistan, and there have been no known instances where a death resulting from small arms fire can be attributable to a failure of issued ceramic body armor. Since hard body armor systems add a significant weight to the burden on the soldier, the testing of body armor has an implied goal of ensuring that survivability standards are met while allowing sufficient soldier mobility and flexibility.

In 1977, a study was performed to correlate the depth that a 200-g, 80-mm hemispherical missile impacting at approximately 55 m/sec penetrated live-animal tissue and other media (Prather et al., 1977). The goal of the Prather study was to develop a simple, readily available backing material for characterizing both the penetration and deformation effects of ballistic impacts on body armor materials and to relate this information to the injury potential of nonpenetrating ballistic impacts.

When there was no penetration of the armor, the researchers noticed that dynamic ballistics forces caused an indent in the recording material behind the point where the bullet struck the front side of the armor. This deformation in the backing material was termed a "backface deformation" (BFD). The depth of the deformation into various media, such as modeling clay or ballistics gelatin, as a function of time was compared to the probability of lethality for an identical degree of deformation inflicted on a live-animal model.

The Prather study observed strong correlations between lethality probability and the deformation of ballistic gelatin[2] and of a modeling clay, Roma Plastilina #1 (RP #1). Ballistic gel required the use of high-speed photography to record the BFDs because the gel was elastic and returned to its original shape immediately after the projectile firing. To avoid the need to use high-speed photography, which was expensive at that time, clay was selected as an alternative and is used today as the medium for recording the BFDs in body armor testing.

RP #1 in its current formulation is the standard recording medium for testing, even though there are imperfect correlations between existing medical data and the BFD testing approach. In a nonpenetrating impact, kinetic energy must be dissipated by the armor through deformation or fragmentation of the armor, bullet, and underlying body wall. The transfer of energy to the body has the potential to cause serious injury or death. Nonpenetrating impact injury is termed "behind-armor blunt trauma" (BABT).

Numerous studies and experiments have been conducted and are ongoing to better determine the relationships among blunt force trauma, human injury, and the body armor testing processes. Since past research was based on smaller and slower bullets, the committee recognized that the *existing research raises concerns regarding the correlation between damage measured in RP #1 and bodily injury at the very high rates typical of BFDs caused by rifle rounds in hard body armor.*

CLAY AND BACKING MATERIALS

The committee assessed the use of clay in testing and described how the variability inherent in the backing material might be incorrectly attributed to variability in the armor. The study investigated the role of the backing material as a recording medium, the properties and limitations of RP #1 clay in body armor testing, and alternatives for future backing materials and systems for testing.

---

[2]Ballistic gelatin is a clear or yellowish gelatin that is the standard medium for evaluating what happens to bullets on impact with soft tissue.

Clay as a Recording Medium

The qualitative assertion that RP #1 exhibits little recovery has been interpreted to mean that the level of elastic recovery is small enough to be safely neglected. This has led to an assumption that the shape of the resultant cavity provides a record of the BFD. Since the relative degree of elastic and plastic deformation will vary as a function of strain rate, the backing material must be characterized under conditions that are relevant to those under which the tests will be performed. The cavity that results from live-fire ballistic testing is indeed related to the deformation on the back face of the armor, but it is not a true record of maximum deflection. It remains unknown how the dimensions of the cavity relate to the true BFD and how such a relationship may depend on the rate at which the cavity is formed.

RP #1 was originally developed as a modeling clay for artists. Over time its composition changed and the clay became stiffer to suit the ceramic arts community's needs. Consequently, testers recognized the need for a method for calibrating the clay. The so-called column drop test was developed in response to this need. Because the oil-based modeling clay is readily softened by heating, ovens are now used on the firing range to warm the clay so that the newer formulations respond in the same way as the older ones.

Experiments conducted by the ATC show that RP #1 exhibits highly variable penetrations under nominally identical conditions. This unambiguously indicates that RP #1 is an inherently imprecise recording medium.

The committee found that *both the spatial and the temporal variations of the modeling clay are significant. Experiments can be conducted to determine the variation due to box geometry and location of the drop in relation to the side of the box.* Also, *the scaling relationship between drop tests and ballistic tests remains mostly unexplored.*

Understanding the structure-property relationships of oil-based modeling clay as they pertain to mechanical working, thermal processing, friction, and how the various ingredients of the clay modify behavior could lead to alternative clay systems with more favorable properties. A clay working group consisting of interested government and civilian experts from the body armor testing community is working to develop a near-term replacement clay that can meet the calibration specification of the column drop test at ambient temperature and whose properties are little affected by temperature.

Recommendation 4-1: The Office of the Director, Operational Test and Evaluation, and the Army should continue to expedite the development of a replacement for the current Roma Plastilina #1 oil-based modeling clay that can be used at room temperature.

Clay Conditioning and Handling

Interim opportunities for improvements in clay conditioning and handling were recommended in Phase II of the study, because *in the short term, testing will continue to be conducted with available RP #1. As long as heating the clay is necessary, cooling will take place, and a post-test calibration drop test, as recommended in the Phase II Report (NRC, 2010), will continue to be an urgent requirement for the Army test operating procedure (TOP).*

There is also a continuing need for detailed and systematic characterization of both the medium and the testing process. *The comprehensive thermomechanical characterization of RP #1 that was recommended in the Phase II Report (NRC, 2010) will quantify the effect of shear history and thermal history on the storage and dissipative components of mechanical deformation. Such a characterization will also quantify the times associated with recovery of properties as well as the thermal properties, including thermal expansion, thermal conductivity, thermal diffusivity, heat capacity, and thermal arrests associated with phase changes.*

In the drop test, the strain rate experienced by the clay is qualitatively lower than the rate experienced in the live-fire ballistic test of armor, and there is little information on clay behavior in these two strain-rate domains. Further, the volumes of cavities formed in the drop tests and the live-fire tests differ significantly. *The testing community would benefit greatly from devising an alternative to the column drop test and certifying the validity of the current drop tests for calibration.*

Medium-Term and Long-Term Replacements for Modeling Clay

There are two broad classes of backing material replacements for consideration in the medium and longer terms: (1) elastic materials that recover their original shape after unloading and (2) plastic materials that preserve a permanent cavity whose dimensions can be correlated to lethality probability. *There is no compelling rationale for expending resources to achieve an interim solution using an elastic material such as ballistic gelatin.* The committee also found that *for the foreseeable future, plastically deforming recording media appear to be the proper choice of backing material for production testing of body armor.*

The committee assessed the potential of the anthropomorphic test module (ATM) technology currently used by the Army for ballistics injury research. *The committee concluded that the use of the ATM represents a transition to a challenging methodology with only limited ability to extend results to injury prediction. Also, it is too costly to be used as a production testing alternative to RP #1 at this time. The ATM is judged a research tool that is not practical or appropriate for widespread deployment in ballistic testing ranges.*

There are several other test devices that are potentially suitable for use in the development of a test methodology for ballistic BABT, but they all need significant development and validation experimentation. Much depends on the degree to which it is desirable to rank armor or predict injury probability, which would have to be addressed. *Overall, instrumented electronic sensor response elements are in a primitive state for the evaluation and assessment by medical researchers of ballistic BABT with rifle round threats. They also are too costly to be used in high-volume production testing. More research and detailed validation is necessary before electronic sensors can be considered as a practical medium- or long-term alternative to the use of RP #1.*

The report describes near-term actions, medium-term needs, and long-term goals that are consistent with earlier recommendations of the Phase II study (NRC, 2010).

Recommendation 4-2: The Office of the Director, Operational Test and Evaluation, and the Army should provide resources and execute the road map described in this chapter and graphically shown in Figure S-1 with the objective of developing a standard ballistics backing material for testing body armor. The properties and behaviors of the material should be well understood. It should exhibit minimal variability due to temperature, working, and aging and require simple calibration techniques and equipment, and it should enable reliable and accurate recording of body armor test results.

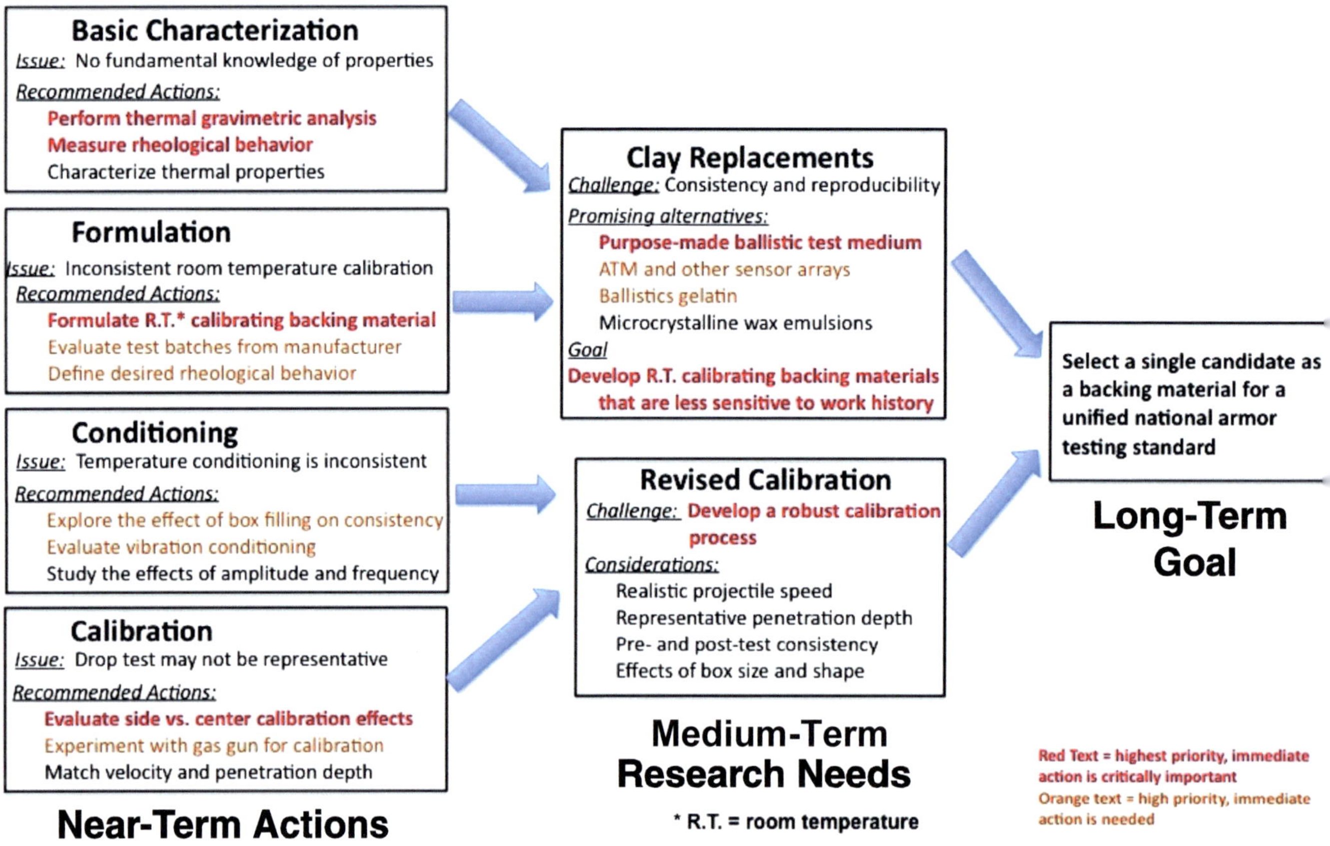

FIGURE S-1 Road map showing suggested near-term actions, medium-term research needs, and a long-term goal to develop a more consistent backing material and a more reliable process for evaluating hard armor. The color coding shows “highest priority” items in red text with “high priority” actions in orange.

INSTRUMENTATION AND PROCEDURES FOR MEASURING AN INDENT IN THE BACKING MATERIALS

The committee was tasked to determine the best instrumentation and procedures for measuring BFD (see Box S-1). To do this, it reviewed technical specifications, viewed demonstrations of the operation and use of current and prospective systems, and evaluated factors such as human handling variability, process transparency, and software variability judgment.

The committee found that *given the current clay variation, a measurement precision (standard deviation) of 0.5 mm is sufficient; instruments featuring greater precision add little practical value to the testing process. Future improvements in the inherent variability of the backing material will require instruments that are correspondingly more precise.* It is important that quantified data from actual tests be obtained for all instruments and measurement scenarios in order to make valid comparisons of instrumentation for different applications.

In evaluating the instrumentation methods, the committee noted that there is unknown variability associated with the software smoothing algorithm used by the Faro laser scanner system.

Recommendation 5-1: An organization such as the National Institute of Standards and Technology should conduct a controlled study to determine the most reasonable and consistent Faro smoothing settings to be used while measuring backface deformations (BFDs) in body armor testing. Similarly, any other software selections that could cause relevant changes to BFD measurements should be studied. Corresponding values for the precision and accuracy of each software setting will need to be quantified.

It is possible that a standard BFD cavity artifact could be used by testers to help to ensure that all measuring devices provide standard measures of accuracy and precision at different locations.

Recommendation 5-2: An organization such as the National Institute of Standards and Technology should develop a standard backface deformation artifact system and procedures to allow operators to ensure that different measurement devices at different locations are able to meet specified levels of accuracy and precision.

Finally, the committee derived criteria for a “best utility” measuring instrument based on its assessment of the characteristics of instrumentation systems presently used by military and commercial testers.

Recommendation 5-3: In anticipation of future test measurement requirements, the Office of the Director, Operational Test and Evaluation, and/or the Army should charter an organization such as the National Institute of Standards and Technology to conduct an analysis of available candidate commercial instruments with inputs from vest users, manufacturers, testers, policy makers, and others. The goal is to identify one or more devices meeting the characteristics of "best utility" measuring instruments as defined in this study to the government, industry, and private testing labs.

The list of best utility instruments should be shared with the National Institute of Justice (NIJ), international allies, and others, as appropriate, to promote measuring instrument standardization for body armor testing nationally and internationally. A formal gauge or "artifact standard" repeatability and reproducibility study is required to quantify accuracy and precision as inputs to the best utility analysis.

STATISTICAL CONSIDERATIONS IN BODY ARMOR TESTING

The Phase II committee was asked to review a statistically based protocol that had been developed by DOT&E with assistance from Army statisticians and testers, and the Phase II report (NRC, 2010) provided initial insights on statistics-related issues. The committee reviewed historical test protocols as well as the new DOT&E first article testing (FAT) protocol and a proposed lot acceptance testing (LAT) protocol with regard to the assumptions underlying the statistical methods and design trade-offs.

The committee found that *because of their differences, and as demonstrated in the DoD Inspector General calculations, neither the historical Army protocols nor the U.S. Special Operations Command (USSOCOM) protocols met the key protocol design requirement as a common standard DoD-wide. In addition, the historical Army protocol did not meet the key design requirement as a statistically principled test.*

During the course of the committee's research and deliberations, the DOT&E, Army, and USSOCOM have endeavored to establish statistically principled test standards that are realistically achievable with the current body armor designs. The committee found these collaborative efforts to be commendable.

The new DOT&E protocol meets both key protocol design requirements; it is statistically principled and it provides a minimum DoD-wide body armor test standard. However, *since the distribution for some combinations of vendor, threat, and design may not be normally distributed, the tolerance-bound calculation that is specified by the protocol may not be appropriate in all cases.*

The committee found that *use of the Clopper-Pearson method for calculating the lower confidence limit is conservative, resulting in actual confidence levels that are at least as great as, and often greater than, the confidence level specified in the standard. The actual confidence level varies substantially as a function of the probability of no penetration [Pr(nP)] of the plates, and it can be quite different for small changes. For most lot sizes, and over the higher levels of Pr(nP), the S-4 inspection level*[3] *results in a greater probability that a lot will pass the LAT.*

The committee concluded that *using a statistically principled protocol enables decision makers to explicitly address the necessary and inherently unavoidable risk trade-offs that must be faced in testing. Furthermore, while additional research and coordination may be necessary to finalize the protocol design, and continuing review will likely be required as manufacturing conditions and plate designs change over time, a statistically principled protocol ensures that decision makers have sound information about body armor performance in order to ensure the quality of a critical soldier safety item.*

Recommendation 6-1: The Office of the Director, Operational Test and Evaluation (DOT&E) should continue to conduct due diligence to carefully and completely assess the effects, large and small, of its statistical protocol as it is adopted across the body armor testing community. In particular, DOT&E should continue to

- Collaborate with the Army and the United States Special Operations Command (USSOCOM) to revise the test protocol as necessary, based on the results of Army and USSOCOM "for government reference" first article testing test results and other empirical evidence, to ensure that currently acceptable plate designs are not eliminated under the new protocol; and
- Regularly assess the impact or impacts of the new protocol on plate design, particularly plate weight, to ensure the test protocol results in body armor that achieves the requisite soldier safety while not negatively, inappropriately, or inadvertently affecting plate design.

Recommendation 6-2: The Office of the Director, Operational Test and Evaluation, should consider modifying the first article testing protocol to

- Generalize the description of the backface deformation (BFD) upper tolerance interval calculation to allow for nonnormal BFD distributions;
- Specify a confidence interval calculation methodology that has better coverage properties, such as the Agresti-Coull interval recommended by Brown et al. (2001) and described in detail in Agresti and Coull (1998); and
- Specify guidelines that will accommodate deviations in environmental conditions and/or plate size from the current 60-plate design matrix.

---

[3]Sample sizes in the protocol are based on special inspection level S-4 of ANSI/ASQ Z1.4-2008 (American Society for Quality, 2008).

Recommendation 6-3: The Office of the Director, Operational Test and Evaluation, and the Army should continue to consult and engage statisticians throughout the process of assessing and revising protocols, comparing the performance of the new and old protocols, assessing the effects of the new protocols, and considering possible changes.

Testers and statisticians should continue to work together as a team to (1) quantify in a statistically rigorous manner the amount of variation in BFD attributable to the testing process and that attributable to the plates and (2) ensure these results are appropriately reflected in an updated protocol. In particular, the statisticians involved with developing and implementing the statistically principled protocol should be involved with the clay experimentation discussed and recommended in the study.

Over the course of the committee's research and deliberations, the DOT&E, the Army, and USSOCOM have endeavored to establish statistically principled test standards that are realistically achievable with the current body armor designs.

Recommendation 6-4: The Office of the Director, Operational Test and Evaluation, the Army, and the United States Special Operations Command should work together to arrive at an acceptable set of test standards for lot acceptance testing that is both statistically principled and is realistically achievable with current body armor designs.

HELMET TESTING

A specific tasking for Phase III of the study was to provide ideas for future improvement of helmet testing. Helmet testing follows a methodology similar to that for the testing of body armor plates. Head forms filled with the same RP #1 modeling clay are heated and subjected to drop tests to assure uniformity. The helmet to be tested is placed over a head form and a test round is fired into the front and side of the helmet. Ballistic forces from the bullet cause an indent in the clay similar to the BFD behind the armor plate, and the indent must be within specifications for it to pass the test.

The committee found that *existing helmet test methodologies, including the current Army test methodology, do not relate directly enough to human injury to confidently assess injury risk from back-face trauma to the head. Improving the link between test methodology and human injury is an urgent matter in light of the newer helmet systems with lower areal densities and increased threat velocities.* Also, *it is uncertain how clay response correlates with human head/skull/brain response. Yet, clay response serves as the basis for current clay-based helmet methodologies.* From a broader systems perspective the same problem exists with body armor plate methodologies. That is, it is uncertain how clay response is correlated with human injury in the thorax.

Recommendation 7-1: The Army should perform research to define the link between human injury and the testing methodology for head behind-armor blunt trauma.

Recommendation 7-2: The Aberdeen Test Center should ensure the following:

1. Dynamic mechanical strain/deformation response of the head surrogate is similar for both types of loading at loading rates typical of behind-helmet response;
2. Response of the head surrogate is similar to that of the human head;
3. Required head quality control calibration is either performed on the head surrogate itself or is shown to be demonstrably represented by a surrogate for the head itself (i.e., by a sample box filled with clay) in controlled testing using a standard test procedure; and,
4. Response of the clay for the low-rate calibration tests is shown to be similar or scalable to the high-rate backface deformation response of the surrogate in controlled testing using a standard test procedure.

The Army Research Laboratory has developed what is referred to as the "Peepsite" head form to deal with some of the shortcomings of existing test head forms. The committee found that the *Peepsite head form reduces or eliminates several potential problems with the NIJ head form that is used in the current clay test methodology.*

A potentially important aspect of ballistic protective helmet design is the suspension system that provides helmet stand-off from the head, an important factor in ballistic protection. This complicates any analysis of injury risk due to deformation of the helmet.

Recommendation 7-3: The Army should investigate use of the Peepsite headform currently in development by the Army Research Laboratory with room-temperature clay. This headform and procedure has potential as a near-term alternative to testing using the National Institute of Justice clay head form tested at elevated clay temperatures.

MEDICAL BASIS FOR FUTURE BODY ARMOR TESTING

Much is to be gained by applying medical knowledge to body armor design and test processes. The committee reviewed applicable advances in medicine and biomechanics since the Prather study and concluded that the researchers at the time made good use of the data that were available (Prather et al., 1977). However, advances in imaging and measurement technology since then could facilitate a better understanding of the injury mechanisms, which will help to identify different and more appropriate engineering tests for armor qualification.

Thoracic Ballistic Test Methodologies

As previously noted, injuries to the thorax due to deformation of the armor are often termed BABT. Dynamic pressures transmitted to the thorax can cause local and remote fractures, contusions, and hemorrhage, as has been demonstrated in numerous animal studies. The committee found that *carried mass, such as that associated with body armor, may decrease a soldier's mobility and lead to fatigue. Further, body armor can prevent high-velocity bullets from penetrating the body but may not protect personnel from the shock wave from the initial projectile impact and the trauma induced by the BFD.*

The committee found that the *details surrounding the force that is transmitted from the body armor to the person wearing the armor, including the amount, the timing, and the immediate and long-term consequences of this force, are unknown.* Techniques are needed not only to identify and treat BABT injuries, but also to assess the risk of BABT injury to those who wear the body armor. An instrumented surrogate (dummy) has been used effectively in many fields of injury biomechanics to evaluate the risk of injury from blunt trauma. Elements of this technique include a biofidelic surrogate, an engineering measurement system, an injury risk evaluation, and validation by physical injury model (such as by tests on animals or cadavers). Development of a relationship between a robust surrogate for injury and a validated injury model is crucial for success of this approach.

The body armor plates were designed to resist penetration by threat projectiles as detailed in the performance specifications. As a consequence, the plates are tested primarily on their ability to defeat the threat projectiles. In combat, the vests and plates also may provide warfighters with an unknown degree of protection against other battle hazards, including blast effects. *The design for future body armor vests should consider blast effects as well as trade-offs between bulk, weight, and protection.* Discrepancies between published measurements of changes in intrathoracic pressure for human subjects exposed to blasts from explosives with and without vests need to be resolved.

Recommendation 8-1: The Army medical and scientific testing communities should adequately fund and expedite the research necessary to experimentally and epidemiologically quantify the physiologic and medical impact of blunt force trauma on the body from both ballistic and blast threats to soldiers.

Cadaveric Experiments for Behind-Armor Blunt Trauma

Although there are several studies using animal and cadaveric experiments to study BABT injuries for hard body armor, the committee found that *the current work does not allow the development of a thoracic BABT injury criterion from existing studies. Additional animal and/or cadaveric experimentation is necessary to develop a BABT injury criterion.* Also, *there is a need for a robust and widely used ballistic trauma injury classification scale.*

Although there are a number of existing injury scales, including a widely used scale for automobile injuries, the abbreviated injury scale promulgated by the Association for the Advancement of Automotive Medicine, none is well suited to ballistic trauma. Data on which to base a satisfactory injury scale will require the collection of military epidemiological data on a large scale.

Models used by blunt trauma researchers do not reflect realistic battlefield threats, and *the fidelity of anatomical, physical, and mathematical finite-element models simulating the human thorax, heart, lungs, liver, and kidneys, is limited at the present time. Thus, damage from transmitted pressures associated with blunt trauma to such organs as the intestines, spinal cord, brain, or vascular system cannot be predicted.*

Recommendation 8-2: The Army should perform high-speed ballistic tests using human cadavers and large animal cadavers to provide responses to deforming hard armor impacted by velocities likely to be encountered in combat. These tests should be extensively instrumented to determine dynamic deformation characteristics in the human and animal torsos to provide data that can be correlated with clay response at the same rates (or with alternative media or other test methodology) and with epidemiology and medical outcomes in the soldier. The studies should ensure that velocity and backface deformation regimes replicate those for current and future desired body armor testing protocols.

The observations and data needed for large animal studies are far more extensive than data collected in the past. As described in Appendix J, studies will require extensive use of pressure transducers, cineradiography, metabolic imaging and neurochemical cerebral spinal fluid and blood assays.

Recommendation 8-3: The Army should perform live large-animal, live-fire tests to simulate the behavior of current and proposed new body armor against expected threats.

Instrumented Alternatives to Determine BABT

Technologies developed for research to evaluate injury effects, such as the ATM and clay sensors, have been considered by the Army for use in developing alternative testing methodologies. The committee found that *instrumented response elements are in a primitive state for the evaluation of ballistic BABT for hard body armor against rifle round threats. Although several devices have associated instrument response and injury criteria that have been validated against a small range of loading conditions, there is no test device suitable for use without further development and validation.* Also, *instrumented anatomical surrogates are not detailed enough to assess ballistic BABT for hard body armor with rifle round threats.*

Recommendation 8-4: The Army should develop finite-element simulation models of human and live-animal thoracic response to behind-armor blunt impact.

The validation of this simulation should be hierarchical from the small scale to the large scale. This includes the dynamic local response of constituent materials such as skin, bone, muscle, lung, liver, and other tissues; the regional response of the tissues under loading; and the global response of the whole torso. It should also include deformations from soft and hard body armor impacted with appropriate threats.

Recommendation 8-5: The Army medical community should enhance the current trauma registries to provide a program of injury epidemiology for ballistic impact, including behind-armor blunt trauma. This should include collection of both injury and noninjury events and should be similar to the federal crash databases used by the Department of Transportation—for example, the Fatality Analysis Reporting System and the National Automotive Sampling System for traffic injuries/fatalities, including injuries induced by both penetrations and backface deformations.

Recommendation 8-6: Using experimentally determined links to injury, response, and epidemiology, the Army should ensure that the clay or other alternative test methodology for hard body armor has humanlike dynamic response and is suitable for the development of behind-armor blunt trauma injury criteria.

Recommendation 8-7: To achieve improvements in behind-armor blunt trauma (BABT) research methodology in the medium term, the Army should develop instrumented thoracic simulators as response elements (sensors). Necessary preludes to this effort include the following:

- Establishing BABT phenomenology and injury criteria using human cadavers, animal models, and field injury epidemiology coupled with well-validated finite-element simulations.
- Establishing human BABT mechanical response for the range of design conditions for personal protective body armor. This should include impact on soft and hard body armor of anticipated threats.

Recommendation 8-8: In the long term, beyond simple clay torso surrogates and one-layer torso simulants, the Army should use the road map in Figure S-2 to investigate the use of detailed anatomical surrogates (such as cadavers, instrumented models, etc.) as research devices to evaluate behind-armor blunt trauma.

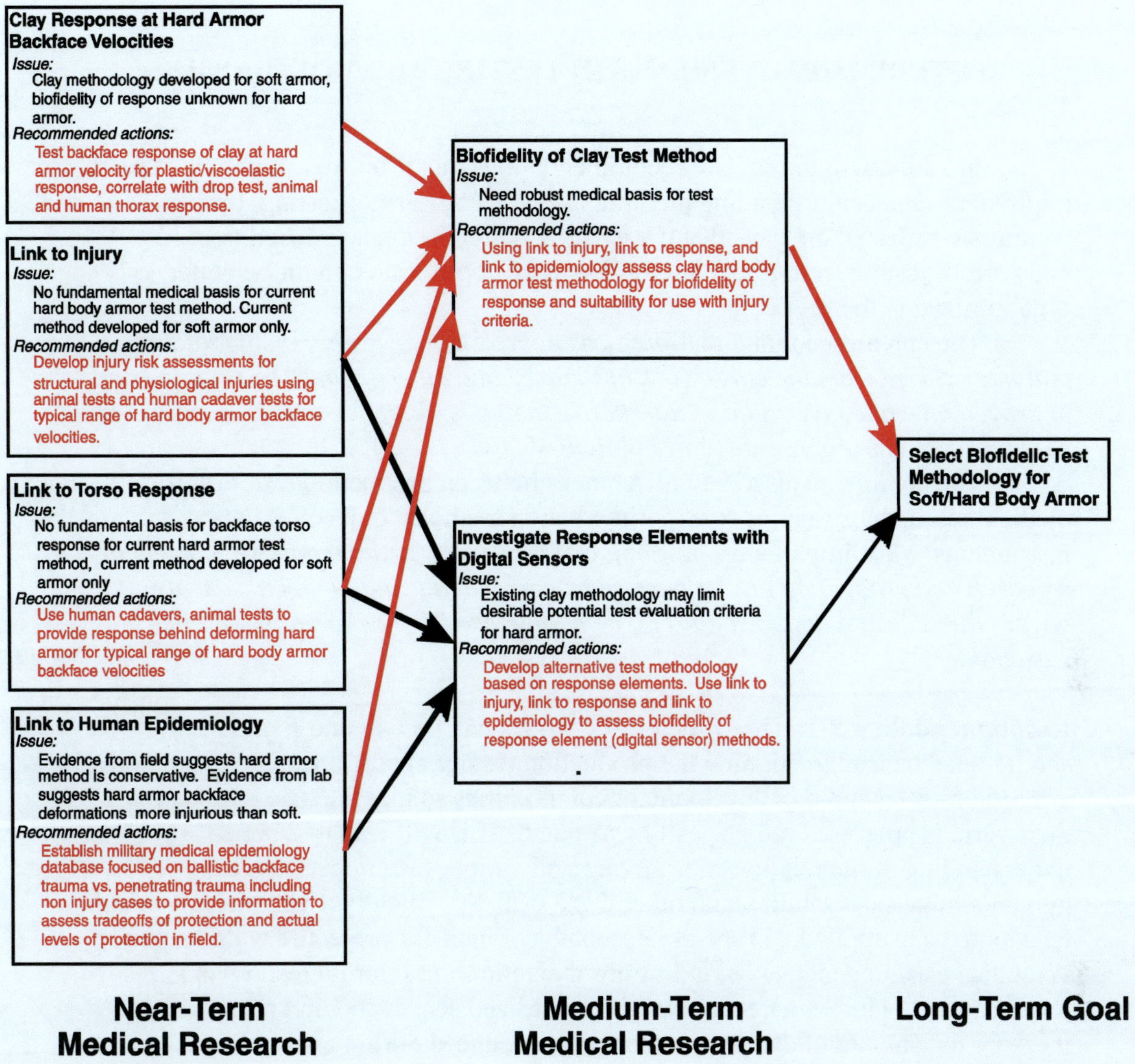

FIGURE S-2 Road map showing suggested near-term and medium-term research needs, and a long-term goal to provide the fundamental medical basis for injury risk assessment behind helmets and hard body armor.

FUTURE IMPROVEMENTS IN TESTING METHODOLOGY

In addition to the several recommendations that propose refinements to improve or replace the standing methodology for body armor testing, the committee reflected on ways that the existing national standards used to guide body armor testing for military and police force applications might be better synchronized in the future.

The committee found that *the current body armor testing methodology that has evolved from the early work of Prather and others should be retained and improved on while investigating alternative methods.* Recommended priorities for near-term actions are illustrated in Figure S-1.

The committee discerned differences between production testers and medical researchers relating to experimentation methods, objectives, and test instruments. Most importantly, it found that *recording medium data from medical research and production testing need to be correlated using identical sensors having the requisite time resolution. The results need to be shared among the stakeholders.*

Recommendation 9-1: The Director of Operational Testing and Evaluation should take the lead in aligning the production testing, medical research, and body armor/helmet technology development communities so that the data outputs from their various processes can be easily correlated. This will lead to a better understanding of the relationships among body armor testing performance, human/animal survivability, and other trade-offs. Specifically, two policies should be adopted and applied: (1) specify acceptable ranges for projectile weights and velocities used to generate behind-armor dynamic forces during testing and research and (2) investigate the use of standardized sensors behind armor to measure the amount of dynamic force that is produced during testing and research.

The overall need is for a coordinating committee to provide oversight and facilitate the exchange of information between stakeholder groups. The committee believes that *the nationally recognized coordination committee recommended in the Phase II report is needed to align and accelerate efforts of technologists, production testers, and biomedical researchers in BABT/BFD-related research for both body armor and helmets.* As an important step in this process, the ad hoc clay working group approach that was started by and is currently chaired by DOT&E offers an organizational nucleus for a way ahead for DoD. The committee agreed that *the original ad hoc clay working group could be expanded to form DoD's portion of the national body armor testing standardization committee recommended in the Phase II report.*

The committee's last recommendation is conceptually the same as Recommendation 15 in the Phase II report (NRC, 2010) but has been expanded to include helmet testing. Helmets and body armor plates have different requirements, and there will likely be different testing standards for them for the foreseeable future.

Recommendation 9-2: The Office of the Director, Operational Test and Evaluation and the National Institute of Justice (NIJ), in collaboration with the military services, unified commands, government testing organizations, NIJ-certified testing laboratories, medical researchers and governmental and commercial material developers should convene a national body armor testing standard committee to review all appropriate considerations and develop recommendations that could lead to updated national body armor configurations and testing standards for body armor and helmet testing.

REFERENCES

AAA (U.S. Army Audit Agency). 2009. Body Armor Testing. A-2009-0086-ALA. Alexandria, Va.: U.S. Army Audit Agency.

Agresti, A., and B. Coull. 1998. Approximate Is Better Than "Exact" for Interval Estimation of Binomial Proportions. The American Statistician 52(2):119-126.

American Society for Quality. 2008. American National Standard Sampling Procedures and Tables for Inspection by Attributes. ANSI/ASQ Z1.4-2008. Milwaukee, W.I.: American Society for Quality.

Brown, L., T. Cai, and A. DasGupta. 2001. Interval Estimation for Binomial Proportions. Statistical Science 16(2):101-117.

GAO (United States Government Accountability Office). 2009. Independent Expert Assessment of Army Body Armor Test Results and Procedures Needed Before Fielding. GAO-10-119. Washington, D.C.: Government Accountability Office.

NRC (National Research Council). 2009. Phase I Report on Review of the Testing of Body Armor Materials for Use by the U.S. Army: Letter Report. Washington, D.C.: National Academies Press.

NRC. 2010. Phase II Report on Review of the Testing of Body Armor Materials for Use by the U.S. Army. Washington, D.C.: National Academies Press.

Prather, R., C. Swann and C. Hawkins. 1977. Backface Signatures of Soft Body Armors and the Associated Trauma Effects. ARCSL-TR-77055. Aberdeen Proving Ground, MD: U.S. Army Armament Research and Development Command Technology Center.

1

Introduction

This chapter provides the study context and task statement. It also describes the purpose and scope of the study and how the report is organized.

BACKGROUND

The National Research Council (NRC) consists of several boards and their associated committees that bring the complexities of protection materials science research and applications into focus for Department of Defense (DoD) sponsors. These include the Board on Army Science and Technology, the National Materials Advisory Board, and the Army Research Laboratory Technical Assessment Board of the NRC's Division on Engineering and Physical Sciences and the Committee on National Statistics of the NRC's Division on Behavior and Social Sciences and Education.

In 2009, the Government Accountability Office (GAO) released a report that commented on the conduct of the test procedures governing acceptance of body armor vest-plate inserts worn by military service members (GAO, 2009). The GAO report, as well as other observations (for example, the Army Audit Agency report to the Program Executive Officer Soldier on Body Armor Testing (AAA, 2009), led the DoD Office of the Director, Operational Test and Evaluation (DOT&E) to request that the Division on Engineering and Physical Sciences conduct an ad hoc study to investigate issues related to the testing of body armor materials for use by the U.S. Army and other military departments.

Study Tasks

Box 1-1 contains the statement of task for the three-phase study. Phases I and II were completed in 2009 and 2010 respectively and resulted in two NRC letter reports (NRC, 2009 and 2010).[4] This report is the Phase III report. To ensure wide dissemination, no classified or restricted information is contained in the reports. The sponsor also specifically requested that the NRC report emphasize the science rather than the policy aspects of the body armor testing issues.

---

[4]Findings and recommendations from the Phase I and Phase II reports are in Appendixes K and L respectively.

Box S-1 Statement of Task

The National Research Council will convene specialists in committee form to consider the technical issues relating to the testing of body armor. To do this the National Research Council shall conduct a 3-phase study:

In Phase I a committee will comment on the validity of using laser-profilometry/ laser-interferometry techniques to determine the contours of an indent made by a ballistic test in a non-transparent clay material at the level of precision established in the Army's procedures for testing personal body armor. If laser-profilometry / laser-interferometry is not a valid method, the committee will consider whether a digital caliper can be used instead to collect valid data. The Committee will also provide interim observations regarding the column drop performance test described by the Army for assessing the part to part consistency of a clay body used in testing body armor. The committee will prepare a letter report documenting the findings from its Phase I considerations. This is a six week effort beginning November 1 2009 and ending mid December 2009.

In Phase II a committee will consider in greater detail the validity of using the column drop performance test described by the Army for assessing the part-to-part consistency of a clay body within the level of precision that is identified by the Army test procedures. The committee will prepare a letter report documenting the findings from its Phase II considerations. This is a three months effort beginning November 1 2009 and ending early February 2010.

In Phase III a committee will consider test materials, protocols and standards that should be used for future testing of personal armor by the Army. The committee will also consider any other issues associated with body armor testing that the committee considers relevant, including issues raised in the Government Accountability Office Report---Warfighter Support, Independent Expert Assessment of Body Armor Test Results and Procedures Needed Before Fielding (GAO-10-119).The committee will prepare a final report. This is a 14-months effort beginning November 1 2009 and ending January 2011.

The final report will document the committee's findings pertaining to the following issues that are of particular immediate concern to DOT&E including the following:

- The best methods for obtaining consistency of the clay, and of conditioning and calibrating the clay backing used currently to test armor.
- The best instrumentation (e.g., laser scanning system, digital caliper, etc.) and procedures to use to measure the back face deformation (BFD) in the clay.
- The appropriate use of statistical techniques (e.g., rounding numbers, choosing sample sizes, or test designs) in gathering the data.
- The appropriate criteria to apply to determine whether body armor plates can provide needed protection to soldiers; this includes the proper prescription for determining whether a test results in a partial or complete penetration of body armor, including, as appropriate, the soft armor underlying hard armor.

The final report will also document the committee's findings regarding any other issues regarding body armor testing that the committee found relevant. The study team will have access to all data with respect to body armor testing that the team needs for the conduct of the study.

The last task for Phase III of the study was to document in its final report any other issues regarding body armor testing that the committee found relevant. In response, this Phase III report also attempts to do the following:

- Provide a roadmap to reduce the variability of clay processes and show how to migrate from clay to future solutions.
- Consider the use of statistics to permit a more scientific determination of sample sizes to be used in body armor testing.
- Develop ideas for revising or replacing the Prather study methodology;
- Review and comment on methodologies and technical approaches to military helmet testing.
- Consider the possibility of combining various national body armor testing standards.

Appendix C contains the specific tasks that were identified for the Phase III portion of the study. A matrix relating the statement of task to specific sections of the report is included as Appendix D.

STUDY CONTEXT

From the outset, the committee recognized that the body armor testing community exists in a charged environment where the lives of service members and law enforcers are at risk. Efforts to improve body armor testing processes should lead to the fielding of more effective body armor and helmets for our servicemen and women. To be most effective, body armor and helmets should be in the "sweet spot" where there is a balance between survivability and light weight. The broad purposes of the study were to verify and validate current test procedures for body armor plates, to investigate longstanding issues related to the testing processes, and to recommend approaches that will improve future testing methodologies and procedures.

Study Implementation

As directed by the task statement, the study was divided into three phases.

- The Phase I letter report focused on the validity of using laser-based measuring techniques to determine the contours of an indent made in a nontransparent clay material by a ballistic test. The report offered interim observations on the column drop performance calibration technique being used by the Army's Aberdeen Test Center for assessing the part-to-part consistency of a clay body used in testing body armor. It also provided immediate feedback on issues raised by the GAO report (GAO, 2009). The specific findings contained in the Phase I letter report are listed in Appendix K.

- The Phase II letter report focused on the use of clay as a backing material during body armor testing. It examined in detail the validity of using the column drop performance calibration and recommended alternatives for future backing materials. Additionally, the report discussed use of statistically based protocols in body armor testing and described an approach to developing a single national body armor testing standard. The recommendations of the Phase II committee are contained in Appendix L.
- In this final report, the Phase III committee has carried out the Phase III tasks and consolidated and expanded on information contained in the two earlier letter reports. This final report provides a road map to reduce the variability of clay processes and eventually migrate from methods based on clay to other methodologies. It also develops ideas for revising the medical basis for testing procedures and addresses technical approaches to military helmet testing.

To complete the study, the Phase III committee conducted data-gathering sessions at the U.S. Army Aberdeen Test Center in Maryland and at the National Academy of Sciences Keck Center in Washington, D.C. The chair assigned committee members to working groups in the following task areas: clay and instrumentation; body armor testing methodologies; statistics; and helmet testing. To facilitate the study, the separate working groups conducted individual data-gathering sessions, teleconferences, and, in two instances, site visits. The leaders of the working groups coordinated the gathering of data and consolidated written inputs into chapters for the overall report. Appendix B provides a list of the committee briefings and activities.

Report Organization

Chapter 1 (Introduction) provides background and context for the study, and Chapter 2 (Overview of Body Armor) provides a detailed description of the body armor testing processes and facilities.

Chapter 3 (Historical Basis for Current Body Armor Testing) reviews the foundational basis for the testing methodology that has been used since the late 1970s. Chapter 4 (Clay and Backing Materials) assesses the use of clay in testing and describes how the variability inherent in the backing material may be incorrectly attributed to variability in the armor. Chapter 5 (Instrumentation and Procedures for Measuring an Indent in the Backing Material) offers insights into measuring devices.

Chapter 6 (Statistical Considerations in Body Armor Testing) discusses findings on the statistical aspects of body armor testing with a focus on body armor plate testing, and Chapter 7 (Helmet Testing) extends the discussion of testing to the testing of combat helmets and provides ideas for future improvement.

Chapter 8 (A Medical Basis for Future Body Armor Testing) describes the current lack of a medical basis for body armor testing and recommends a direction for future medical research and analyses. Finally, Chapter 9 (Future Improvements in Testing Methodology) looks to the future of body armor testing and describes what is needed to improve or replace the methodology that has for decades underpinned body armor testing.

The report includes several appendixes as described in the chapters and listed in the contents.

REFERENCES

AAA (U.S. Army Audit Agency). 2009. Body Armor Testing. A-2009-0086-ALA. Alexandria, Va.: U.S. Army Audit Agency.

GAO (Government Accountability Office). 2009. Warfighter Support: Independent expert assessment of army body armor test results and procedures needed before fielding. GAO-10-119. Washington, D.C.: Government Accountability Office.

NRC. (National Research Council). 2009. Phase I Report on Review of the Testing of Body Armor Materials for Use by the U.S. Army: Letter Report. Washington, D.C.: National Academies Press.

NRC. 2010. Phase II Report on Review of the Testing of Body Armor Materials for Use by the U.S. Army. Washington, D.C.: National Academies Press.

2

Overview of Body Armor

The purpose of this overview is to provide a broad introduction to the nature of ceramic body armor plates and helmets as used by many U.S. servicemen and women; the medical basis for determining the relationships among body armor, blunt force trauma in humans and the testing of body armor; and techniques used by the U.S. Army to test the effectiveness of body armor.

BACKGROUND

The evolution of body armor in the United States dates from a series of inventions in 1861 when thin steel plates were enclosed in military jacket materials to protect against saber attacks and bullets (Peterson, 1950). Whether or not to use protective armor was a personal choice and depended on cost ($5-$7), weight (2 kg), and appearance (too unmanly). The use of vests ceased after the American Civil War and did not reappear in earnest until 100 years later, when the U.S. government began to supply law enforcement and public officials with protection from small-arms bullets. During the Vietnam War, U.S. military forces widely began wearing soft Kevlar-based protective vests. During the wars in the Middle East, U.S. personnel—both military and civilian—in the combat zone were required to wear protective vests containing hard body armor plates.

Currently, there are two major types of personal body armor, soft and hard. Soft armor vests are designed to protect against shrapnel resulting from explosions and against low-velocity, low-energy bullets (e.g., 9 mm, or .38 caliber). Hard armor and bullet-proof vests incorporate hard-plate inserts made of polyethylene or ceramic composite material in soft armor vests to defeat high-velocity threats such as 7.62 mm (.30 caliber) and 12.7 mm (.50 caliber) rifle bullets. The original work in standardizing body armor testing (discussed in Chapter 3) focused on soft armor, but the threats to both warfighters and law enforcement personnel are currently from shrapnel and projectiles of higher energy and higher velocity than anticipated 35 years ago. Therefore, much of the current research is on improving hard body armor.

Modern hard body armor can defeat incoming pistol and rifle rounds, trading energy and momentum deposition into the armor for deformation of the armor. This deformation includes direct deformation of the body armor in the case of soft body armors and deformation with fracture in hard body armors. A technical explanation of how ceramic body armor is able to defeat a threat projectile and protect the wearer of body armor is contained in Appendix E. This deformation, however, has the potential for creating injuries in the thorax behind the armor that may generally be characterized as blunt trauma. These injuries are often termed "behind-armor blunt trauma."

This study focuses on hard body armor (referred to as "body armor") and helmets.

Ceramic Plates in Body Armor

The Phase II report (NRC, 2010, p. 4) described the use of ceramic materials in body armor as follows:

> Ceramic materials have been used successfully in personal armor systems to defeat small-arms threats. They are preferred for personal armor systems because they are lighter than more traditional armor made of metallic alloys. Properties that contribute to the performance of ceramic armor include superior hardness, low density, favorable elastic constants, and high compressive strength. However, as stand-alone items, ceramics would not be particularly good because of their low tensile strength, brittle response, and sensitivity to small mechanical defects such as pores and cracks. Hence, ceramics are used in combination with other materials, such as polymers and metals, to form laminar composites that provide excellent properties for body protection. A typical insert (also referred to as a "plate") of body armor consists of a layer of dense boron carbide or silicon carbide backed by a layer of metal or polymer composite; the entire plate is wrapped in tightly woven ballistic fabric. The ceramic layer breaks up an incoming projectile and dissipates its kinetic energy. The layer of polymer composite and/or metallic alloy provides ductility and structural integrity and spreads the forces resulting from the impact of a projectile over a larger area.

The use of ceramic materials has been successful. The military collects data on casualties resulting from possible penetrations of body armor by enemy rounds, and there have been no known soldier deaths due to small arms that were attributable to a failure of issued ceramic body armor (NRC, 2010).

Fiber and Resin Composites in Helmets

Like body armor, current ballistic protective helmets employ a passive momentum defeat mechanism in which a bullet with a small mass and high velocity progressively engages a larger mass of high-performance fiber/resin composite, decreasing the bullet velocity and locally transferring momentum to the helmet. This process continues until all the momentum of the incoming round is deposited into the helmet or the helmet is defeated and penetrated by the incoming round. Even if the incoming round does not penetrate the helmet, there is still potential for substantial local head contact from sufficient helmet "backface deformation" (BFD) into the head and resultant traumatic brain injuries.

Survivability vs. Mobility

All hard body armor systems currently add a significant burden of weight on the soldier. Armor testing therefore has implicit goals of ensuring that body armor meets survivability standards while allowing sufficient soldier mobility and flexibility. To provide soldiers with more weight than necessary to defeat a specified threat can lead to unintended consequences such as premature exhaustion and restricted ability to move rapidly and react appropriately in life-threatening situations (NRC, 2010).

MEDICAL STUDY BASIS FOR TESTING BODY ARMOR

In 1977 a study was performed to correlate the depth that a 200-g, 80-mm hemispherical missile impacting at approximately 55 m per second penetrated live animal tissue and other media (Prather et al., 1977). The goal of the study was to develop a simple, readily available backing material for characterizing both the penetration and deformation effects of ballistic impacts on soft body armor materials and to relate this information to the injury potential of nonpenetrating ballistic impacts. When there was no penetration of the armor the researchers noticed that dynamic ballistics forces caused a deformation in the recording material directly behind the point where the bullet struck the front side of the armor. This deformation in the backing material was the BFD mentioned above. The depth of penetration into various media such as modeling clay and ballistics gelatin as a function of time was compared to the probability of lethality for the same penetrator entering into a live animal model (goats were used as models) (Clare et al., 1975).

Prather et al. (1977) observed strong correlations between lethality probability and penetration into ballistic gelatin[5] and also into the modeling clay Roma Plastilina (RP #1) . The ballistic gel required the use of high-speed photography to record BFDs because the gel was elastic and returned to its original shape after the projectile firing. To avoid the necessity of using expensive high-speed photography, an alternative material was sought that would retain its deformation and be easily measured using inexpensive manual calipers. The first conclusion of the Prather report had a profound effect on testing over the next 30 years. It reads as follows (Prather et al., 1977, p. 11):

> A readily available, easy-to-use backing material, Roma Plastilina 1, has been found which can be correlated to tissue response for use in characterizing both the penetration and deformation effects of ballistic impacts on soft body armor materials.

---

[5]Ballistic gelatin is a clear or yellowish gelatin that is the standard medium for seeing and evaluating what happens to bullets on impact with soft tissue.

RP #1, a commercially available artists' modeling clay, has since been adopted as a recording medium to assess the likelihood of injury or death from ballistics, and its use has been extended from assessing soft armor, such as Kevlar vests, to assessing hard armor plates and helmets, knife wounds, industrial injuries, such as one to a drop-forge operator, and nonlethal projectiles (Lyon, 1997; Chadwick et al., 1999; O'Callaghan et al., 2001; Vaughan, 2001; and Karahan, 2008). RP #1 appears to have become an industry standard despite being a questionable correlative of the human body. The deformation left in the modeling clay has until recently been measured using manual calipers. Over the past few years more technologically advanced laser profilimeters have been mandated for measuring the deformations made during Department of Defense body armor testing.

Since the original Prather effort, a number of studies and experiments have been conducted to better determine the relationships among blunt force trauma, human injury, and body armor testing processes. Even though there is no correlation between medical data and the BFD approach, the committee believes that the current methodology for testing body armor should be continued until it is replaced by a better methodology. As stated earlier, the current approach has allowed the Army to send body armor with adequate survivability characteristics (no known deaths due to penetrations of rounds the armor was designed to defeat) to soldiers in combat.

In this study the committee will offer ideas that may lead to a refinement or replacement of the original Prather methodology.

BODY ARMOR TESTING PROCESS

The Army's procedures for testing hard body armor by measuring the deformations in clay backing from ballistic impacts are documented in "Test Operations Procedure (TOP) 10-2-210: Ballistic Testing of Hard Body Armor Using Clay Backing," dated October 1, 2008 (ATC, 2008). As described in the Phase I report, the approach may be summarized as follows (NRC, 2009, p. 6):

> A clay box [6] and a clay chest plate appliqué [7] [See Figure 2-1] are assembled, appropriately calibrated for part-to-part consistency using the column-drop performance test, and placed upright in the test holder. . . . Independently, a "shoot pack" is prepared. ...To create a shoot pack, the armor plate is placed in a fabric envelope together with multiple layers of Kevlar to replicate the vest worn by the soldier. The dimensions of the armor plate depend on the size of the vest ... and can range from

[6]A plywood-backed aluminum frame (~61 × 61 × 14 cm) filled with modeling clay is subsequently referred to in this report as a "clay box" or as a "part" when discussing part-to-part variations. Since 1977 the modeling clay of choice for the testing community has been Roma Plastilina #1.

[7]As shown in Figure 2-1, the appliqué is an additional layer of clay that has been molded to the shape of the specific armor plate to be tested.

18 × 29 cm to 28 × 36 cm, with a thickness of approximately 2 cm. The vest has a significant nonconstant radius of curvature. Once assembled, the shoot pack is pressed firmly into the surface of the appliqué to ensure a conformal fit. The shoot pack is then removed and the laser scanning system is used to scan the surface of the appliqué in order to provide a reference surface relative to which subsequent deformations caused by the firing of the projectiles can be compared.

. . . The laser scanning system is moved out of the way and the shoot pack is repositioned onto the surface of the clay, with care taken not to disturb the reference surface, and the shoot pack is secured. … The selected projectile is then fired into the shoot pack, after which the shoot pack is removed from the clay and inspected for penetration . . .

FIGURE 2-1 The clay appliqué applied to the clay box. SOURCE: Richard Sayre, Deputy Director, Office of the Secretary of Defense, Director of Operational Test and Evaluation (OSD DOT&E) Live Fire Test and Evaluation, and Tracy Sheppard, Executive Officer and Staff Specialist, OSD DOT&E Live Fire Test and Evaluation, "DoD In-Brief to the National Research Council Study Team," presentation to the Body Armor Testing Phase I committee, Aberdeen, Maryland, November 30, 2009.

During the test, the velocity of the projectile is measured using Oehler Model 57 Ballistic Screens to verify that it was within the desired range. Any test round that penetrates the armor plate (a partial penetration) and continues on to completely break any Kevlar fiber on the back of the shoot pack is considered a complete penetration. A typical displacement or indent in the clay made by the deformation of the armor is shown in Figure 2-2. The nominal design specification is that the maximum depth in the clay relative to the original surface be less than 43 mm. That is, a BFD less than 43 mm deep is considered to indicate

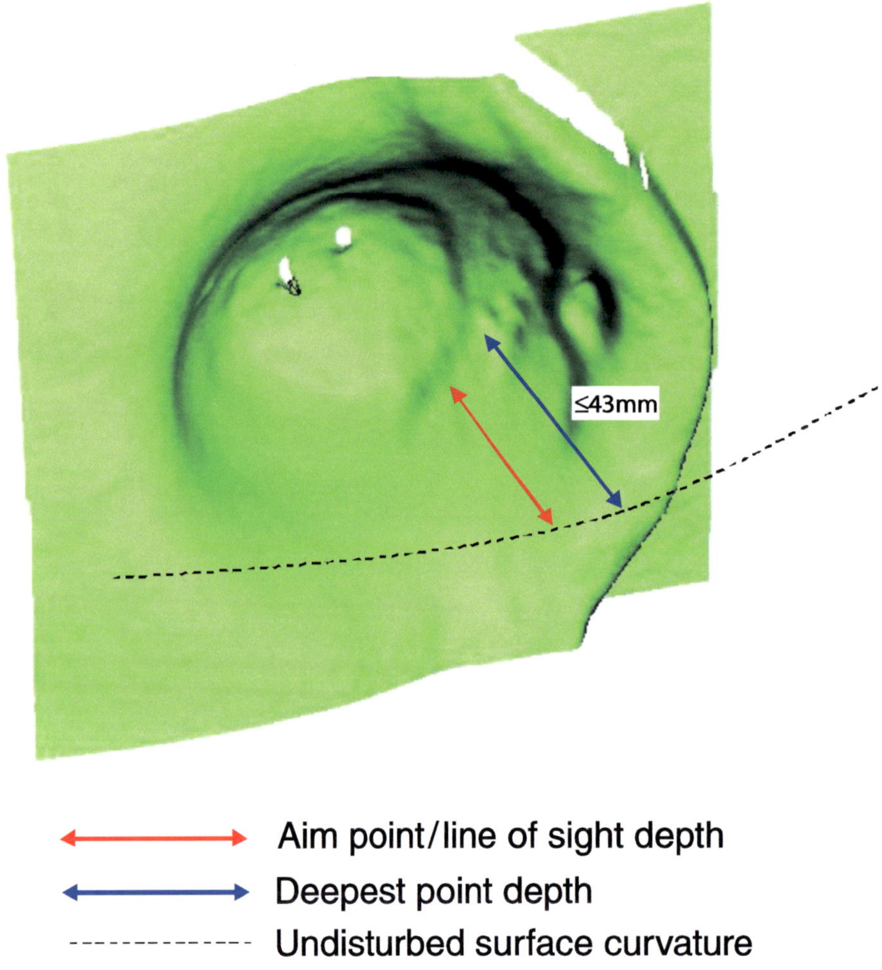

FIGURE 2-2 Surface of the BFD as measured by a laser scanning system. SOURCE: Richard Sayre, Deputy Director, Office of the Secretary of Defense, Director of Operational Test and Evaluation (OSD DOT&E), Live Fire Test and Evaluation, and Tracy Sheppard, Executive Officer and Staff Specialist, OSD DOT&E Live Fire Test and Evaluation, "DoD In-Brief to the National Research Council Study Team," presentation to the Body Armor Testing Phase I committee, Aberdeen, Maryland, November 30, 2009.

acceptable performance of body armor in service.[8] Experimental data collected by the Army indicate that under nominally identical conditions the standard deviation for the maximum depth of the BFD (hard armor) is in the range of 2.5 to 4 mm.[9] The BFD measurements in combination with the penetration data are used to evaluate the armor.

The deformation is measured with the laser scanning system. The data are collected and used to compute the profile (depth distribution) indent. The deformation is analyzed and serves as an indication of the survivability of a soldier subjected to a similar shot and protected by a similar plate in a protective vest.[10]

The testing of protective helmets also involves the measurement of clay deformation. In the helmet test methodology, a helmet is placed over a head form filled with modeling clay. A test round is fired against the helmet on the head form. The ballistic forces from the bullet cause an indent in the clay. The indents must not exceed the specifications for maximum acceptable indents on both the side and front for the helmet to pass the test. (A detailed discussion of the helmet test process with illustrations is contained in Chapter 7).

Body Armor Testing Range

A typical firing range used to test body armor uses a rifle-like device to fire a projectile against an armor plate. An electronic instrument is used to measure the velocity of the projectile before impact. The armor plate being tested is affixed to an oil-based backing of modeling clay, which is left with a crater in its surface as a result of the impact. A laser system is used to measure the before and after surface geometry of the indentation in the clay. The indoor range set up for testing body armor at the Aberdeen Test Center (ATC) is shown in Figure 2-3.

---

[8]As described in the Phase II report, there has been variation over time about the allowable BFD depth (NRC, 2010). Traditionally, the Army has used 43 mm (with penalty points given for BFDs in the 44-47 mm range); the new DOT&E protocol requires that the 90 percent BFD upper tolerance limit be less than 44 mm with 90 percent confidence for the first shot and with 80 percent confidence for the second shot.

[9]James Zheng, Chief Scientist, PEO Soldier, "Ballistic Protection for Warfighters," presentation to the Body Armor Testing Phase I committee, Aberdeen, Maryland, November 30, 2009.

[10]The Prather study showed that for various media, including the modeling clay Roma Plastilina #1, there is a correlation between the depth of penetration as a function of time and the probability of lethality for the same penetrator entering a human surrogate (goat) (Prather et al., 1977). The study addressed depth but did not address volume of the indentation.

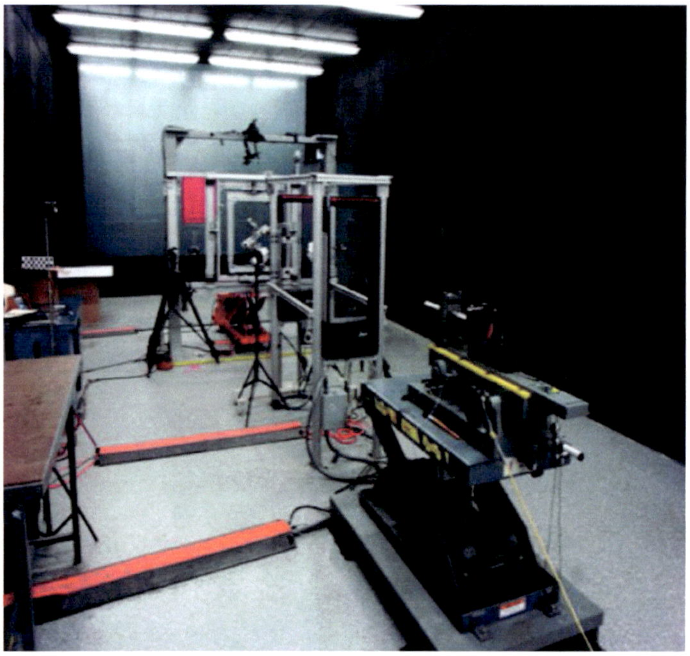

FIGURE 2-3 The body armor test range at ATC. SOURCE: John Wallace, Technical Director, Aberdeen Test Center, "Body Armor Test Capabilities," presentation to the Body Armor Testing Phase II committee, Aberdeen, Maryland, March 10, 2010.

GOVERNMENT ACCOUNTABILITY OFFICE REPORT

A primary motivation for the study was the Government Accountability Office (GAO) report GAO-10-119, "Report to Congressional Requesters, Warfighter Support, Independent Expert Assessment of Army Body Armor Test Results and Procedures Needed Before Fielding" (GAO, 2009). In the report, the GAO recommended that "the Army should provide for an independent ballistics evaluation of the First Article Testing results," that "the Army should assess the need to change its procedures based on the outcome of the independent experts' review and document these and all other key decisions made to clarify or change the testing protocols," and that "the Army provide for an independent external peer review of ATC's body armor testing protocol, facilities and instrumentation" (GAO, 2009, p. ii). The committee has addressed questions raised by the GAO throughout this report, and a summary of the committee's responses to specific issues raised in the GAO report is contained in Appendix F.

While addressing the GAO concerns was of importance, the committee and sponsor have endeavored to focus the study on findings and recommendations that improve current testing processes. In turn, such improvements offer a way to field a lighter, more survivable body armor for our nation's military forces.

REFERENCES

ATC (U.S. Army Aberdeen Test Center). 2008. Test operations procedure (TOP) 10-2-210: Ballistic testing of hard body armor using clay backing. Aberdeen Proving Ground, Md.: Aberdeen Test Center.

Chadwick, E., A. Nicol, J. Lane, and T. Gray. 1999. Biomechanics of knife stab attacks. Forensic Science International 105(1): 35-44.

Clare, V., J. Lewis, A. Mickiewicz, and L. Sturdivan. 1975. Blunt trauma data correlation. EB-TR-75016. Aberdeen Proving Ground, Md.: Edgewood Arsenal.

GAO (Government Accountability Office). 2009. Independent expert assessment of army body armor test results and procedures needed before fielding. GAO-10-119. Washington, D.C.: Government Accountability Office.

Karahan, M. 2008. Comparison of ballistic performance and energy absorption capabilities of woven and unidirectional aramid fabrics. Textile Research Journal 78(8):718-730.

Lyon, D. 1997. Development of a 40mm nonlethal cartridge. ARL-TR-1465. Available online at http://www.defense-technology.com/pdfs/resources/development%20of%20a%2040mm%20nonlethal%20cartridge.pdf. Last accessed March 10, 2011.

NRC (National Research Council). 2009. Phase I Report on Review of the Testing of Body Armor Materials for Use by the U.S. Army: Letter Report. Washington, D.C.: National Academies Press.

NRC. 2010. Phase II Report on Review of the Testing of Body Armor Materials for Use by the U.S. Army. Washington, D.C.: National Academies Press.

O'Callaghan, P., M. Jones, D. James, S. Leadbeatter, S. Evans, and L. Nokes. 2001. A biomechanical reconstruction of a wound caused by a glass shard: A case report. Forensic Science International 117(3): 221-231.

Peterson, H. 1950. Body armor in the Civil War. Ordnance 34:432-433.

Prather, R., C. Swann, and C. Hawkins. 1977. Backface signatures of soft body armors and the associated trauma effects. ARCSL-TR-77055. Aberdeen Proving Ground, Md.: U.S. Army Armament Research and Development Command Technology Center.

Vaughan, N. 2001. Assessment of aprons for protection against drop forging projectiles. Contract Research Report 395/2001. Sheffield, U.K.: Health and Safety Laboratory.

3

Historical Basis for Current Body Armor Testing

This chapter discusses the foundational basis for the current body armor testing methodologies as practiced by military and law enforcement agencies. In the 1970s developments in fiber technology and protective vests along with injury biomechanics investigations led to the study performed by Prather et al. (1977) (called "the Prather study" in subsequent text). That study provides the basis for the current clay-based test methodology for the assessment of blunt trauma risk from backface deformation (BFD). The work focused on protection from low-velocity handgun rounds using soft body armors. It included an injury assessment methodology developed using animal tests and the correlation of animal chest deformation response with the response of simulant materials at velocities that are typical of rounds used to test soft body armors. A diagram of this process is shown in Figure 3-1.

The process can conceptually be separated into two stages. As shown in Figure 3-1, the first stage is a soft body armor evaluation using paired (goat and simulant) tests to look at realistic deformation responses and fatalities behind soft body armor on goats. Typical penetration depths were derived using gelatin as a tissue stimulant for comparison. The second stage is an injury risk assessment using a hard cylindrical impactor into goat chests and correlated depth of penetrations using the same impactor into gelatin and clay. The development of this process is discussed below.

BACKGROUND

In January 1973, the Law Enforcement Assistance Administration, a branch of the U.S. Department of Justice (DOJ), tasked the U.S. Army Land Warfare Laboratory at Aberdeen Proving Ground, Maryland, to develop lightweight, inconspicuous protective garments for public officials in response to an increasing number of armed assaults on public officials. The U.S. Army Land Warfare Laboratory contacted the Biophysics Division of the U.S. Army Biomedical Laboratory for assistance in developing a research program to accomplish this task. Other players included the U.S. Army Natick Laboratories and Aerospace Corporation. In March 1973, the program was expanded to include protection for law enforcement personnel (NIJ, 2001).

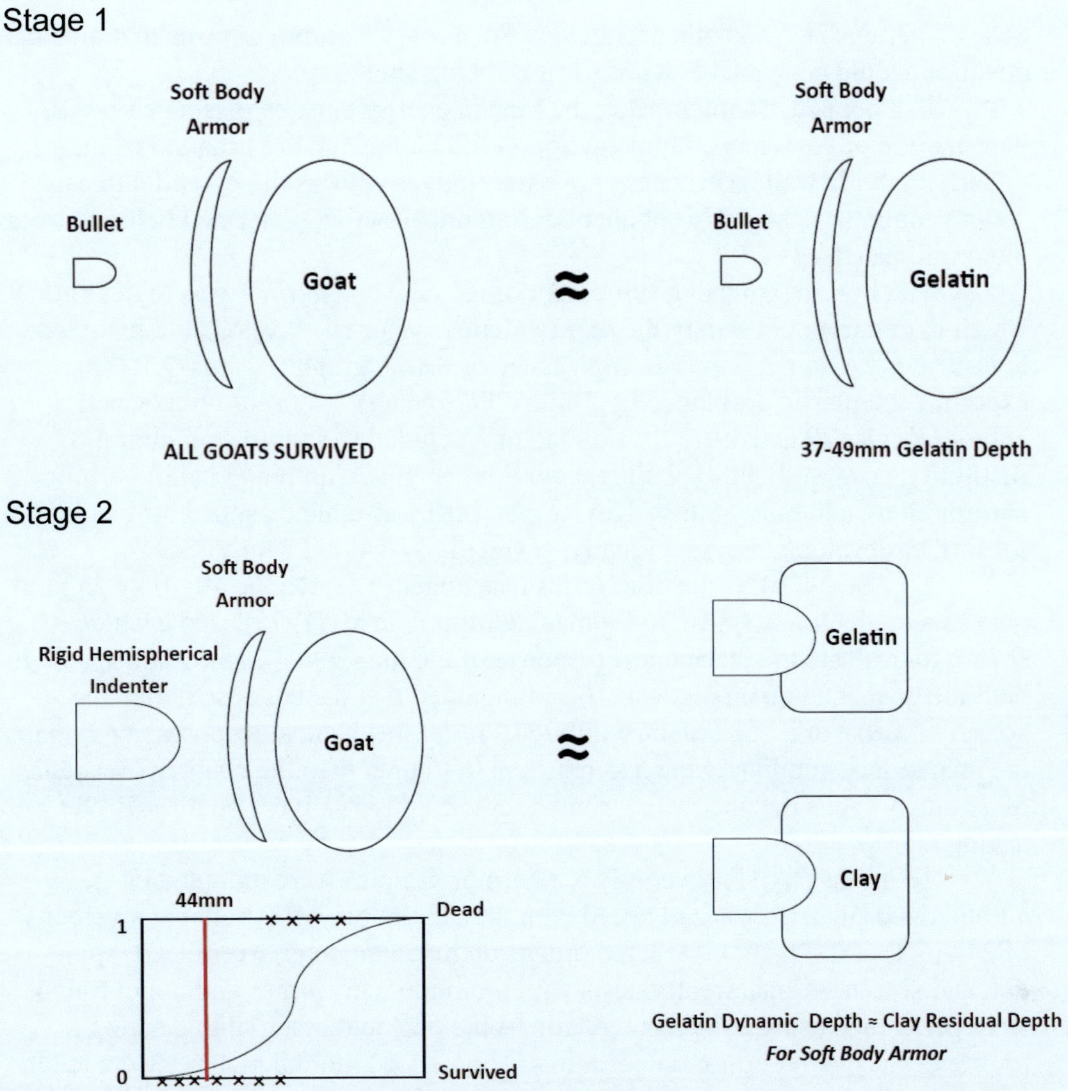

FIGURE 3-1 Overview of development of Prather clay methodology. Stage 1: depth correlation of goats with gelatin in soft body armor. Stage 2: injury assessment with rigid impactor onto goats, gelatin, and clay. SOURCE: Prather et al., 1977.

In the 1970s, 80 percent of the civilian handgun threat comprised .38-caliber and smaller handgun rounds. The primary rounds chosen for this program were the .38-cal, 158-grain lead round nose (LRN) bullet with initial velocity of 244 m/sec (800 ft/sec) and the .22-cal, 40-grain long rifle high-velocity bullet with an initial velocity of 305 m/sec (1,000 ft/sec). The garments developed under this program had to be lightweight, inconspicuous, and wearable. Additional requirements included protection from bullet penetration, blunt trauma mortality risk of less than 10 percent, and sufficient protection to allow the wearer to walk away from any shooting incident. Note that these last two requirements are not necessarily contradictory, because overall mortality risk might involve delayed hemorrhage and other sequelae. So, a major assumption was that medical attention would be available within 1 hr of being shot.

In a nonpenetrating impact, the kinetic energy must be dissipated by the deformation of the armor, fragmentation of the bullet, and deformation of the underlying body wall. This energy transfer to the body has the potential to cause serious injury or death. This nonpenetrating impact injury is termed behind-armor blunt trauma (BABT).

The first step in the development of the new body armor was to determine which materials could satisfy the requirements. Materials investigated included high-tenacity nylon, nylon felts, high-tenacity rayon, graphite yarns, XP (an experimental plastic developed by Phillips Petroleum), Monsanto fibers, and DuPont Kevlar 29 and 49. Selection factors included weight to strength ratio, flexibility, cost, availability, ballistic qualities (ballistic limit and behind-armor deformation), and tailorability. Kevlar 29 (K29) was ranked as the best candidate for further development: specifically, seven plies of 400/2 denier K29.

To assess BABT, the biophysics researchers selected the 40-50 kg angora goat as a model for a typical 70-kg man. Goldfarb et al. (1975) used a waterjet stream to evaluate the mechanical response of the lung, liver, kidney and spleen in both the goat and human organs. They concluded that the collapsed lung and spleen of a goat and of a human exhibited similar mechanical responses, and that the goat kidney and liver were less resistant to trauma than the counterpart organs of a human. Thus the goat was assumed to be a conservative model for BABT testing.

To assess BABT, seven-ply K29 armor samples were mounted on anesthetized angora goats and tested with the .38-caliber LRN bullet at a velocity of 244 m/sec (800 ft/sec). Targeted organs included the lung, liver, heart, spine, gut, and spleen. Concurrently, tests with the same body armor and bullet rounds were performed using 20 percent gelatin as the backing material to develop a profile of the behind-armor deformation. High-speed motion pictures were taken of the impacts on gelatin in order to derive the rate of deformation, as well as the deformation depth, volume, and area. These measures were then correlated to the damage seen in liver, lung, spleen, and heart injuries to goats. The average deformation recorded from the gelatin profiles was 44 mm, as shown in Figure 3-2, and 44 mm was therefore selected as the BABT standard injury reference value. It is important to note that no deaths were seen from the back-face effects in goats with the .22- or .38-cal rounds that corresponded to the 44-mm deformation in gelatin (Goldfarb et al., 1975).

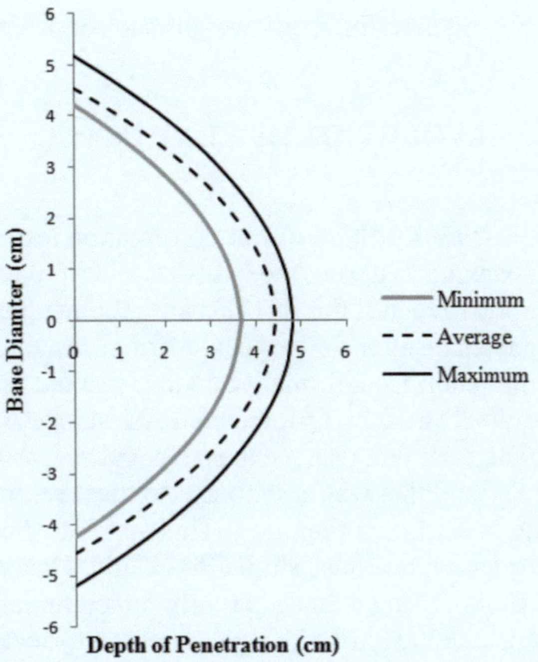

FIGURE 3-2 Blunt deformation profiles into gelatin using seven-ply K29 armor samples mounted on gelatin and tested with the .38-cal LRN bullet at 213 m/sec (800 ft/sec). Comparison experiments with anesthetized angora goats with the same armor and round showed no fatalities. The deformation envelopes shown in this figure were obtained as fits to respective indentation profiles read from the high-speed video film frame exposed at the time of maximum penetration. It was noted by the author that the penetration profiles were not necessarily parabolic and that, in some cases, the fitted curve was not as deep as the deepest part of the uneven surface it approximated. SOURCE: Prather et al., 1977.

Concurrent with the BABT testing, Clare et al.(1975) were developing blunt trauma correlation models formulated from experimental data sets obtained from tests on unarmored animals, where the physical characteristics of the impacting projectile (mass, velocity, diameter) were known. High mass (50-200 g), low-velocity impacts were involved. The first model, a four-parameter discriminant model, accomplishes its discrimination in a plane whose axes, x_1 and x_2, are defined as follows:

$$x_1 = \ln [MV^2]$$

and

$$x_2 = \ln [WD]$$

where M is projectile mass in grams, V is projectile impact velocity in meters per second, W is experimental animal body weight in kilograms, and D is projectile diameter in centimeters.

The discriminant lines establish three zones—low, medium, and high lethality. As the impact dose increases, the probability of lethality should also increase for targets having the same body weight and for projectiles of the same diameter.

EVOLUTION OF CLAY USAGE

Before the mid-1970s, behind-armor deformation testing used 20 percent ballistic gelatin as a backing material; the ballistic gelatin required the use of high-speed photography to record the BFD because the medium was elastic and returned to its original shape after the projectile firing. An alternative material was sought that would retain its deformed state to avoid the use of expensive high-speed photography. The Law Enforcement Assistance Administration requested a backing material that was inexpensive and reusable, that exhibited little material recovery, and that was easy to use so that law enforcement agencies could conduct testing at their own facilities. This material should exhibit a penetration and deformation response similar to gelatin. Data already existed for impacts on the goat thorax using a 200-g, 80-mm noncompliant hemispherical impactor with an impact velocity of 55 m/sec. These deformation-time histories were used to compare the response of various materials under similar impact conditions.

Figure 3-3 shows the response of the most promising materials tested. Although none of the materials duplicated the thoracic response, Roma Plastilina #1 clay had a deformation depth response similar to that of gelatin and was considered to be a suitable tissue simulant that was easy to use, inexpensive, and repeatable and that required no high-speed photography.[11] The clay and the ballistic gelatin were generally softer and less resistant than the goat thorax to the impactor at the testing velocity of 55 m/sec.

Blunt impactor data on goats were used to link this deformation response with fatality using a logistic regression model (Clare et al., 1975). This relationship was derived between deformation and the probability of lethality as shown in Figure 3-4. The displacement levels for goat survival and death are indicated in the figure. The figure shows that a 44-mm deformation in the goats is correlated with ~10 percent probability of death from the impactor, similar to the initial program requirement of less than or equal to 10 percent. Further, because this deformation depth in the goats is less than the depth at which any of the goats died in the impactor tests, it was selected as the injury reference value in the clay.

[11]Gelatin was recommended for consideration as a possible mid-range alternative to clay by the Phase II committee (NRC, 2010), in part because the marginal costs for high-speed photography imaging technologies are now much lower. As discussed in Chapter 4, advances in sensor technology can also provide such time-resolved information.

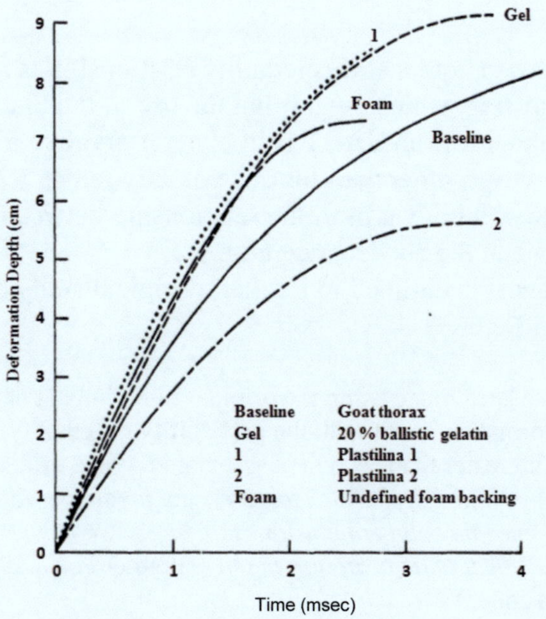

FIGURE 3-3 Deformation depth vs. time of candidate materials in a goat thorax using a blunt impactor at 55 m/sec. SOURCE: Prather et al., 1977.

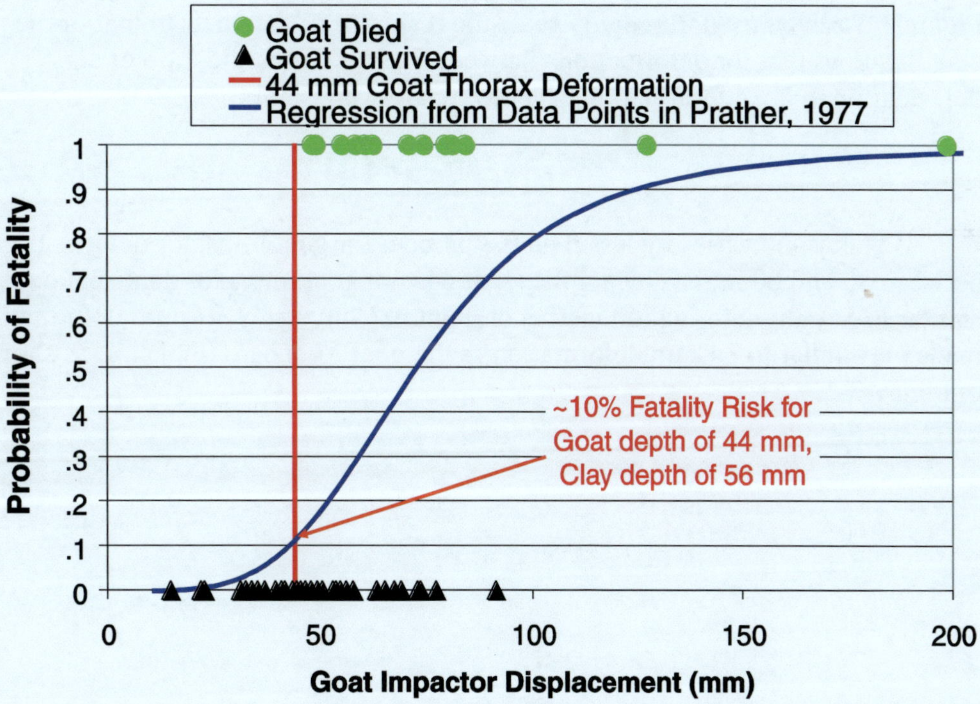

FIGURE 3-4 Logistic regression model of death vs. deformation for blunt impact into goat chests. SOURCE: Based on data from Prather et al., 1977.

It is important to note that this lethality relationship is not derived from actual body armor testing, which means that the use of the blunt impactor tests to model the injury behavior from BABT is likely conservative for this impact velocity range. However, other than for this velocity range, which is typical of handgun rounds into soft body armor, the relationship between injury and deformation response in the clay is uncertain.

A further caveat is outlined in Prather's original study on the injury regression shown in Figure 3-4:

> Attempts have been made using the original blunt impact data to correlate deformation depth with the probability of lethality. A depth of penetration greater than 50 mm is associated with a probability of lethality of approximately 15%. *However, the available data is limited and hence no solid conclusions can be drawn as yet regarding the effect of deformation depth.* (Prather et al., 1977, p. 10, emphasis added)

Thus, the original injury correlation is quite limited, even for soft body armor back-face effects. For a given impact, the clay and gelatin depth of penetration were found to be generally greater than the goat depth of penetration. Figure 3-3 can be used to roughly scale the response of the goats to that of the clay. It shows that for deformations between 3 and 60 mm, the ratio of clay deformation to goat deformation is approximately constant:

$$D_{clay}/D_{goat} = 1.28$$

This value varies by less than 0.3 percent for goat impactor deformations between 30 and 60 mm. This implies that 44-mm goat impactor deformation is similar to 56-mm deformation in clay or gelatin. Conversely, 44-mm deformation in clay is similar to 34-mm deformation in the goat for a hard impactor, as shown in Figure 3-5.

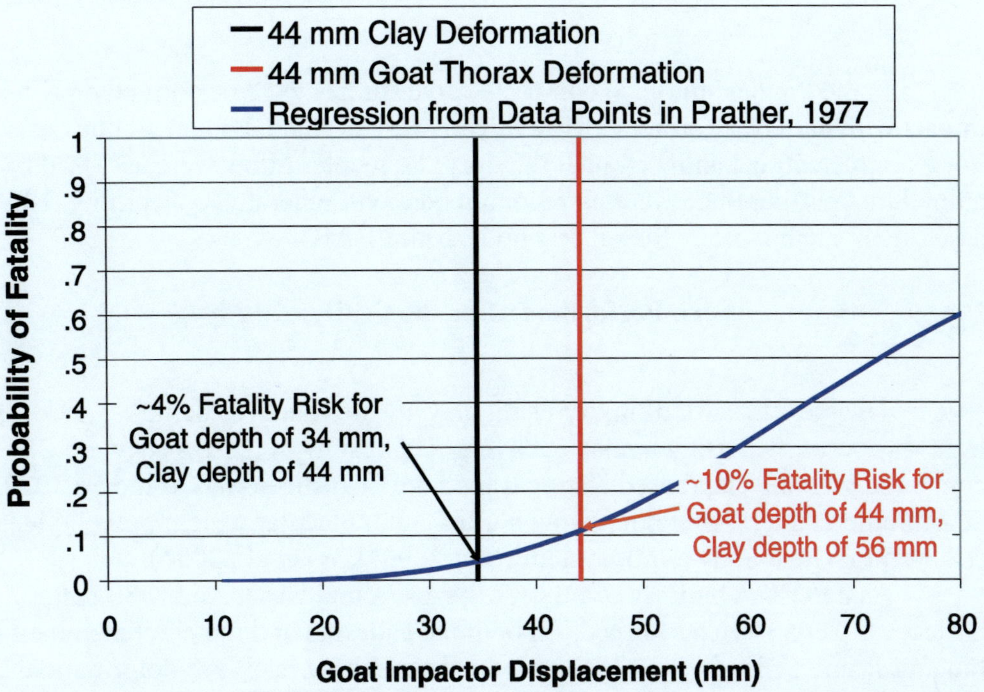

FIGURE 3-5 Logistic regression model of death vs. deformation for blunt impact into clay using deformation response into goat chests and clay. SOURCE: Based on data from Prather et al., 1977.

HIGH-ENERGY THREATS

In the late 1970s, the primary civilian ballistic threat changed to more powerful handgun rounds. Research was initiated to develop body armors to protect against these higher energy threats. The threats investigated included the .357-cal, 158-grain semiwad cutter at 396 m/sec (1,300 ft/sec) and the 9-mm, 124-grain full metal jacket at 350 m/sec (1,150 ft/sec). Investigations determined that these rounds would not penetrate 16 plies of K29. They also showed that the behind-armor deformation profiles for this soft body armor were similar to those derived under the original program. Limited goat studies demonstrated injuries similar to those incurred in the seven-ply tests. The program was terminated before sufficient tests could be conducted to verify these preliminary conclusions. Though there are no existing reports on this work, the studies did not produce deaths in the animal model in tests with actual soft armor.

Rifle Threats for Hard Body Armors

In the Prather study, no comprehensive studies were performed on rifle threats with hard body armors. Rifle threats were evaluated on an ad hoc basis. For example, .50-cal antipersonnel threats to helicopter pilots were assessed in animal models, but there were no recommendations generated concerning risk assessment methodology for generic body armor BABT.

Work Performed after the Prather Study

To assess the risk of injury using clay at rifle round velocities, a series of tests was performed using human cadavers. The results of these tests were compared with the clay-based National Institute of Justice (NIJ) Standard 1010.04 at a commercial test laboratory using a ultrahigh molecular weight polyethylene (UHMWPE) hard body armor system (NIJ, 2000; Bass et al., 2006).

As described by Bass et al.: ". . . the test round was a 7.62 M80 ball projectile. Tests were performed [on both the cadavers and the clay] at velocities ranging from ~670 m/sec to ~800 m/sec. The resulting backface deformations [Figure 3-6] showed a very low correlation of deformation with the range of velocities [Figure 3-7]. In contrast, the human cadaver, over the same velocity range, showed a wide range of injury outcomes that generally scaled well with impact force and velocity."

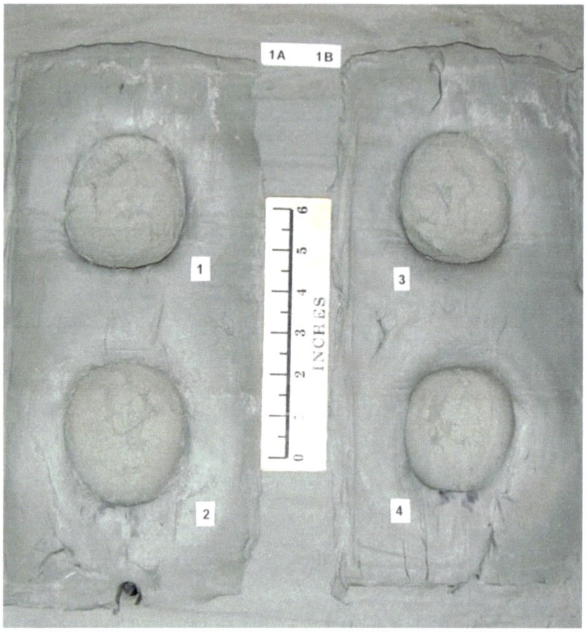

FIGURE 3-6 Clay deformation behind hard armor with rifle round threats. SOURCE: Bass, 2006.

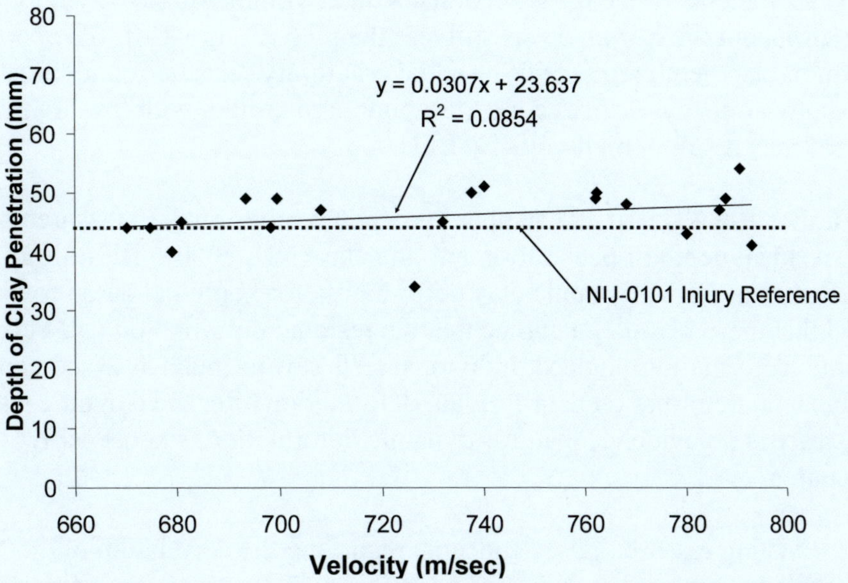

FIGURE 3-7 Variation of clay penetration depth with velocity for behind-body armor deformation (7.62-mm NATO round, UHMWPE body armor). SOURCE: Bass, 2006.

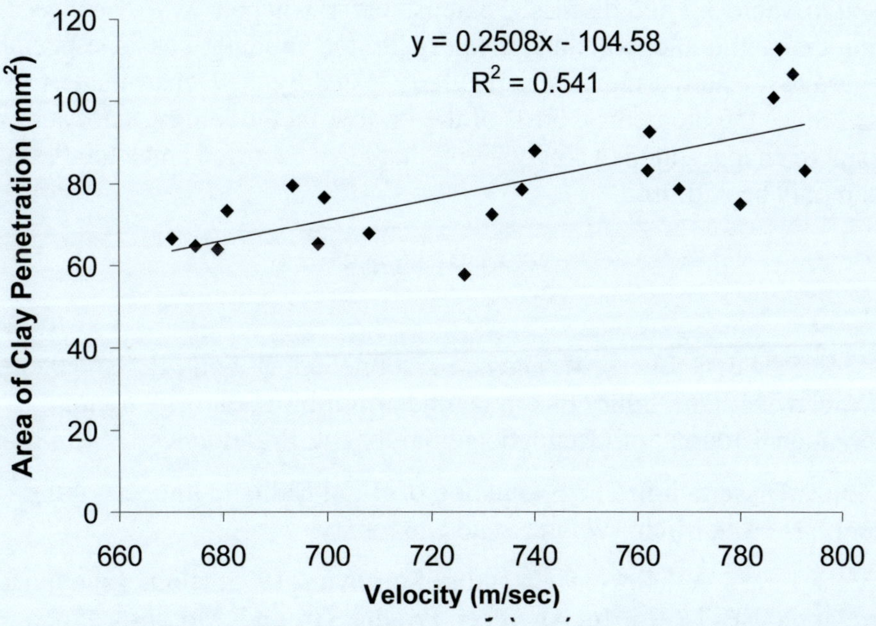

FIGURE 3-8 Variation of clay penetration area with velocity for behind-body armor deformation (7.62-mm NATO round, UHMWPE body armor). SOURCE: Bass, 2006.

Measurements of the cross sectional area or volume of the clay improved the correlation, but the R^2 value was still less than 0.6 (Figure 3-8). The poor correlation of clay depth with resulting cadaveric injury for rifle round threats with this body armor type raises concerns about hard armors with low areal density that may result in high-velocity BFDs.

Further, the NIJ 0101.04 standard procedure relies on the measurement of the static residual depth of penetration into the clay (NIJ, 2000). Bir (2000) performed an analysis of dynamic clay deformation for nonlethal baton rounds and found that there was no guarantee that the residual deformation was equal to the dynamic deformation. Indeed, individual tests saw as much as 20 percent greater dynamic deformation than residual deformation after the dynamic test. In addition, there is no evidence that this dynamic deformation is not sensitive to rate or contact area.

Finding: Existing research raises concerns regarding the correlation of the damage measured in the clay with the bodily injury at the very high rates typical of backface deformations caused by rifle rounds in hard body armor.

CURRENT STANDARD

Strengths and weaknesses of the current Prather methodology are displayed in Table 3-1 and discussed extensively in Chapter 9. Key concerns regarding the methodology include the very limited validation basis, especially with regard to the hard armor plates regularly tested by the Department of Defense. Since the biomedical basis of the Prather methodology is not current, the impact of changes in clay composition that have occurred since the Prather study can only be surmised.

REFERENCES

Bass, C., R. Salzar, S. Lucas, M. Davis, L. Donnellan, B. Folk, E. Sanderson, and S. Waclawik. 2006. Injury risk in behind armor blunt thoracic trauma. International Journal of Occupational Safety and Ergonomics 12(4):429-442.

Bir, C. 2000. Dissertation: The Evaluation of Blunt Ballistic Impacts of the Thorax. Detroit, Mich.: Wayne State University.

Clare, V., J. Lewis, A. Mickiewicz, and L. Sturdivan. 1975. Blunt trauma data correlation. EB-TR-75016. Aberdeen Proving Ground, Md.: Edgewood Arsenal.

Goldfarb, M.A., T.F Ciurej, M.A. Weinstein, L. Metker. 1975. A method for soft body armor evaluation: Medical assessment. EB-TR-74073. Aberdeen Proving Ground, Md.: Edgewood Arsenal.

NIJ (National Institute of Justice). 2000. Ballistic resistance of personal body armor. NIJ Standard 0101.04. Available online at http://www.ncjrs.gov/pdffiles1/nij/183651.pdf. Last accessed March 10, 2011.

NIJ. 2001. Selection and application guide to personal body armor. NIJ Guide 100-01. Available online at http://www.ncjrs.gov/pdffiles1/nij/189633.pdf. Last accessed March 10, 2011.

NRC (National Research Council). 2010. Phase II Report on Review of the Testing of Body Armor Materials for Use by the U.S. Army. Washington, D.C.: National Academies Press.

Prather, R., C. Swann, and C. Hawkins. 1977. Backface signatures of soft body armors and the associated trauma effects. ARCSL-TR-77055. Aberdeen Proving Ground, Md.: Edgewood Arsenal.

TABLE 3-1 Strengths and Weaknesses of the Prather Methodology

Strengths	Weaknesses
Ease of use	Clay constituents have changed considerably since original study
Immediate results	Clay variability (handling, thixotropy, temperature effects, etc)
Relatively low cost	Current methodology requires elevated clay temperatures
Large historical database of results	All variability in testing results is assumed to be design flaws in the armor
Apparent success in field for soft body armor	Method has limited medical validation for soft body armor
Apparent success in field for hard body armor	Method has no medical validation for hard body armor
	Pass/fail criterion

4

Clay and Backing Materials

This chapter discusses the role of the backing material as a recording medium, the properties and use of Roma Plastilina #1 (RP #1) modeling clay in body armor testing, and potential alternative backing materials and systems. It concludes with a road map for the body armor testing community to achieve reductions in the variability of clay as backing material for testing processes.

USE OF BACKING MATERIAL AS A RECORDING MEDIUM

As introduced in Chapters 2 and 3, the RP #1 modeling clay backing material used in armor testing has two important purposes. The first is "to simulate [some aspects of] the tissue response appropriately beneath the point of impact so that . . . ballistic data generated in laboratory tests can be correlated to the effects seen on the human body" (Prather et al., 1977, p. 7). The second purpose of the backing material is to mark the extent of backface deformation (BFD) during ballistic testing. Multiple materials are available to simulate a body; in fact, at the time it was introduced, modeling clay was recognized to only approximate tissue response, and empirical correlations were needed to develop a probability for lethality or injury. The chief advantage of modeling clay over other materials available at the time was that it better served the function of recording the BFD, because when impacted, it deforms plastically and a permanent cavity (also termed indent, impression, or crater) is developed under the point of impact. Correlations were developed between the geometry of the cavity and the probability of lethal injury. These results, however, do not predict a strain-rate dependence for the mechanical response of RP #1 and therefore increase the committee's sense that obtaining direct measurement of the mechanical response of RP #1 in the strain-rate regime, corresponding to the development of the cavity in live-fire testing, should be a high-priority task.

The role of a backing material such as RP #1 is to serve as a recording medium. That is, the backing material must exhibit plastic deformation. Ideal plasticity, illustrated in Figure 4-1a, exhibits no deformation until a critical stress is exceeded, at which point it deforms irreversibly (Fung and Tong, 2001). Thus, a backing made of such a material would serve as a "contour gauge" that would perfectly preserve the locus of points that corresponds to the maximum BFD.

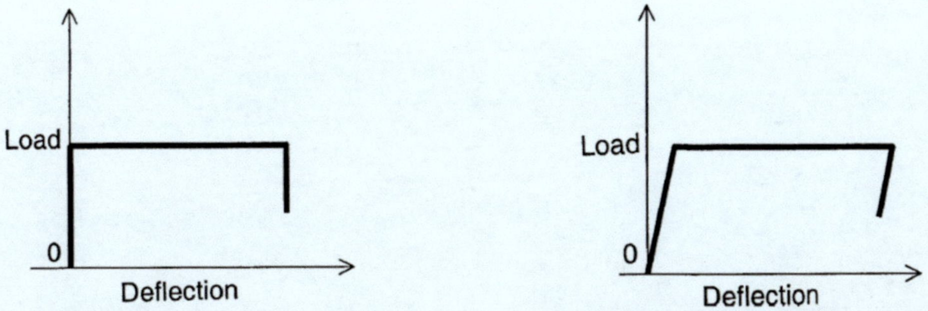

(a) Ideal plasticity **(b) Linear elastic—ideal plastic material**

FIGURE 4-1 A schematic illustration of the stress-strain curves for two idealized solids. The material corresponding to (a) exhibits ideal plasticity, in which there is no deformation until a critical stress (the yield point) is exceeded, at which point the material continues to deform at a constant rate until the stress is removed. The instant the stress falls below the critical value, such a material will stop deforming—that is, it exhibits no recovery. In contrast, linear elastic–ideal plastic material deforms elastically as the stress is applied before the plastic yield point. As before, the material deforms irreversibly when the yield point is exceeded. But in this case, upon removal of the stress, the elastic portion of the deformation is recovered as illustrated in (b). Real materials always exhibit some degree of elastic recovery. SOURCE: Fung and Tong, 2001, Copyright 2001, World Scientific Publishing Co.

A contour gauge is a device familiar to craftsmen. It consists of a linear array of steel pins held parallel by a light clamping force. A typical device is illustrated in Figure 4-2. The pins are held in place with friction and therefore do not move until the application of stress. The relative motion in this case is caused by moving onto a shaped surface, but the principal is the same as in the armor test. In the latter case, the relative motion is the same, but it is the back face of the armor that moves into the backing material. If the backing material exhibited ideal plasticity, the resultant cavity would be a record of the maximum deflection of the BFD of the armor system, but this is manifestly not the case.

As illustrated in Figure 4-1b, the deformation of real materials differs in important ways from ideal plasticity. The first distinction is that all real materials have a finite elastic modulus. The consequence of this is that the material deforms reversibly prior to the onset of yielding and will exhibit elastic recovery when the load is removed. In the context of armor testing, this means that the cavity that remains in the backing material after the armor system has been struck by the projectile will be smaller than the maximum BFD.

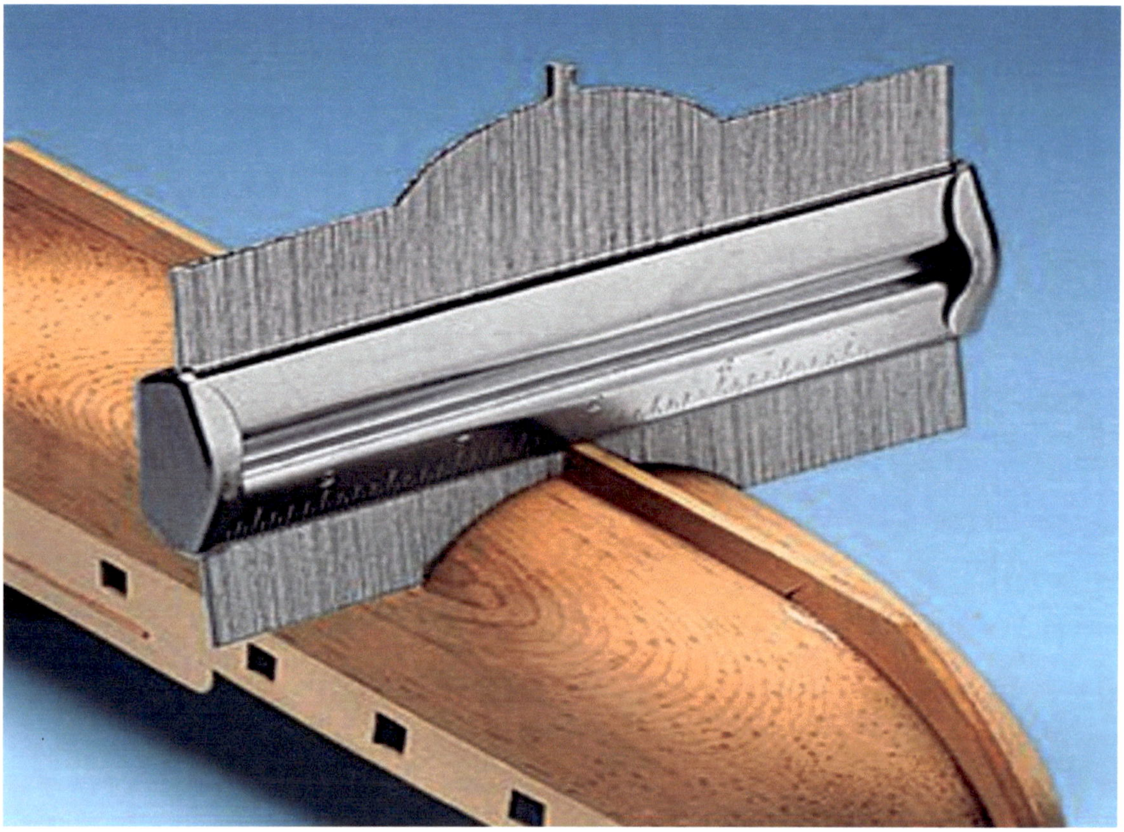

FIGURE 4-2 A contour gauge in use. The parallel metal wires slide under the force that results from pressing the tool onto (or into) a shaped surface. The wires closely approximate plastic behavior in that they do not move until the applied force exceeds the frictional force produced by the clamping force. Given the high elastic modulus for the steel wires relative to the peak stress during sliding, there is effectively no elastic recovery when the tool and the surface separate. As discussed in the text, the backing material used in ballistic testing of armor is meant to serve an analogous role in that it should deform as the back face of the armor system moves and capture a permanent record of this transient event. SOURCE: Micromark, photo of a 5 in. metal contour gauge, found at: http://www.micromark.com/5-Inch-Metal-Contour-Gauge,9335.html.

In some materials elastic recovery is so large that they do not store any memory of the event. Prather et al. (1977) noted that ballistic gelatin, for example, is a highly elastic material and exhibits nearly total recovery. Constraining his choices to low-cost readily available materials, Prather et al. identified an oil-based modeling clay, RP #1, as a material that exhibited sufficient plasticity to evidence post-test cavities with geometries that correlated to lethality probabilities (Prather et al., 1977). It must be noted that in a presentation to the committee, Mr. Prather indicated that the study results should be considered provisional (i.e., not final or fully worked out or agreed upon at the time). He also noted that RP #1 was "convenient," and this attribute seems to have dominated as it rapidly became widely used.[12]

As time passed and a wide range of investigators used RP #1, two sources of confusion emerged. First, many assumed that it was a simulant when it was not. Second, the Prather report's description of RP #1 has been misunderstood. It stated that RP #1 was "a highly plastic material which undergoes viscous flow when deformed and exhibits little recovery, thus providing a readily available cavity formed during impact from which measurements can be taken."

The qualitative assertion that RP #1 exhibits little recovery has been interpreted to mean that the level of elastic recovery is small enough to be safely neglected. This led to the assumption that the shape of the cavity is a record to the BFD. It is not. As early as 1974, measurements of elastic springback were made using a modified Charpy impact tester (Aerospace Corporation, 1974). (A Charpy tester consists of a pendulum fitted with a weighted hammer that is allowed to swing into the sample material from a prescribed height, i.e., a given potential energy.) The difference in distance between the maximum point of the penetrator during its swing was compared to the size of the cavity in the RP #1, with this difference being the measure of displacement of the modeling clay during unloading. The results indicate that elastic recovery is in excess of 40 per cent and in some cases more than 70 per cent. That is, the differences are very large. Results from the Aerospace Corporation final report are shown in Table 4-1.

TABLE 4-1 Elastic Recovery in Modified Charpy Testing of Oil-Based Modeling Clay

# of Plies of Kevlar-29	Peak Load, N	Max. Depth of Indentor (mm)	Depth of Cavity (mm)	Difference (apparent recovery) (mm)	Difference (apparent recovery)	Expected* Elastic Recovery (mm)
3	4671	37.34	18.54	18.8	50%	12.45
3	5449	39.37	10.16	29.21	74%	18.8
5	10453	41.66	24.13	17.53	42%	20.07
5	10787	47.24	24.38	22.86	48%	20.83

*Calculated by Aerospace using "punch formula" $d_e=[(1-\upsilon)P_i a]/G$.

SOURCE: Committee-generated, derived from data in Table II, p. A38 (Aerospace, 1974).

[12] Russell Prather, Survice Engineering Co., "Prather Study Results" presentation to the committee on August 11, 2010.

Results of drop tests conducted by H.P. White Laboratory, Inc., also were consistent with significant elastic recovery.[13] Furthermore, low-rate indentation experiments on plasticine, which is the same class of material as RP #1, indicate that recovery would be expected at high rates (Huang et al., 2002). Thus, the cavity in the RP #1 is not a record of the BFD. It is, as originally stated by Prather, "a readily available cavity formed during impact from which measurements can be taken" and to which correlations can be made (Prather et al., 1977). This is a critical point to recognize when considering either a replacement or the potential for improving the backing material performance by adjusting the formulation to produce a "ballistic grade."

Another very important point is that the relative degree of elastic and plastic deformation will be expected to vary as a function of strain rate. That is, the material must be characterized under conditions that are relevant to those under which tests will be performed. To the knowledge of the committee this has never been done.

Although the properties of RP #1 have not been reported as a function of the strain rate, those of other candidate backing materials have been. For instance, the compressive properties of 20 per cent ballistic gelatin measured at 10ºC using a modified split Hopkinson bar as a function of strain rate over a range comparable to the range of interest (hundreds to thousands of reciprocal seconds) (Salisbury and Cronin, 2009). The compliance is observed to change by a substantial amount, with the gel perhaps 10 times stiffer at the high strain rate. Also showing the dependence on strain rate is a study that compared ballistic gelatin with physically associated styrene-isoprene triblock copolymer gels (Juliano et al., 2006).

In sum, RP #1 was selected as a material of convenience rather than on the basis of well-determined engineering properties. It serves as a recording medium rather than a body simulant. The cavity that results from live-fire ballistic testing is related to the BFD of the armor, but it is not a true record of the maximum deflection. It remains unknown, therefore, how the dimensions of the cavity relate to the true BFD (and how such a relationship depends on the rate at which the cavity is formed).

---

[13]Don Dunn, H.P. White Laboratory, Inc., "Commercial Body Armor Testing Perspectives," presentation to the committee, August 9, 2010.

CHARACTERISTICS AND PROPERTIES OF RP #1

Behavior in Testing

Column-Drop Test

As standards have evolved, column-drop tests have been introduced to ensure that the modeling clay used for each test has well-defined behavior. The drop test consists of dropping a cylindrical steel mass with a hemispherical cap (44.5 mm in diameter) of defined mass (1 kg) from a height of 2 m. The mass is then removed, and penetration is quantified by measuring the distance between the original flat clay surface and the deepest point in the indent. As the deepest point lies on a highly regular hemisphere, it can be readily and reliably located by an operator using a digital caliper. The Phase I committee letter report (NRC, 2009) found that a digital caliper is adequate for this measurement because of the well-defined planar reference, the smooth and shallow indentation, and the ease of locating the center of the indentation.[14, 15]

To assess the appropriate methodology for measuring the dimensions of deformed RP #1, it is useful to review the general characteristics of prior observations of its deformation, and this is best done by reviewing the results of so-called column-drop tests.

The introduction of the column drop test is another consequence of widespread adoption of Prather's originally preliminary recommendation. "Roma Plastilina #1" is a trade name and as such does not embody a set of technical specifications. This was not an issue at the time as the recommendation was not expected to become a standard. However, it has been confirmed that the formulation of RP #1 has evolved over time. In part the evolution was in response to the primary customer base for clay (artists) making performance requests and in part it was due to the shifting availability of raw materials from different suppliers. While this may be commonplace for commercial products, it has had profound effects on the use of RP #1 as a backing material for live-fire testing of ceramic body armor. To quote Aberdeen Test Center (ATC) personnel, "The mechanical properties of Roma Plastilina #1 are dramatically different from the clay that was used in 1977."[16]

---

[14]Finding 3 of the Phase 1 letter report stated that "the digital caliper is adequate for measurements of displacements created in clay by the column-drop performance test: there is a well-defined reference plane, and one can visually see the surface of the clay, given that the depression is relatively shallow (approximately 22 to 28 mm) and fairly smooth" (NRC, 2009).

[15]Finding 4 of the Phase 1 letter report stated that "The column-drop performance test (including the testing protocols, facilities, and instrumentation) is a valid method for assessing the part-to-part consistency of clay boxes used in body armor testing" (NRC, 2009).

[16]Scott Walton and Shane Esola, Aberdeen Test Center, "ATC Perspective on Clay used for Body Armor Testing," presentation to the Body Armor Testing Phase II committee, March 10, 2010.

One consequence of shifting composition of the clay has been that a need was recognized to find a way to calibrate the modeling clay that was compatible with use on a ballistic firing range. This led to the development of the so-called column-drop test. Although it is not possible to trace the history of the test in published documents, it appears to have been developed in response to testers noting that newer versions of RP #1 were stiffer than older versions. Given that fact and the fact that oil-based modeling clay is readily softened by heating led to the use of ovens to warm the clay so that it behaved similarly to the older (de facto reference) formulation. The column-drop test developed to assess the similarity of clay behaviors.

Several variants of the drop test are currently employed. At ATC, the drop test consists of dropping a cylindrical steel mass with a hemispherical cap (44.5 mm in diameter) of defined mass (1 kg) from a height of 2 m onto RP #1 contained in a clay box. The mass is then removed, and penetration is defined by measuring the distance between the original flat clay surface and the deepest point in the indent. As the deepest point is determined by a highly regular hemisphere, it can be readily and reliably located by an operator using a digital caliper, and the depth at this point can be measured by any of a number of techniques. As noted earlier, the digital caliper is adequate for this because of the well-defined planar reference, the smooth shallow indentation, and the ease of locating the center of the indention.

The three photographs in Figure 4-3 illustrate the drop test. The cavity resulting from the drop test is of a volume and shape that is qualitatively similar to the cavity from an armor test. Both craters are tens of millimeters in depth and width, and both are smooth, regular shapes. However, the deformation rate experienced by the clay is markedly different. As demonstrated to the committee, the weight impacts the surface of the clay slightly faster than 6 m/sec, whereas the back face of the armor system moves at a velocity nearly an order of magnitude greater, just over 50 m/sec.[17]

[17]Ibid.

FIGURE 4-3 Column-drop test as performed at ATC. The overall setup is shown in (A). The weight, shown up close in (C), is held in place by an electromagnet at the top of the antiyaw tube. Upon release the weight accelerates under gravity and is implanted into the surface of the modeling clay. The weight is manually removed and the depth of the cavity (two are visible) is measured. Also visible are two thermometer probes used to track the temperature of the modeling clay. The results of a typical drop are shown in (B). Notable is the significant yaw (inclination with respect to the normal of the clay surface). SOURCE: ATC, 2008.

Nonetheless, the column-drop test is what is used to determine if the clay box is what is termed "within calibration" and therefore can be used to test the hard armor plates. The criterion for test/no test is that the cavity resulting in this test is 25 ± 3 mm (ATC, 2008).

Drop test results reviewed by the committee were all obtained using the standard clay box on which the clay appliqué is mounted for the live-fire testing of hard armor, as described in Chapter 2. Four characteristics typify the results:

1. Drop test results exhibit scatter even under nominally identical conditions;
2. The flow of RP #1 in response to load (rheology) depends on thermal history or heating;
3. The rheology of RP #1 also depends on prior working (shear history); and

4. Drop test results depend weakly on location in the clay box.

The effects of temperature have been systematically studied by both Army ATC personnel and an independent lab. The Army study employed the standard drop test. A clay box was thermally equilibrated at 40°C (104°F) and subjected to serial drop tests over time as the clay box was allowed to cool, approaching room temperature. Although temperature measurements were taken, they were not reported; instead, the variation with time was presented. These data (Figure 4-4) reveal two very important characteristics of the modeling clay with respect to this application: (1) there is substantial lot-to-lot variation (under nominally identical conditions different boxes yielded penetrations that varied systematically from 1 to 2 mm) and (2) the drift with time is significant compared to the allowed range for "calibration," that is, ±3 mm. Over the 45 min of the test the average penetration in all cases was reduced by more than 4 mm. One implication of the latter characteristic is that the majority of clay boxes that are within calibration when removed from the oven can be predicted to fall out of calibration during the 45-min time window.

A second result from the same study is given in Figure 4-5. In this figure drop tests results using weights of different geometries are presented. The information implies that there is no particular advantage of any one shape. The three different geometries that are tested reveal equally useful information. However, the results do make startlingly evident the magnitude of the scatter associated with drop test results; it is disturbingly large compared to the allowed calibration range.

A qualitatively similar degree of scatter was observed in a study of drop test penetration as a function of radial position measured from the center of the box (see Figure 4-6) (Esola et al., 2010). In this study, there was a large box-to-box variation in drop-test penetration and substantial scatter under nominally identical conditions. Significantly, there was not a systematic trend with respect to radial distance from the center. In most boxes there was only a weak variation with distance, but there were some tests in which the edges were significantly less deeply penetrated and some in which the penetration was deeper near the extremities (see Figure 4-7). The results of this study can be summarized as follows:

- There was only weak correlation between radial position and average penetration depth;
- Variability of the average penetration depth under nominally identical conditions was significant;
- Variance increased as a function of distance from the center; and
- Well-used clay boxes exhibited behavior different from "dormant" boxes or from new boxes until they had been used for a while; however, the time constants for changes in behavior were not determined.

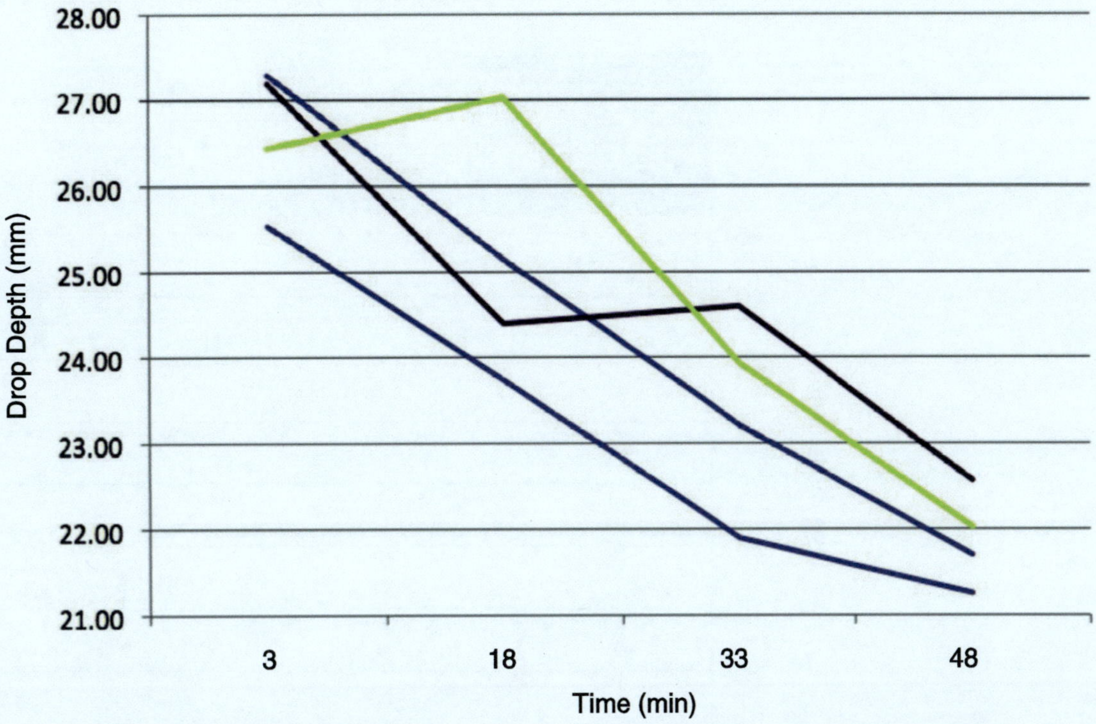

FIGURE 4-4 The results of drop tests on clay boxes allowed to naturally cool from 40°C to normal room temperature (roughly 23°C). Drops were made in a randomized 4 × 4 grid. The surface was not repaired between drops, and drops were intentionally separated to minimize potential interference. Four separate clay boxes were used, each represented by a different line on the graph. Each point on the graph is the average of two drops. Initial pairs of drops were made 3 min after removal from the oven, and subsequent data were taken in 15-min intervals. Although there is scatter, over the range investigated the slopes of the curves are all consistent with a decrease in average cavity depth of 1.5 mm every 15 min. The difference in the absolute values of the cavities resulting from the drop tests is attributed to lot-to-lot variation in the modeling clay and differing lengths of time in service. SOURCE: Scott Walton and Shane Esola, Aberdeen Test Center, "ATC Perspective on Clay used for Body Armor Testing," presentation to the committee, March 10, 2010.

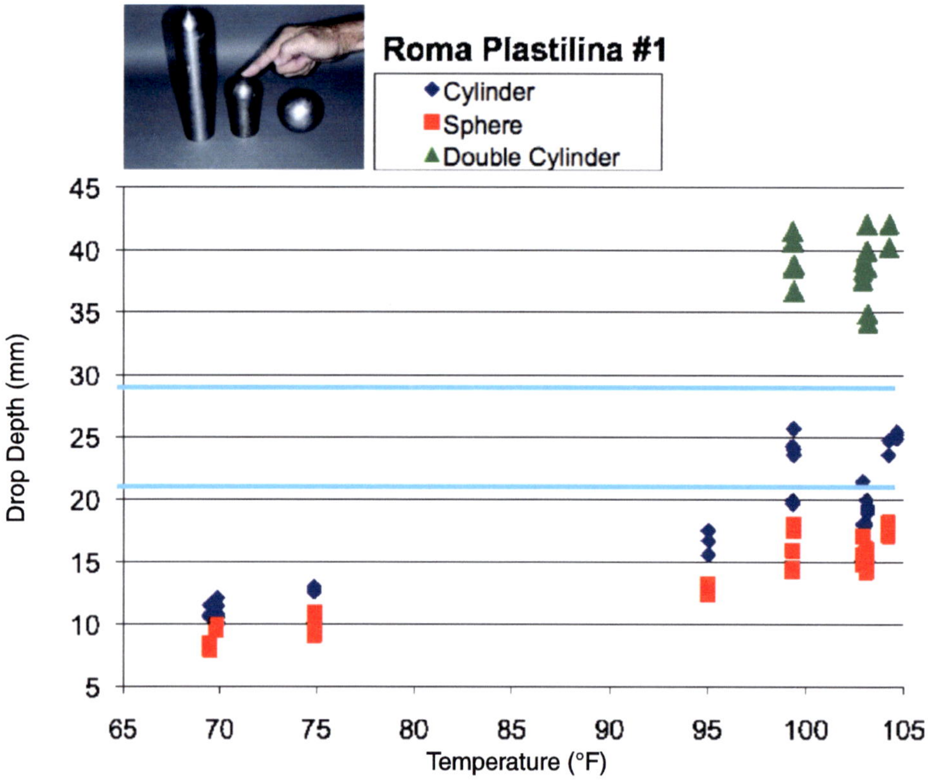

FIGURE 4-5 Drop test results using the standard Army right-circular cylinder with a solid hemispherical cap (44.5 mm [1.75 in.] in diameter with a mass of 1 kg [2.2 lb]), a similar non-standard double-length cylinder of the same diameter with the same type of hemispherical cap, and sphere with the diameter specified in the National Institute of Justice Standard (NIJ), 63.5 mm (2.5 in.) in diameter. The two horizontal blue lines represent the upper and lower limit of the calibration range. The most striking feature of the results is the observed scatter – which appears similar for all three classes of weight geometry. Under nominally identical conditions, the scatter is a substantial fraction of the allowable range! This is particularly so when the temperature is in the range of that seen in typical practice. SOURCE: Scott Walton and Shane Esola, Aberdeen Test Center, "ATC Perspective on Clay used for Body Armor Testing," presentation to the committee, March 10, 2010.

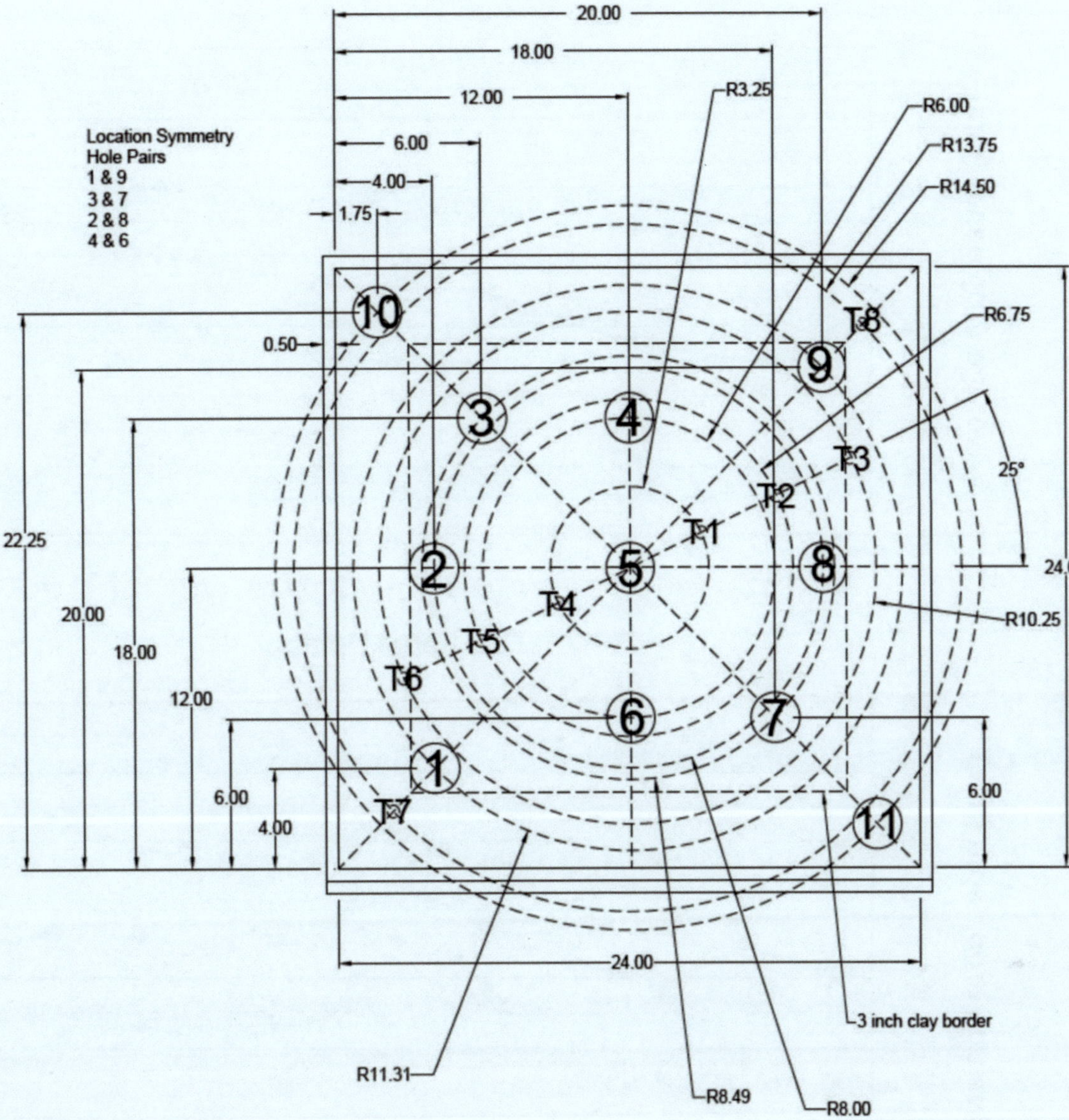

FIGURE 4-6 Spatial pattern used in a series of experiments to determine the effect of position on the size of the cavity produced during a drop test. The positions of the drops are given by the circled numbers (all dimensions are in inches). Thermocouples were placed at the locations given by numbers preceded by the letter T. (Temperature data were tabulated but not used in the analysis.) SOURCE: Esola et al., 2010.

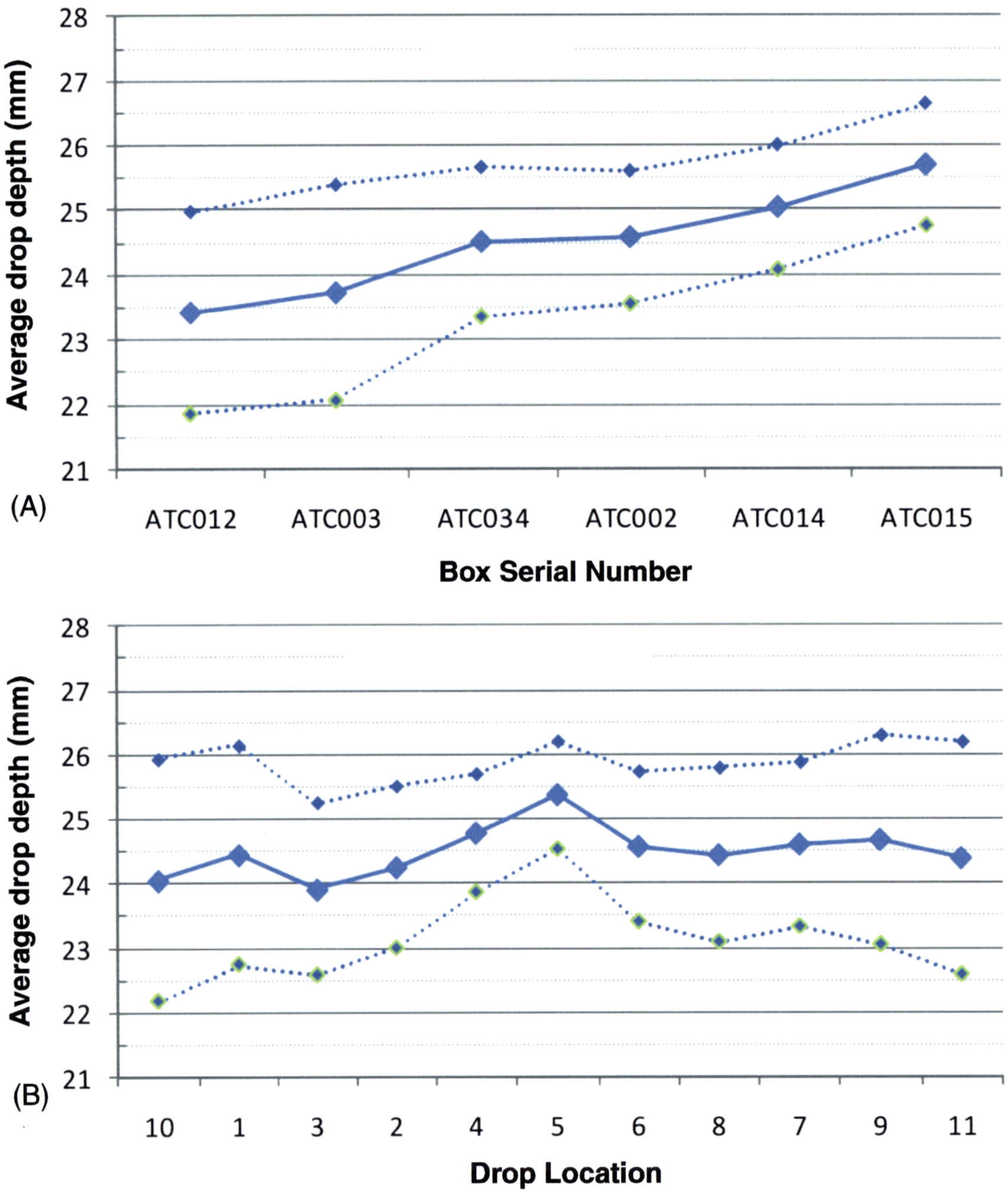

FIGURE 4-7 Illustrative results from a study on the effect of radial position on depth of penetration during a drop calibration test. The data presented in (A) represent the average and standard deviation for drops in different serial-numbered boxes. The ordering of the data is arbitrary—it is not intended to show a trend but rather the magnitude of the difference that can be expected from one clay box to another (a difference of 2 mm could be associated with a box switch). The scatter is very large: The upper and lower curves represent not the total observed range but the ± 1 standard deviation (about 68 per cent of observations fall within this interval when a data set is normal). The data presented in (B) show not only that the average penetration is shallower the farther from the center (the indexes are those from Figure 4-6), but also that the scatter was substantially greater

(some boxes saw deeper penetrations farther from the center). The increase in variance was the largest effect, with only two of the six clay boxes exhibiting a moderate linear correlation to distance from center where drop depth decreased with radial distance; all other boxes exhibited weak or no correlation. SOURCE: Esola et al., 2010.

Similar results are found in a separate study conducted in an independent laboratory (Roberson et al., 2010). Some of the data in that study were obtained using a capillary rheometer. Such data are notoriously hard to interpret for systems such as this (a highly filled wax) owing to effects such as plug flow, wall slip, and flow instabilities associated with the axial migration of the low viscosity binder, and it does not appear that appropriate precautions were taken (Suwardie et al., 1998). Nonetheless, the study reported important results using calibration-drop tests (performed to the NIJ standard rather than the Army standard). In the NIJ-standard drop test, the weight is a sphere with a diameter of 63.5 mm (2.5 in.). The standards for penetration are 19 ± 3 mm for each drop, 19 ± 2 mm for the average of the prescribed five drops).

Roberson et al. (2010) reported similar results both with and without working the surface of the clay box through manual use of a rubber mallet to deliver repeated blows across a well-defined area. The calibration drop tests were performed to NIJ standards, which employ a sphere similar to that illustrated in the inset of Figure 4-5. Like the independent tests at the ATC, Roberson et al. observed substantial box-to-box variation that depends on shear history and temperature. There was significant scattering of penetration depths on the order of the range of allowed values ±2 mm. Under all conditions, the same temperature for the same box led to deeper penetrations after the modeling clay had been worked. The data from (Roberson et al., 2010) seems to support—at least in the case of the calibration drop test—the long-standing practice of correcting for box-to-box variation or differences in work history by changing temperature.

Another result in Roberson et al. is that the softening associated with mechanical working is reversible. That is, RP #1 is observed to stiffen when allowed to rest undisturbed for long periods of time. The results trace the behavior of oil-based modeling clay immediately after the requisite deformation to form the as-received 2-lb blocks into a clay box (soft); then after ageing undisturbed for 6 months (stiff); and, finally, after working the surface of the clay box (once again soft).

In sum, under all drop-testing conditions reported, RP #1 shows highly variable penetrations under nominally identical conditions. Typically, repeated drop tests show results covering a large fraction of the allowable (wide) range. This shows clearly that RP #1 is an inherently imprecise recording medium.

Finding: Both the spatial and the temporal variations of the modeling clay are significant. The response of the clay depends on temperature, shear history, and, perhaps, proximity to the edge of the box in which it is contained. Column drop experiments can be conducted to determine the variation due to box geometry and location of the drop in relation to the side of the box. Analysis of these results should lead to uniform and evidence-based procedures for use by the body armor community.

High-Velocity Tests

Although there is a paucity of high-velocity data on strain rates approaching those associated with the motion of the back face of an armor system, there are two sets of results that can be considered.

The first set is an extension of the study by Roberson et al. (2010) on the effects of ageing, heating, and working. In that study, soft armor rather than ceramic hard armor was tested, but a lower threat round was used so that penetration depths were comparable to what would be expected in a test of hard armor against a higher threat round. The work systematically compared the depth of penetration during live-fire testing of soft armor in worked and unworked clay. The interesting result is that working the clay surface typically resulted in penetration that was 4 mm deeper. This is very close to the results of the drop test after the modeling clay had been worked even though the penetrations in the ballistic tests were three to four times deeper. However, additional empirical data is needed as the existing data do not reveal an obvious scaling relationship.

The second set of high-speed data comes from a nonlethal weapons development program (Weber, 2000). In this study, .32-cal (7.62-mm) rubber sting balls were shot from a high-velocity gas gun over a wide range of velocities into a clay box that was within calibration using the NIJ standard.[18] The box was a square of the same edge length as a clay box used in the Army's test operating procedure (TOP), but it was roughly one-third as deep—that is, 50 mm (2 in.) rather than 140 mm (5.5 in.). Figure 4-8 shows the configuration that was tested and the indentations from calibration drop tests and sting ball experiments. The results, presented in Figure 4-9, do not reveal any reason to expect a qualitative change in deformation behavior of modeling clay up to strain rates that approach those experienced in armor testing. Neither do they suggest that variance is lower at high velocities.

One point that is stressed by Weber is that the modeling clay is a relatively insensitive medium—that is, small differences in penetration depth correspond to large differences in sting ball velocity (and therefore energy). This means that accurate measurement of penetration depth is essential and that scatter in the data reflects a significant error in apparent projectile velocity.

Finding: Although the data are suggestive, the scaling relationship between drop tests and ballistic tests remains mostly unexplored.

[18]The sting ball hits the clay at a velocity three times that typical of the armor backface, but the diameter of a sting ball crater is less than a third as large as a typical BFD.

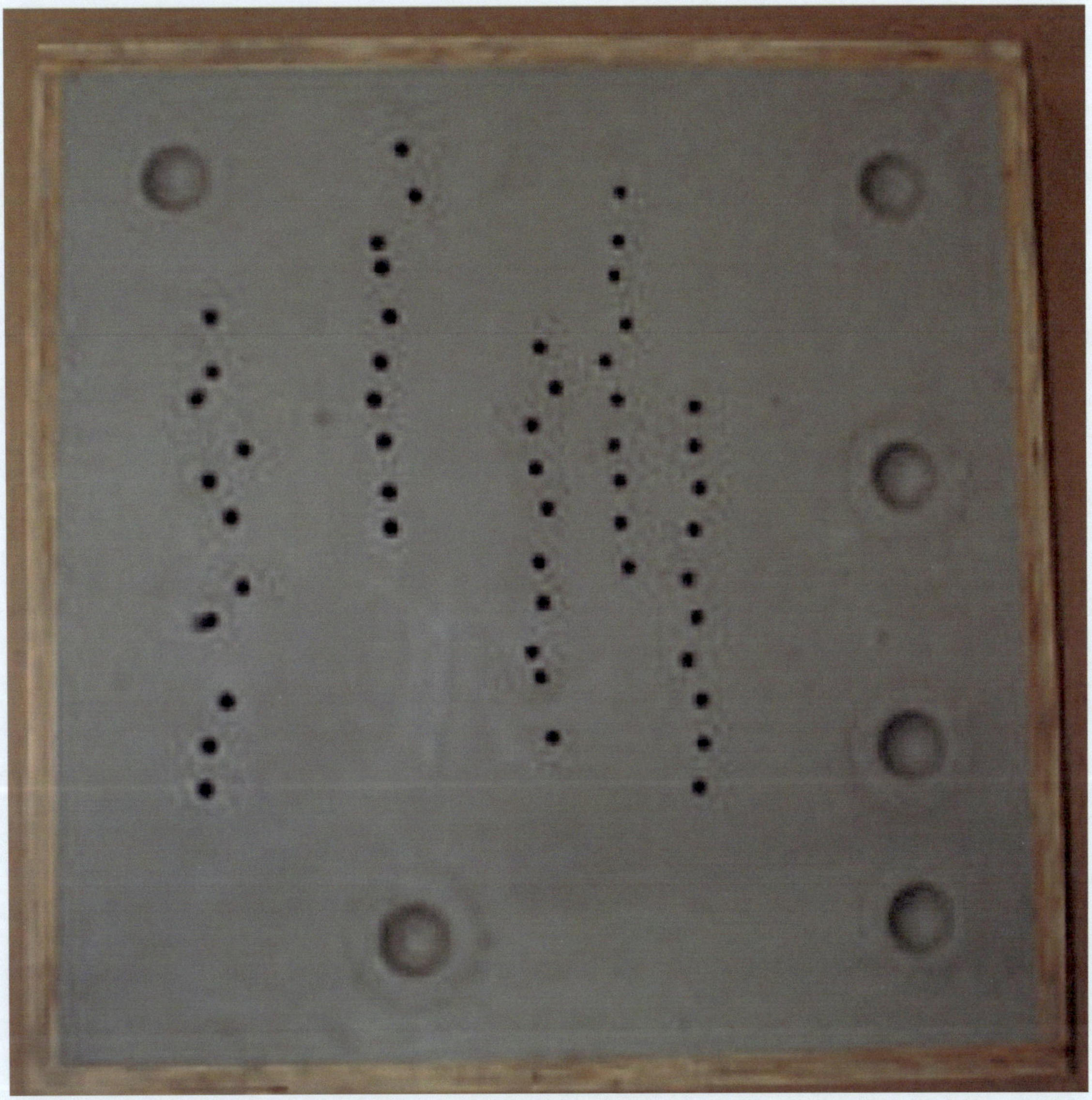

FIGURE 4-8 A clay box used for .32-cal rubber sting ball testing. The large diameter impressions represent NIJ standard calibration drops. SOURCE: Weber, 2000.

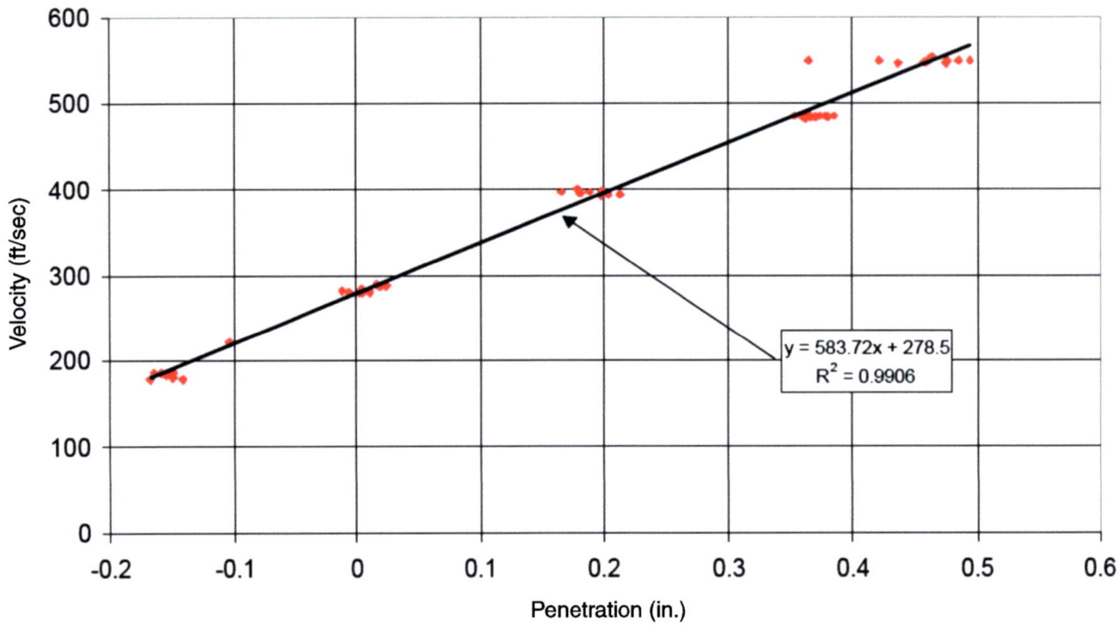

FIGURE 4-9 Results of sting ball experiments with projectiles ranging from 55 m/sec (180 ft/sec) to 168 m/sec (550 ft/sec). Penetration depths are measured from the aft end of the sting ball (they were not removed before measurements were taken). Converting to the tip of the sting ball (i.e., adding 0.32 in. to the recorded value and converting to mm) gives roughly 21 mm for the deepest penetration. Two observations may be made. The first is that there is no break in the curve. Over the wide range tested there is a linear relationship between the depth of penetration and the speed of the projectile at impact. This gives some measure of confidence in the drop test as a measure of consistency. The second observation is that a significant amount of scatter appears in the data; this is more true when penetration depths approach those that would be seen in ballistic testing. SOURCE: Daniel Weber, Edgewood Chemical Biological Center, "Measuring Impact Velocities of Non-Lethal Weapons," presentation to NDIA - Non-Lethal Defense IV, March 21-22, 2000.

Taken as a whole, the work discussed above indicates that when considering the methodology, preparing and working the modeling clay, or interpreting the results of ballistic tests, the following points must be taken into account:

- As a recording medium RP #1 is inherently highly variable, giving noisy results;
- The response of the oil-based modeling clay RP #1 is dependent on both shear history (or working) and thermal history;
- There is a paucity of data at strain rates approaching those experienced in real armor test conditions; and

- Using thermal conditioning to correct for differences in shear history and ageing appears justified based on drop-test results.

Furthermore, when considering a road map for the development of an improved ballistics-grade clay or for the selection of an alternate backing material, the degree to which the first three points can be addressed must be rigorously evaluated.

Influence of Structure on Properties of Oil-Base Modeling Clay

Thixotropy as a Phenomenon

Understanding the behavior of an apparently complex material such as RP #1 or designing from scratch a backing material that offers desirable, predictable, and controllable properties will benefit from an understanding of microstructure. That is, if the phases present, their spatial distribution, and their relative amounts are known, the system has a much better chance of serving the intended function. Mechanical properties can be related to the structure, allowing a meaningful prediction of strain rate, shear history, and thermal history. Furthermore, properties such as thermal conductivity, diffusivity, and thermal expansion all can be deduced or modeled using appropriate mixing rules once the phase distribution has been well determined.

Although the specific formulation of RP #1 must, understandably, remain proprietary for commercial reasons, the general composition is understood. Combined with information in the open literature, this allows the relationship between internal structure (or microstructure) and properties to be anticipated to a degree that is helpful to understanding how they might be improved. The two properties of most interest are the shear- and temperature-dependence of the RP #1.

The RP #1 formulation is known to include several multicomponent organic phase(s), a kaolinite filler, and two other inorganic fillers, sulfur and zinc stearate; this places RP #1 in a family of well-known systems. For example, a published formulation in a patent assigned to 3M Corporation specifies 40 to 60 percent inorganics (by weight) in a plasticized wax medium. The inorganic phase contains 15-30 per cent powdered sulfur, 40-70 per cent clay, and 15-30 per cent zinc stearate.

Such multiphase materials often exhibit a finite yield strength and thixotropy; that is, the material does not flow until it experiences a stress in excess of a critical value (the yield stress), and the material properties following the initiation of flow depend on time (thixotropy). Before they yield, the materials appear to respond as viscoelastic solids. The probable cause of shear sensitivity and thixotropy is shear-induced modification of the microstructure, which changes the modulus, yield stress, and viscosity among other things. Accordingly, these materials can exhibit aging, shear conditioning, and even "avalanche" (runaway) behavior (Bonn and Denn, 2009).

Thixotropy can arise for a number of reasons and in a wide variety of materials. One classic example of such a system is a dilute suspension of clay minerals in a fluid such as water, but this is an illustrative system only. Indeed, quite varied and distinct systems exhibit thixotropy.

As outlined in the comprehensive review by Barnes (1997, p. 2) the term thixotropy was coined in the 1920s:

> In 1923, Schalek and Szegvari found that aqueous iron oxide gels have the remarkable property of becoming completely liquid through gentle shaking alone, to such an extent that the liquified gel is hardly distinguishable from the original sol. These sols were liquified by shaking [and] solidified again after a period of time . . . . The change of state process could be repeated a number of times without any visible change in the system . . . . The term thixotropy was then coined by Peterfi in 1927 . . . in the first paper that properly described the phenomenon. The work combines the Greek words thixis (stirring or shaking) and trepo (turning or changing). (Barnes, 1997, p. 2)

Solid particles suspended in a liquid medium comprise one classic system that exhibits thixotropy when there is a driving force for the establishment of a space-filling agglomerate (a gel) due to net attractive interactions between particles. The attractive interaction has to be weak enough to be readily broken through the application of mechanical energy. If it is very weak the gel can be readily and reversibly reduced to a liquid state by heating as the thermal energy of the particles becomes greater than the interparticle bond strength. Such a system is perhaps the most intuitive class of thixotropic materials (see Figures 4-10 and 4-11).

However, many complex fluids exhibit thixotropic behavior—or, more generally, rheology that is dependent on thermal and shear history. All that is necessary is for a weakly bonded network of a relatively rigid dispersed phase to exist in a matrix of relatively flowable matter. Examples include many foods and (of direct relevance to the use of RP #1 for measuring BFD) ointments and cosmetics (Borwankar, 1992; Barnes, 1997; and Abu-Jdayil, 2003). In these all-organic systems, a three-dimensional network forms of a phase that is relatively rigid because, for example, it is crystalline or the intramolecular bonding is stronger due to polarity.

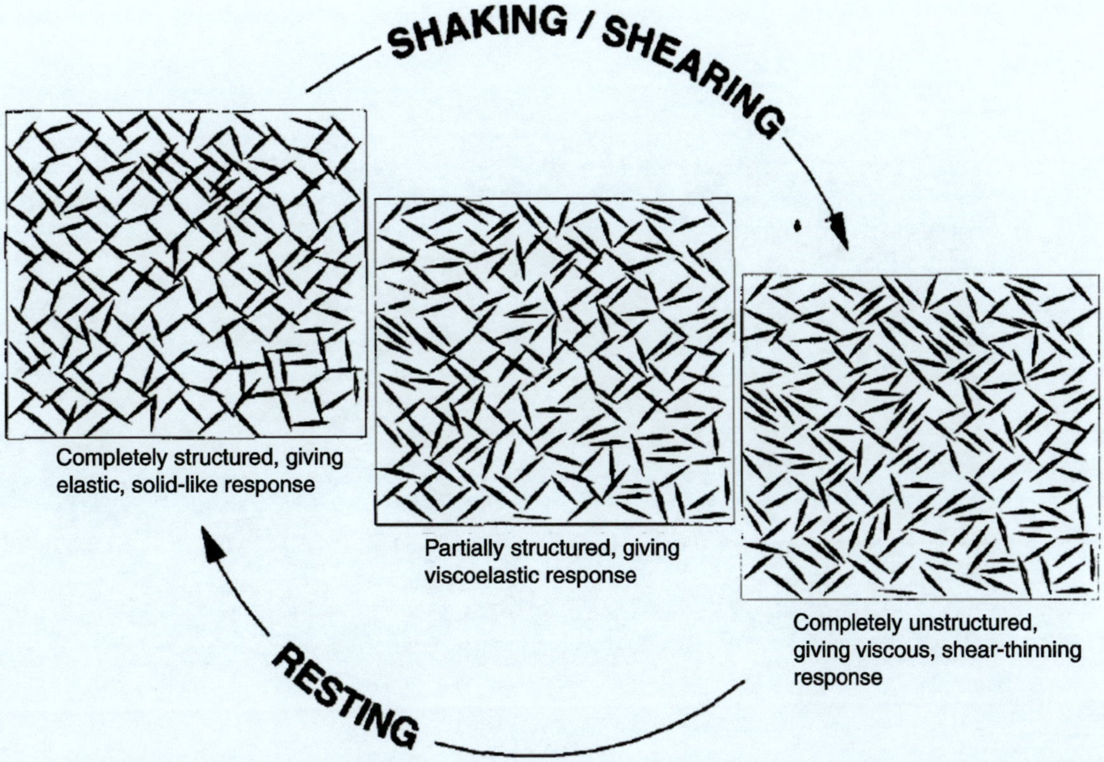

FIGURE 4-10 A schematic illustration of the "thixotropic cycle" of a two-phase system. The system at rest is a metastable structure that consists of a three-dimensional network of a rigid dispersed phase in a flowable matrix. In this schematic the dispersed phase is a nonspherical (platy) particle surrounded by a liquid. The interparticle potential is attractive between edges and faces, leading to the so-called "house of cards" structure as evidenced, for example, by the montmorillonite clays in Benna et al. (2001). It is significant that the schematics of structure are not dramatically different in the completely structured, partly structured, and completely unstructured states. This is consistent with the notion that the topological connectedness of the structure is critical, not the volumetric density of particles, which changes only at local levels. The upper arrow indicates the destruction of the network caused by the application of mechanical forces. This is generally considered an effectively instantaneous response. The lower arrow represents the reformation of the three-dimensional network when the forces are removed. Significantly, the regeneration of the structure takes significant time. The forces that drive the reformation are the interparticle forces (a combination of van der Waals, hydrogen, and weak electrostatic forces). The rate of reformation can be altered in some circumstances by heating or small-amplitude vibration, as these increase the mobility of the particles. If the structure does not reform, then the material is not thixotropic: It was in an unstable initial state. If the structure reforms so quickly that the lag is not observable, the material also is not thixotropic: Here, the term "structural viscosity" is used. SOURCE: Reprinted from Barnes, H., Thixotropy—A review, Journal of Non-Newtonian Fluid Mechanics, 70/1-2:1-33. Copyright 1997, with permission from Elsevier.

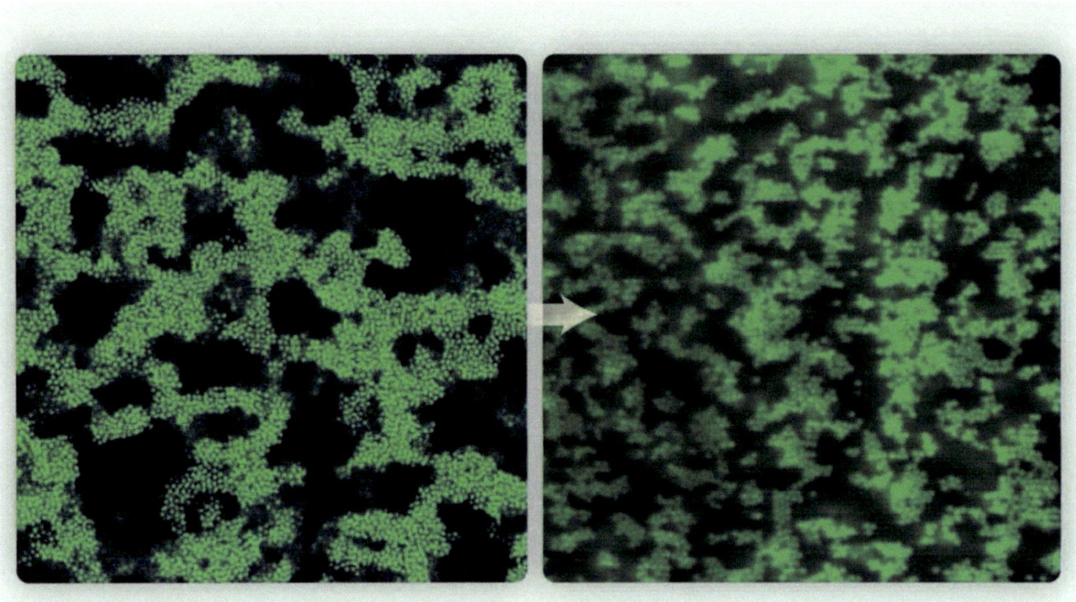

FIGURE 4-11 Optical micrographs of a three-dimensional network of spherical latex particles. The same colloidal gel is shown at rest (A) and just after flow (B). In (A), a clearly percolating structure (i.e., connected in three dimensions) can be observed. This material has a yield stress of ~5 Pa. In (B), the gel has broken into many smaller flocs. The material no longer has a measurable yield stress. Particle diameter, 1.3 μm. SOURCE: Bonn and Denn, 2009. Reprinted with permission from AAAS.

Phase Makeup

The manufacturer describes RP #1 as an oil-based, nonhardening modeling clay. In this context the word "clay" refers to a class of tactile and rheological behavior. That is, RP #1 looks and feels like a clay-water system. In this sense it is like a "polymer clay" such a Fimo or Sclupey. Polymer clay has been described as "a sculptable material based on the polymer polyvinyl chloride."[19] It usually contains no clay minerals, and is called "clay" only because its texture and working properties resemble those of mineral clay.

The following information about the composition and processing of RP #1 is available from the manufacturer's Web site and may therefore be regarded as public:[20]

---

[19]Available online at <http://en.wikipedia.org/wiki/Polymer_clay>. Last accessed December 21, 2010.

[20]Available at <http://www.sculpturehouse.com/plastilina_info.aspx>. Last accessed December 21, 2010.

> The main ingredients are wax, oil, and clay flour that is used as a binder. All plastilina is produced hot, and then cooled and extruded into the shape that will eventually be available for sale in art supply stores. There are basically three groups of plastilina: professional grade, school grade and industrial grade.
>
> The professional grade plastilina commonly contains sulfur, to make the smoother, more homogeneous texture required by professionals. ROMA Plastilina is an oil- and wax-base modeling material preferred by professional sculptors.

The following description is from a competitor:[21]

> Plastilina (or plasticine) is an oil-based, nonhardening modeling compound. Because it contains wax instead of water, plastilina remains pliable and can be used over and over again. It is smooth and does not stick to the hands and fingers, unless it is very warm. [This] clay is made by combining various waxes with different properties—ranging from soft to hard and from plastic to brittle—and melting them until well blended in a steel vat. A powdered filler and pigments are added to the melted wax and then mixed together to form a finished product.

This is sufficient information when combined with information in the open literature to guide an engineering understanding of the material when used as a backing material. In the following discussion, attention is first focused on the behavior of the organic constituents and then on how two inorganic dispersed phases can be expected to modify that behavior.

<u>Organic Phases.</u> To begin with it is useful to consider the behavior of a mixture of microcrystalline wax, grease, and oil and in this context to make use of the structure-property discussion about the rheology of a model ointment that is a mixture of petrolatum, mineral oil, and microcrystalline wax (Pena et al., 1994). The first point to be made is that all three constituents are obtained from the distillation of petroleum. Therefore, it is correct to view the creation of the ointment, and most likely the plasticized wax base of modeling clay, as a selective reconstitution of mutually soluble fractions of petroleum to achieve a desired set of properties.

All of these materials are natural products comprising a mixture of straight-chain, branched-chain, and cyclic hydrocarbons. Yet, they are distinct classes of material. Mineral oil is composed mainly of liquid hydrocarbons at room temperature and is a distillation product, whereas both petrolatum and microcrystalline wax are derived from the residue of distillation. In fact, petrolatum is considered to be a soft microcrystalline wax with a high oil content (or, conversely, microcrystalline wax is a de-oiled petrolatum). Both petrolatum

[21] Available online at http://www.vanaken.com/howclay.htm. Last accessed December 21, 2010.

and microcrystalline wax contain small irregular crystals whose size is believed to be limited by the presence of oil.

Other work has determined that petrolatum is a two-phase colloidal gel of liquid, microcrystalline, and crystalline hydrocarbons. When cooled, the wax crystals form a three-dimensional network with the liquid present in the interstices of that network (Barry and Grace, 1971). Because the crystals have a relatively high coordination number, the system develops a finite yield point. The rheological properties are dependent, of course, on the relative fractions of crystalline solid and liquid elements of the microstructure.

Of particular note in the context of the observed shear history dependence of modeling clay, a model all-organic ointment exhibits rheology that is dependent on both thermal history and shear history. In particular, the effect of increasing the liquid fraction (by increasing the mineral oil content) and the microcrystalline fraction (by adding microcrystalline wax) has been demonstrated (Pena et al., 1994). Increasing the microcrystalline fraction builds up the three-dimension network and increases the static yield value. During shear the three-dimension network is broken down and the material shows a marked softening that is not recovered even after a week at rest.

The effect of temperature in this system is complex because there are qualitatively distinct phases that have distinct melting and softening points. Notably, multiple inflections and thermal arrests occur in the ranges associated with thermal treatment of RP #1: 37°C-39°C, 46°C-47°C, and 57 °C -58°C (98 °F-102°F, 115°F-117°F, and 135°F-136°F).

These observations are important as they bear directly on the presumption that shear history and thermal history produce comparable changes in the response of modeling clay. While the empirical data discussed in the previous section showed an apparent equivalence between heating and working, a study by the National Institute of Standards and Technology (NIST) on the rheological properties in torsional shear of an earlier formulation of RP #1 found that the material was highly nonlinear and time dependent and that the shear properties of kneaded and "melted" (raised to a temperature of 90°C) clay at a fixed test temperature were different (NIST, 1994).

In considering the development of material systems and associated testing protocols, potential backing materials are unlikely to be anisotropic on a macro scale, since any organic polymers that may be contained will not be of sufficiently high molecular weight to induce orientation. Plasticity is not dependent on the presence of polymers; it normally requires the presence of weak bonds, but these may be colloidal interactions.

Inorganic Phases (Fillers). Turning to the role of the inorganic filler ("clay flour"), the first question is whether the presence of a particulate induces shear-dependent (thixotropic) behavior. The answer is clearly no. For example, in a study on the dispersion of the ferrofluid γ-Fe_2O_3 in paraffin, no thixotropy is observed with or without a magnetic field (Hosseini et al., 2010). These results are in keeping with the more general observation that ceramic particles are readily dispersed in wax matrices and form the basis of low-pressure injection molding of ceramics (Peltsman and Peltsman, 1984; Zorzi et al., 2003).

In such molding formulations it is important to have engineering control over the interparticle forces. Generally, it is desired to produce a stable dispersion in which the particles do not agglomerate. This is readily accomplished through the use of surfactants, including stearic and oleic acids for oxides and fatty amines for nonoxides (Peltsman and Peltsman, 1984; Lenk and Krivoshchepov, 2000; Zorzi et al., 2003).

The three filler components control the stiffness of the modeling clay and its surface finish. One material, clay mineral, is a low-cost platy particulate often used to stiffen thermoplastic organic systems (Rothon, 1999). Both zinc stearate and powdered sulfur are used as solid lubricants, the former in ceramic processing and the latter in powder metallurgy (Reed, 1988; Blagin, 1966). Presumably the main role of these agents is to modify the friction resulting from contact with the tool working the surface. (It is unclear that such surface properties affect the performance of the oil-based modeling clay used in ballistic testing.)

In sum, there is sufficient information in the engineering literature to provide a basis for the understanding the structure-property relationships of RP #1 oil-based modeling clay that pertain to mechanical working, thermal processing, and friction and how the various ingredients modify behavior. Furthermore, the principles of design are sufficiently clear that an alternative system with more favorable properties can be expected to be successfully developed.

SHORT-TERM DEVELOPMENT OF AN INTERIM STANDARD CLAY FORMULATION FOR BALLISTIC TESTING

The Army's protocol for ballistic testing of soft and hard body armor specifies RP #1 as the backing material (DoD, 2008). Since the initial validation studies, the formulation of RP #1 has changed, and this has changed its properties. Whereas historically calibration and testing could be performed at room temperature, the clay must now be above 100°F to pass the column-drop test (described in the section on clay behavior). The committee was informed that the thermal conditioning temperature has increased about 1°F every year.

In response to these known deficiencies of the current backing material, the Director of Operational Test and Evaluation (DOT&E) established an ad hoc clay working group whose members are technical clay experts from the Department of Defense (DoD), NIST, NIJ, private laboratories certified for testing body armor, and others. The group's purpose is to pursue short-term improvements in clay formulation and processing, and short- and long-term alternatives to clay. A short-term goal is to develop a replacement for RP #1.

Based on their experience, members of the clay working group have developed the following desirable characteristics of clays for ballistic testing:

1. Known, controlled, and consistent change in properties as temperature is changed.
2. A long useful life for repeated testing at room temperature.
3. Known, controlled, and consistent change in properties due to cold working (thixotropic effect).
4. Excellent dimensional stability.
5. Minimal stickiness to the target (i.e., the clay must not peel away when the target is removed) but high stickiness of clay to clay.
6. Easy moldability, so that clay blocks can be formed with no voids, air bubbles, or gaps.
7. Long shelf life (more than 1 year).
8. Nontoxic, minimum odor, and reasonable price.
9. Specifiable and controllable mechanical properties: density, seismic velocity, elastic modulus, shear modulus, grain size, hardness, etc.

Because its properties depend on shear history, time, and temperature, RP #1 appears to meet only some of these criteria. For example, it is typically heated to over 100°F to meet the calibration specification, which limits its useful life for testing at room temperature to less than 45 minutes. The current formulation also requires a complex preparation and packing procedure to produce boxes with uniform, reproducible properties that can pass the calibration test described in MIL-STD-3027 (DoD, 2008). The goal of the clay working group is to develop a short-term replacement that can meet the calibration specification at ambient temperature and minimize the sensitivity of the properties of the clay in the box to cold working.

In addition to the criteria developed by the clay working group, two additional considerations could facilitate development of a clay replacement in the short term. First, the formulation could be simplified by minimizing the number of ingredients. For example, as previously noted, RP #1 contains sulfur, which has an unknown effect on its performance in ballistic testing. Minimizing the number of ingredients should reduce variability in performance over time and simplify attempts to characterize and model performance. Second, the current RP #1 formulation of microcrystalline wax, oil, and grease includes clay as an inorganic filler. The inherent anisotropic (i.e., platy) nature of the clay particles may complicate the behavior of the RP #1. Eliminating the clay particles or replacing them with an inorganic filler that has an equiaxed particle morphology may provide properties that are less dependent on work history and time.

Two approaches are possible for the procurement of a standard ballistic clay from an industrial supplier. One would be to develop a material specification that uses a precise composition formulated with particular raw materials that are called out in the specification. This approach would guarantee a consistent product as long as the raw materials do not change but would not allow the supplier to adjust the formula in the event that properties change because raw materials are no longer available or that the properties of the raw materials themselves change over time. This approach could cause the properties of the standard ballistic clay to evolve, as happened with RP #1. The second approach would be to develop a performance specification. This approach would allow the supplier to continually evaluate and adjust the composition to produce a consistent product. It would put the burden on the Army to specify the properties that are most important to the application but would seem to be the best approach to meet the need for a consistent backing material.

The presence of an organic phase in a clay-like modeling material is important for two main reasons. First, it provides a continuous phase that is highly viscous and possesses a substantial modulus, which cannot be achieved with a water-based system. Second, the organic phase is not volatile, which gives the backing material a long usable lifetime with consistent performance.

Temperature sensitivity in the material is much less an issue than is the fact that the current RP #1 must be used at elevated temperature, which causes the temperature and temperature-dependent properties of the material to change substantially during testing. Such a problem would not arise with a temperature-sensitive material that could be used at room temperature, which is why the focus for a replacement modeling clay is to develop a room-temperature material.

Recommendation 4-1: The Office of the Director, Operational Test and Evaluation, and the Army should continue to expedite the development of a replacement for the current Roma Plastilina #1 oil-based modeling clay that can be used at room temperature.

Conditioning and Handling of Clay

The conditioning and handling procedures associated with the use of modeling clay in live-fire ballistic testing of body armor can be reviewed in light of the above information.

As described in the NIJ standard and the Army TOP, and as observed at ATC and during tours of three commercial testing laboratories, standard practices for conditioning and handling clay demonstrate operational awareness of the influence of both shear and thermal history on the rheology of modeling clay.

First, the modeling clay is heavily worked using mallets or tampers when the standard clay box is filled. The box surface is scraped before calibration to create a flat surface of precisely known elevation. The manner of preparing and working the clay is appropriate given the thixotropic nature of the clay composition. That is, the procedure described is consistent with standard practice to remove behavioral artifacts in the material that are due to the manufacturing process and to erase any differences that might be associated with the length of time the modeling clay has been stored. Thus, this procedure represents good practice.

The clay boxes are thermally equilibrated for at least 3 hr at 40°C (104°F) prior to use. During drop calibration testing, standard practice is to insert two thermometers into the clay mass. The thermometers are inserted at points approximately along the box face diagonal one-quarter of the way from the opposing corners. They are placed about halfway down through the clay. These measurements are used to confirm that thermal equilibrium has been established.

The personnel who repair and recondition clay boxes follow procedures that are consistent with standard practice by artists and others for filling space without entrapping air. That is, small additions are made sequentially and each is heavily sheared by hand to express any entrapped air. This procedure represents good practice. Army personnel related that periodic X-rays of clay boxes only very rarely indicate the presence of entrapped air. The same rebuilding procedure appears to be used both in rebuilding indentations produced during calibration and after an armor test to restore the box. However, at the conclusion, test clay containing any debris (such as fabric, fragments of body armor, or projectiles) is removed prior to rebuilding the surface.

Improvements to Mechanical Working

Given that mechanical working is known to break down the internal structure of the modeling clay, changing its flow behavior, and that mechanical working is routine and unavoidable given the need to assemble and rebuild the clay box, it is highly desirable to ensure that the entirety of the box can be worked in situ. It is standard practice in mechanical testing of thixotropic materials to establish a standard state, typically by preshearing.

A partial solution is to work the exposed surface of the clay box either manually or with a mechanism such as powered concrete tamper. This could be improved by removing the plywood backing and working both sides.

An alternative is to devise a means of working the entire volume of clay in place in the box. Examples of in situ mixing of soils exist in the open literature (Topolnicki, 2004). These might be adapted to the present purpose, but it is recognized that doing so would be a tough engineering challenge.

The committee was shown results of a preliminary experiment in which a clay box was placed on a vibration table and indentation experiments were carried out before starting the vibration and after 20, 30, and 40 min of vibration. The vibration frequency and amplitude, however, were not available. The average indentation increased monotonically from about 12 mm to just under 16 mm, with an approach to asymptotic behavior evident at 30 minutes. This result suggests that in situ mechanical working to obtain a uniform consistency should be feasible, and this work should be continued. The importance of such an effort is manifest in a statement by an ATC Protective Equipment Division science officer, who stated that in his opinion perhaps some of the most significant improvements to reduce variation in the testing process could come from the hand processing that goes into filling the clay boxes and working the clay before and after test firing.[22]

In-box mechanical conditioning alternatives were recommended in the Phase II report and offer the possibility of generating a mass that responds more predictably and uniformly (NRC, 2010). Further, such in-box conditioning might

[22]Shane Esola, Aberdeen Test Center, "Clay Calibration Study: Radial Dependence of Calibration Drop Depths," presentation to the committee, October 13, 2010.

reduce the need for elevated temperature, with its concomitant need for precise control. Testing of such in-box mechanical working methods should continue.

Thermal Conditioning

Temperature changes affect the consistency, probably in large measure by changing the viscosity of the organic phase and, possibly, at sufficiently high temperatures, causing a phase change of one or more microstructural elements; the temperature must be maintained within a fixed interval for the clay to exhibit acceptable mechanical properties. Temperature drift with time has a measurable effect on drop test penetration, for example, and can move a block out of specification during the course of a test.

The current practice of elevating the temperature of the clay box and yet using it at room temperature means that an inescapable drift is superimposed on the experiment. This is a fundamentally flawed approach and leads to undesired laboratory practice. For example, when conforming to the NIJ standard, which requires a post-test calibration drop, operators will be predisposed to heating the clay to the highest temperature that yields drops that are within the permitted range, increasing the likelihood of satisfactory post-drops after the block has had some (perhaps variable) time to cool. This is contrary to the spirit of the 19 ± 3 mm NIJ standard even though it is formally compliant to the standard. The Army TOP does not require or allow a post-test drop but constrains the test time to 45 min in duration. However, it is known that the behavior of that clay is substantially modified over this period. Not permitting a post-test drop creates a situation in which the block is used under conditions when it will almost certainly be out of specification.

There appear to be only two plausible solutions to this dilemma. The first is to heat the clay throughout the test. However, there is no obvious engineering approach to this. The alternative is to reformulate the backing material to a ballistics grade so as to achieve a material that can be used at room temperature and does not require heating to calibrate.

Finding: In the short term, testing will continue to be conducted using the existing Roma Plastilina #1. As long as heating the clay is necessary, cooling will take place, and the post-test calibration drop test, as recommended in the Phase II Report (NRC, 2010), will continue to be an urgent requirement for the Army Test Operating Procedure.

Systematic Characterization

Fundamental thermomechanical information about the clay formulation appears to be lacking. Plasticine rheology has been widely studied over the decades due to the technical importance of modeling materials in a number of scientific and technical applications. It is a complex material with a response that has been shown to depend on strain, strain rate, and thermal and mechanical history. The committee is unaware, however, of any linear viscoelastic measurements at low or high frequency for either worked or unworked RP #1 to determine the relative recoverable (elastic) and dissipative (viscous) fractions of the response; linear viscoelastic measurement, even at frequencies well below operational time scales, can be sensitive indicators of structural change. Nor have there been shear measurements at a range of temperatures to determine the viscoelastic solid response prior to yielding, the equilibrium yield stress, aging, the thixotropic response, or the apparent equilibrium viscosity and shear modulus as functions of shear rate. These are properties that are likely to affect (and hence correlate with) clay response during calibration and testing. Such measurements are standard practice in other industries that use similar materials, including oil well drilling, personal products, etc. These properties are also required for any simulation intended to relate indirect measurements to the mechanics of body armor deformation. In particular, viscoelastic recovery will give a measured BFD that is less than the maximum experienced dynamically during the test.

A priori calculation of temperature change and straightforward calculation of temperature variations within the box as a function of time require knowledge of the thermal diffusivity of the clay, which has also not been measured for the materials in use. Thermogravimetric measurements to measure weight loss and components that may be eluted over time at a fixed temperature have likewise not been carried out.

The committee notes that the Army has a contractual relationship with Rutgers University to carry out rheological and thermophysical measurements, but neither the scope of the effort nor the measurement data were available.

Finding: The comprehensive thermomechanical characterization of Roma Plastilina #1 that was recommended in the Phase II Report (NRC, 2010) will quantify the effect of shear history and thermal history on the storage and dissipative components of mechanical deformation. Such a characterization will also quantify the times associated with recovery of properties, as well as, the thermal properties, including thermal expansion, thermal conductivity, thermal diffusivity, heat capacity, and thermal arrests, associated with phase changes.

Calibration Drop Test

The current implementation of the column-drop test to calibrate RP #1 is clearly able to reveal changes in the resistance to penetration associated with box-to-box variation, heating or cooling, and working (or mechanical conditioning). It also has proved useful in quantitative assessment of the spatial variation in penetration resistance in the plane of the top surface of the box. As such, the column-drop performance test (including the testing protocols, facilities, and instrumentation) is a valid method for assessing the part-to-part consistency of the clay boxes used in body armor testing.

The noise inherent in the results of the drop test is also present in the results of other tests, including the Charpy impact test and high-speed ballistic impacts of small-diameter sting balls, and is therefore inherent to the material and not a variance associated with the impactor.

The key limitation is that in the drop test the strain rate experienced by the clay is qualitatively lower than that experienced in the live-fire ballistic test of armor. However, very little information is available to unambiguously relate the behavior in these two domains.

Another discrepancy between the cavities formed in drop tests and those formed in live-fire testing is the volume of material deformed. The typical depth of penetration of the drop-test indenter, 25 mm, is roughly half the distance of a BFD that would pass under the Army's previous protocol (43 mm). However, given the scaling results from using indenters of different size and scale (see Figure 4-5 inset), this is unlikely to be a big effect.

Several opportunities exist to improve the calibration drop test. These include establishing correlations to standard desktop laboratory instruments such as the penetrometer, which is used by soil scientists and food rheologists. The chief advantages here are the portability and convenience of the device. However, the greatest benefit would come from the creation and use of a high-velocity impactor such as a gas gun. As pointed out in the section "Characteristics of Clay Behavior Observed in Drop Tests," at the time of impact, the velocity of the indenter used in the current calibration test is just over 6 m/sec. However, ATC personnel told the committee that the back face of the armor moves at >50 m/sec after being impacted by threat projectiles.[23] That is, the current calibration occurs under significantly lower strain rates than those experienced in the armor test.

The Army is developing a gas gun capable of directing a penetrator onto the surface of a clay box at ballistic velocities. Although the gas gun is only one approach to achieving high velocities, it is the approach used in the following discussion for illustrative purposes.

The first advantage of employing high speeds is that impactors will penetrate to depths comparable to the BFD in ballistic tests. In addition, a gas gun will, at least in principle, be able to deliver penetrators ranging from spheres to other specialty shapes. It would allow choosing steel spheres to reproduce a cavity in the clay that approximates the dimensions associated with the BFD in a ballistic test. Then, too, it could reproduce the impacter of the original Prather study (1977), which will permit directly comparing the original work with work using modern oil-based modeling clay formulations and conditions.

Shaped impactors can be designed to reproduce the force distributions expected when a blunt trauma occurs as a projectile strikes hard armor. Such shaped impactors are commonly used in injury simulation to induce specific and reproducible forces over a well-defined area. This option is particularly appealing as work progresses to measure the force distribution associated with armor testing (Raftenberg, 2006). Shaped indentors of different radii at constant impact velocities might be used to probe the backing material at a variety of depths up to and perhaps exceeding the expected BFD. An additional appealing possibility is using small-diameter spheres because this allows a high-density matrix of small

[23]Scott Walton, and Shane Esola, Aberdeen Test Center, "ATC Perspective on Clay used for Body Armor Testing," presentation to the Body Armor Phase II committee, March 10, 2010.

impacts that might permit direct measurement of clay homogeneity (Weber, 2000).

It is important to stress there are two different functions of an improved calibration test. The first is to characterize the variability of clay within a given box at a given time in a manner that is directly relatable to the BFD. The second, equally important role is to use such a system to estimate the variation of BFD measurements both within a given box and between boxes, under realistic testing conditions using existing test protocols. The latter would help to provide information of use in the statistical analysis of armor testing results.

The Phase II study (NRC, 2010) recommended that the Army should quickly develop and experiment with a gas gun calibrator, or an equivalent device, capable of delivering impactors to the surface of clay boxes and determining the local variations within a clay box at speeds and depths corresponding to BFD (NRC, 2010). The experiments should be used to estimate as accurately as possible the variation of BFD measurements both within a given box and between boxes, under realistic testing conditions using the existing test protocols. Such experiments can form the basis for refinements in calibrating the clay for testing.

Finding: The testing community would benefit greatly from devising an alternative to the column drop test and certifying the validity of the current drop tests for calibration.

ALTERNATIVE BACKING MATERIALS AND SYSTEMS

As stated earlier, the committee strongly supports DOT&E and the Army in developing a standard ballistic backing material that would be an improvement on the current RP #1 for body armor testing. The Phase II study was interested in learning more about possible mid- and long-term replacements for modeling clay. Briefings from the Army and the personal knowledge of committee members led to suggestions of ballistic gelatin and microcrystalline waxes as possible alternatives.

Medium-Term and Long-Term Replacements for Modeling Clay

There are two broad classes of backing material for live-fire ballistic testing of armor when BFD is the quantity is of interest; both classes were considered in the original Prather study. The first is elastic materials that recover their original shape after unloading. In this case, the signal that is used to determine BFD is a set of images collected, for example, by high-speed video. The prototype for this type of material is ballistic gelatin. The second class of backing material is plastic material that preserves a permanent cavity whose dimensions can be correlated to lethality probability.

The issue with the first class of materials is the cost and complication of recording images. One such complication is devising and standardizing calibration procedures for the image gathering. The problem with the second class is assuring that the material is capable of quantifying elastic and plastic deformation as the recording medium, and then assuring minimal variability and reproducibility from test to test and box to box.

Ballistic Gelatin

At the time of the Prather study, the cost and availability of high-speed video were limiting factors, so the decision was made to recommend a plastic recording medium. In the modern era, video recording has become low cost and readily available. This might suggest a move to a transparent material such as ballistic gelatin in order to gain, for example, information about the dynamics of the event as armor deforms into a transparent body simulant such as ballistic gelatin. However, it is likely that the window for this strategy has both opened and closed, because advancements in sensor technology, discussed below, offer the benefit of time-resolved information in a solid-state reusable device. Such sensor systems typically incorporate a very tough transparent polymeric material behind the armor and in front of the sensor array.

While gelatin is the preferred material for penetrating injury studies, it cannot be shown that it offers qualitative advantages when the subject of interest is trauma associated with nonpenetrating injury (Harvey et al., 1962; Metker et al., 1975; Fackler and Malinkowski, 1985). Gelatin must be prepared as freestanding blocks, is labor intensive, typically is not reusable, has the potential for spoilage, is temperature sensitive, and requires specialized storage facilities. Nevertheless, it has been used in this application with appropriate imaging technology.

In addition to ballistic gelatin, other transparent elastically deforming materials are available (Uzar et al., 2003; Juliano et al., 2006; Moy et al., 2006). These materials are claimed to have advantages over ballistic gelatin because they are less sensitive to processing history, less affected by temperature, have a longer shelf life, and exhibit increased durability during repeated use. Moreover, because they are a synthetic rather than natural product, they are less variable in molecular structure.

In (NRC, 2010), the Phase II committee suggested in its Recommendations 11 and 12 (see Appendix L) that the Army should consider experimenting with ballistic gelatin and/or microcrystalline wax as a medium-term solution to the issues associated with the current recording medium, RP #1. However, the information detailed above, coupled with an improved understanding of the microstructure-property relationships in oil-based modeling-clay, calls for a general shift in emphasis.

If unlimited resources (time and money) were available, it would remain the case that there is potential benefit to exploring, through experiment, qualitatively different media with different means of recording BFD. However, such resources are limited, so, based on preliminary results—that is, small-batch samples obtained by the clay working group—it appears that qualitatively new classes of materials are not an appropriate focus for development work.

Finding: There is no compelling rationale for expending resources to achieve an interim solution using an elastic material such as ballistic gelatin.

Ballistic-Grade Clay

In view of the finding above, attention must be directed at developing an improved ballistic-grade clay or an alternative plastic material to serve as a recording medium. As stated earlier the committee strongly supports DOT&E and the Army as they work to improve the current RP #1 modeling clay as a backing material to test body armor.

In fact, if it is understood that the term "modeling clay" is meant to broadly encompass materials that exhibit the working characteristics and feel of traditional clay- mineral formulations, then ballistic-grade modeling clay may meet two objectives. That is, it might serve as an incremental (or substantial) adjustment of the existing formulation (RP #1) or it might be based, for example, on an unfilled microcrystalline wax that is appropriately plasticized (with grease, oil, or a synthetic polymer). As a matter of course, but not as a research thrust, other materials (such as soap or paraffin) should be considered but investigated in any detail only if it can be determined that they offer the promise of dramatically reducing variability and dependence on shear and temperature history compared to modeling clay.

Finding: Plastically deforming recording media appear to be the proper choice of backing material for production testing of body armor for the foreseeable future.

A road map for improving the performance of a ballistic-grade modeling clay or other plastically deforming medium is presented in detail in the last section of this chapter.

Electronic Recording Systems

As will be discussed in Chapter 8, a variety of electronic systems have been used by medical researchers to assess behind-armor blunt trauma (BABT) injuries. As a consequence, ATC has considered the possible use of electronic sensor-based systems in production testing of body armor in the medium or long term. Conceptually, having the same electronic sensor recording media used by medical researchers is intriguing since it could facilitate the correlation and interchange of data produced by the two communities.

Two examples of such systems are the anthropomoprhic test module (ATM) developed by the Army Medical Research and Materiel Command and the BABT Rig developed by the U.K. Defence Evaluation and Research Agency. The ATM and the BABT Rig are examples of electronic- sensor-based systems that may ultimately help to synchronize research by the medical and armor testing communities. However, as is discussed in detail in Chapter 8, the transition to such a system presents a variety of scientific and logistic challenges, not to mention a significant increase in cost. Further, the discussion there calls into question the predictability of injuries from sensor data regardless of the fidelity of the data collection system. In considering the ATM and BABT Rig systems for the production testing of hard body armor, the committee drew two conclusions that are relevant to the potential future replacement of RP #1 for backing material:

Finding: The use of the anthropomorphic test module (ATM) represents a transition to a challenging methodology with only limited ability to extend results to injury prediction. Also, it is too costly to be used as a production testing alternative to Roma Plastilina #1 at this time. The ATM is judged a research tool that is not practical or appropriate for widespread deployment in ballistic testing ranges.

Finding: Overall, instrumented electronic sensor response elements are in a primitive state for the evaluation and assessment by medical researchers of ballistic behind-armor blunt trauma with rifle round threats. They also are too costly to be used in high-volume production testing. More research and detailed validation is necessary before electronic sensors can be considered as a practical medium or long-term alternative to use of Roma Plastilina #1.

ROADMAP FOR IMPROVING THE TESTING PROCESS

The Army's protocol for ballistic testing of soft and hard body armor specifies RP #1 as the backing material (DoD, 2008). However, it would not be an exaggeration to say that the move of RP #1 from being a material of convenience to becoming an industry standard was an accident of history, because it is a far from ideal backing material for the testing of hard body armor. It is complex and highly dependent on its shear and temperature history; its chief limitation, moreover, is that it exhibits inherently high variability during service in this capacity. While RP #1 fails, in multiple ways, to meet the criteria of the ad hoc clay working group,[24] its central drawback is not mentioned—namely, the high level of variance associated with its deformation under nominally identical conditions.

This variability of RP #1, discussed earlier in this chapter, is reported in all systematic studies of drop-test calibration and in the limited studies on high strain rate. Unless and until a backing material can be identified or developed with low variability during plastic deformation, the disadvantages identified by the ad hoc clay working group remain secondary. Furthermore, a substantial reduction in variance would be required before the use of high-precision measurement techniques can be justified.

[24]Shane Esola, Aberdeen Test Center (ATC), "ATC Perspective on Clay used for Body Armor Testing," presentation to the Body Armor Phase II committee, March 19, 2010.

Based on the desirable characteristics of a backing material and the needs of the testing community, the road map below describes one possible path to develop an improved backing material and, equally importantly, a uniform protocol for calibrating the backing material and testing hard armor panels. This is intended to guide both the ad-hoc clay working group and agencies that would fund the activities that are needed to reach the ultimate goal of having a single, uniform, reproducible process for evaluating hard armor. The activities are divided into near-, medium-, and long-term. A very brief description of the activities is provided below, and the road map itself is shown in Figure 4-12.

In both the Phase I and Phase II reports, the committee made recommendations that were intended to guide the development of a more consistent backing medium, but no priority was assigned to the recommendations (NRC, 2009; NRC, 2010). The purpose of this section is to propose a road map to guide development of a more consistent backing medium for ballistic testing of hard armor plates. The improved medium should be able to meet calibration specifications at room temperature while minimizing other problems encountered with clay, specifically RP# 1. Given the realities of limited funds and human resources, the road map provides a prioritized, time-phased way ahead for accomplishing the tasks.

Based on briefings, demonstrations, and other interactions, the committee recognizes the tremendous effort that ATC has invested to understand and mitigate the consistency and calibration problems associated with the current backing material. The committee was briefed numerous times by ATC personnel and has found these dedicated professionals to be forthcoming and truly interested in finding the root causes of the issues associated with the current backing material. More important, ATC has been aggressively pursuing long-term solutions to the known issues with the current backing material.

However, ATC is not a research organization. As a consequence, it must rely on others for fundamental studies. In particular, the Army Research Laboratory has been a strong partner for ATC in the research and development of replacement materials. This close partnership will likely be required for successful implementation of this road map.

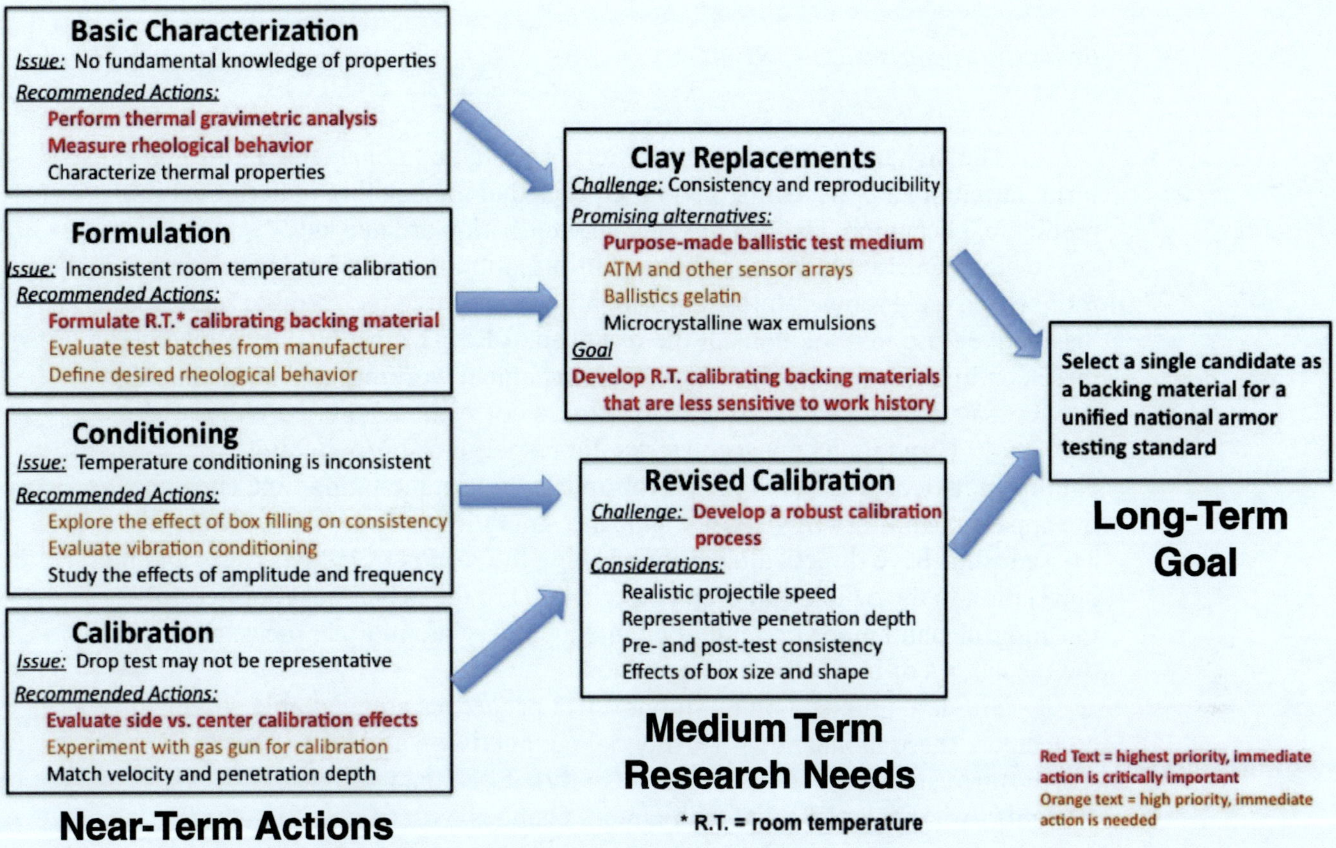

FIGURE 4-12 Road map showing suggested near-term actions, medium-term research needs, and a long-term goal to develop a more consistent backing material and a more reliable process for evaluating hard armor. The color coding shows "highest priority" items in red text with "high priority" actions in orange.

Near-Term Actions

The critical near-term need is for the development of a backing material that can be calibrated at room temperature. To reach that goal, actions are needed in four areas:

1. Characterization of the current backing material, RP #1;
2. Lab trials with alternative formulations from the current supplier;
3. Exploration of in-box conditioning methods; and
4. Study of improved calibration procedures.

Activities suggested for each action are prioritized in sections below and in Figure 4-12 to help guide efforts.

Basic Characterization of RP #1

The highest priority in the near term should be to characterize the current formulation of RP #1. Armor testing requires that the backing material exhibit predictable behavior. Two classes of characterization are needed.

Fundamental thermomechanical information for RP #1 is lacking either at room or elevated temperatures. Specifically, the relative recoverable (elastic) and dissipative (viscous) fractions of the response of RP #1 must be determined under three conditions: as manufactured; after mechanical working; and after extended service, say, 6 months. Measurements should cover the temperature range of 23°C to 40°C and include viscoelastic solid response prior to yielding; the equilibrium yield stress; and the thixotropic response, including time constants or the apparent equilibrium viscosity and shear modulus as functions of shear rate. These results have direct value in interpreting the results of ballistic tests and the correlation to the calibration drop tests. These are the properties required for meaningful simulations that might be intended to relate indirect measurements to the mechanics of body armor deformation.

Similarly, no data on thermophysical properties are available and must be collected. These include density, thermal conductivity, and heat capacity (permitting calculation of thermal diffusivity). Such data permit the straightforward calculation of temperature changes with respect to both position in the clay box and time after removal from the oven. They must be known in order to quantitatively predict the impact of clay-box cooling during service on the range.

Thermogravimetric measurements are needed to measure weight loss due to selective evaporation of components during service. The obvious need for this is underscored by the on-range observation of sulfur evaporation and condensation associated with annealing the clay boxes in the ovens. It also is plausible that low molecular weight hydrocarbons are lost. Differential thermal analysis should be done in parallel with thermogravimetric analysis to help determine the relative contributions of evaporation or chemical reactions (e.g., oxidation) to observed weight changes.[25]

Formulation

In the near term, incremental improvements in modeling clay to allow it to serve as a backing material need to focus on a reformulation to permit calibration and live-fire testing at room temperature. One benefit of this is to simplify range practice, but the chief, and vital, benefit is to remove the substantial drift of clay

[25]The committee noted that the ATC has a contractual relationship with Rutgers University to carry out rheological and thermophysical measurements, but the scope of measurements and the data were not available to the study group. Successful conclusion of this effort must be a high priority, along with any additional work to generate the requisite data set.

properties observed with time after removal from the oven (see Figure 4-4). A room-temperature system would eliminate the need for a post-test drop test.

The committee is aware that Chavant, the manufacturer of RP #1, has been contracted to develop a replacement for RP #1 that will meet the historical calibration specifications at room temperature. As material is provided to the government, rapid turnaround in assessment is required to justify this investment of resources.

The committee has offered two considerations that apply to the reformulation effort. The first is that the formulation be simplified by minimizing the number of ingredients, in particular, the fillers. As previously noted, RP #1 contains both zinc stearate and sulfur in addition to clay. It is unclear how this combination affects its performance as a backing material. The goal of simplifying the formulation, with concomitant tightening of specifications on raw materials, is to reduce the lot-to-lot (and therefore box-to-box) variability of reformulated backing material.

The second consideration is related to the characteristics of the filler phase. The filler employed in the formulation should be equiaxed (i.e., roughly spherical) rather than platy or otherwise anisotropic. The goal of this selection is to reduce variation of mechanical response to shear history and direction or location in the box.

Conditioning

Clay conditioning includes both thermal history and shear history. As discussed earlier in this chapter, a clay box that has been heated to 40ºC cools significantly during the time associated with ballistic testing and causes “drift,” in the engineering sense of the word, in the results.

Given that the preliminary results suggest the feasibility of mechanical vibration as a method to achieve in-box working of RP #1, this work should be systematically extended. In particular, experiments should be conducted with controlled, well- characterized vibration conditions to determine the effects of frequency and amplitude on properties so that reasonable limits can be established for the conditioning treatments. The assessment of flow should be extended to the standard drop tests as well as to the high-rate tests discussed in the next section.

Further, the relaxation behavior of clay after vibration needs to be systematically investigated. The time required for the clay to return to its previbration state needs to be characterized to determine if vibration will produce a state that is stable enough to allow for testing before relaxation becomes significant.

If necessary, ATC should engage the research community to conduct further evaluations. In addition to the vibration conditioning treatment, ATC is advised to explore alternative methods for box filling and clay processing prior to box filling as well as alternative methods for in-box clay working.

Calibration

It is important to stress there are two different functions of an improved calibration test. The first is to characterize the variability of clay within a given box at a given time in a manner that is directly relatable to the BFD. The second, very important role is to use such a system to estimate the variation of BFD measurements both within a given box and between boxes, under realistic testing conditions using existing test protocols. The latter will help to provide information of use for the statistical analysis of armor testing results. Specifically, statistical analyses of the test protocols require quantification of how much of the observed variation in BFD is due to the clay medium (and the test protocol in general) and how much is due to variation in the armor plates. The actual plates cannot be used to answer this question because of the destructive nature of the tests, and the results could be confounded with variation in manufacture. Clearly, the conditions of the column-drop performance test are very different from those experienced by the modeling clay during the actual ballistic test of the armor.

The Army has been developing a gas gun capable of directing a penetrator onto the surface of a clay box at velocities comparable to those of a BFD. The committee strongly feels that the gas gun needs to be deployed to probe the strain rate effect. The first advantage of employing high speeds is that impactors will penetrate to depths comparable to the BFDs in ballistic tests. In addition, a gas gun will in principle be able to deliver penetrators ranging from spheres to other specialty shapes.

In particular, the committee suggests that the feasibility be assessed of using shaped impactors designed to reproduce the force distributions expected when a blunt trauma occurs as a projectile strikes hard armor. This option is particularly appealing as work progresses to measure the force distribution associated with armor testing (Raftenberg, 2006).

An additional interesting possibility is to use small diameter spheres, because this would allow a high-density matrix of small impacts that may permit direct measurement of clay homogeneity (Weber, 2000).

Priorities for Near-Term Actions

Highest priority should be given to the following:

- Determining the rheological properties of RP #1;
- Determining the thermal properties of RP #1; and,
- Formulating a replacement backing material derived from RP #1 that calibrates at room temperature. Other small batch clays should continue to be studied.

High priority should be given to the following:

- Defining the desired rheological behavior;

- Evaluating conditioning with vibration; and,
- Deploying gas gun calibration.

Finally, medium priority should be given to the following:

- Determining the effect of mechanical compounding before filling the box;
  —Exploring alternative box-filling by pressing;
  —Exploring alternative means of in-box working;
- Studying the effects of amplitude and frequency; and
- Matching velocity and penetration depth in calibration.

Medium-Term Research Needs

Completion of the near-term actions will have several outcomes. First, the data gathered from these studies will enable selection of a short-term replacement for RP #1 as a backing material for ballistic testing of hard armor panels. Analysis of results from the near-term activities will also define some fundamental research tasks that are necessary for the development of a long-term replacement for RP #1 as well as (possibly) a revised set of calibration procedures.

Clay Replacements

Assuming that room-temperature use has been achieved in the short term, the overarching goal of medium-term research is to produce a backing material that retains the room-temperature characteristic and exhibits batch-to-batch consistency while delivering both substantially lower variance during plastic deformation and less sensitivity to shear history.

Three possible replacements were discussed in the Phase II study (NRC, 2010). These include a revised formulation similar to RP #1; microcrystalline wax emulsions without inorganic fillers; and ballistics gelatin. The first two possibilities would be plastically deforming materials that serve as recording media; ballistics gelatin, by comparison, is a transparent, fully elastic medium through which transient deformation can be imaged. Each has advantages.

The original rationale for selecting an oil-based modeling clay remains valid. Such clays are inexpensive and straightforward to implement in a ballistics range setting. The ability to image through an elastic material offers the possibility of collecting dynamic information, including true measures of the maximum extent of transient deformation.

Two paramount issues need to drive the development of improved or new backing materials. First is the reliability of the data, which means the material must exhibit a deformation behavior that is characterized by low variance and by minimal dependence on temperature and processing history. The second key issue is practicality. This includes cost, service life, ease of processing and handling, and cost and complexity of associated equipment (notably the cost of equipment to measure deformation).

Plastically Deforming Media. As discussed previously in this chapter (see "Influence of Structure on Properties of Oil-Base Modeling Clay"), the principal relationships between composition, phase distribution, microstructure, and macroscopic mechanical properties are understood. Therefore the scientific basis for a development program is in hand. However, the gap between principle and practice may be large and may require uncommon expertise to bridge successfully.

The committee believes that the Army should engage an industrial firm rather than start an in-house development program to develop, and possibly supply, a formulation. Expertise resides at large chemical companies such as those involved in the development of cosmetics, lubricants, drilling fluids, molding waxes, and the like. However, it is recognized that the small potential market may make this an unattractive research program for many companies. If a market survey proves this to be the case, the committee suggests that a highly focused university-based program may offer the best potential for development of a new class of plastically deforming recording media.

The committee feels the Army should develop a set of criteria to be used to guide such a development process at this time. Further, the criteria should include the ability to restore the material to a well-defined standard state through a combination of working and heating. This is properly regarded as the equivalent of solution-annealing, working, and tempering of metals to achieve a well-defined condition.

Elastically Deforming Media. Several transparent elastically deforming media are available in addition to ballistic gelatin, including other polymer systems (Uzar et al., 2003; Juliano et al., 2006; Moy et al., 2006), and the committee concludes that such materials may become of interest under three conditions. The first is if a low-variance, plastically deforming material has not been, or cannot be devised. The second is if the equipment to record images through a transparent medium is less expensive than the equipment to measure the geometry of a cavity at required resolution. And the third is if dynamic information becomes important for assessing injury probability. Therefore this will be an area of interest over the longer term, and the degree of interest will ultimately depend on the success of other ongoing and planned effects.

In the event that elastically deforming materials become a high priority, the committee recommends connecting to ongoing research efforts on gelatin and its replacements. It should be noted that although ballistic gelatin is widely used, it too suffers from gaps in knowledge about its fundamental structure–processing–property relationships.

Solid-State Instrumentation. The focus of the development of solid-state instrumentation appears to be biofidelity. The range of these developments is discussed in Chapter 8. The committee believes the Army should critically examine whether or not biofidelity is important to, or even desired for, production range testing of armor. In the event that it is judged to be so, ATC should maintain an awareness of developments in the Army-funded programs and others and be prepared to rapidly adopt technologies for range practice as they become mature technologies.

In the event that biofidelity is not held to be a part of range testing, the Army should look to migrate sensor arrays in a manner that is compatible with range practice. One hypothetical example would be to consider using a planar sensor array covered with a thick planar slab of polymer behind a modeling clay appliqué. Evaluation criteria for such an approach should include cost, reliability, needed electronics, and availability of suppliers of modules. Such incremental inclusion of sensors would offer the best of both worlds—easy processing of appliqué with highly workable clay, but none of the processing, handing, boundary constraint issues, and history effects of the clay box. However, the committee wishes to reiterate that sensor development must supplement rather than replace the near-term actions outlined above.

Revised Calibration

In parallel with the clay replacement, improved calibration procedures are needed. Combining information attained by studying gas gun and other alternative calibration methods with a more consistent conditioning process (including the elimination of any conditioning steps) will enable development of a more robust calibration procedure that can be more readily and reliably extrapolated to ballistic testing conditions.

ATC must work with other testing organizations and industrial practitioners to devise a calibration methodology that balances the needs of the various constituencies. A manifest goal to this collaboration should be the definition and adoption of a single testing methodology. If promising backing materials other than clay are identified in the future, then calibration procedures for those materials that balance the needs of the various constituencies will also have to be developed.

Summary of Medium-Term Research Needs

The highest medium-term priorities are to develop backing materials that can be calibrated at room temperature and are less sensitive to work history and cost effective and to develop a robust calibration process that is well suited to the alternative backing materials.

Initial research actions include these:

- Accomplish basic research on clay alternatives
  - —Improved or new plastically deforming backing material(s),
  - —Ballistics gelatin or other transparent elastic media, and
  - —ATM and/or sensor arrays.

- Have potential suppliers formulate test materials.
- Direct basic research on alternative calibration processes.

The desired outcomes are to develop replacement candidates for RP #1 and issue a revised calibration procedure.

Long-Term Goal

The ultimate goal is to develop a consistent, robust protocol for ballistic testing of hard armor plates that is also cost-effective for manufacturers and other testing laboratories. Selecting or developing a well-characterized high-performing replacement for RP #1 and devising an improved calibration procedure will enable development of a standard testing configuration and procedure. The success of the near- and medium-term activities could lead to a single, uniform test standard that could be used by all of the members of the testing community.

Recommendation 4-2: The Office of the Director, Operational Test and Evaluation, and the Army should provide resources and execute the road map described in this chapter and graphically shown in Figure 4-12 with the objective of developing a standard ballistics backing material for testing body armor. The properties and behaviors of the material should be well understood; it should exhibit minimal variability due to temperature, working, and aging and require simple calibration techniques and equipment; and it should enable reliable and accurate recording of body armor test results.

REFERENCES

Abu-Jdayil, B. 2003. Modelling the time-dependent rheological behavior of semisolid foodstuffs. Journal of Food Engineering 57(1):97-102.

Aerospace Corporation. 1974. Equipment Systems Improvement Program–Protective Armor Development Program–Final Report, volume 3. ATR-75(7906)-1. El Segundo, Calif.: Aerospace Corporation.

ATC (U.S. Army Aberdeen Test Center). 2008. Test Operations Procedure (TOP) 10-2-210 Ballistic Testing of Hard Body Armor Using Clay Backing. Aberdeen Proving Ground, Md.: Aberdeen Test Center.

Barnes, H. 1997. Thixotropy—a review. Journal of Non-Newtonian Fluid Mechanics 70(1-2):1-33.

Barry, B., and A. Grace. 1971. Structural, rheological and textural properties of soft paraffins. Journal of Texture Studies 2(3):259-279.

Benna, M., N. Kbir-Ariguib, C. Clinard, and F. Bergaya. 2001. Card-House microstructure of purified sodium montmorillonite gels evidenced by filtration properties at different pH. Progress in Colloid and Polymer Science 117:204-210.

Blagin, V.I. 1966. Compacting properties of powder mixtures. Powder Metallurgy and Metal Ceramics 5(6):454-457.

Bonn, D., and M. Denn. 2009. Yield stress fluids slowly yield to analysis. Science 324(5933):1401-1402.

Borwankar, R. 1992. Food texture and rheology: A tutorial review. Journal of Food Engineering 16(1-2):1-16.

DoD (U.S. Department of Defense). 2008. DoD Test Method Standard for Performance Requirements and Testing of Body Armor. MIL-STD-3027. Arlington, Va.: Department of Defense.

Esola, S., B. Gillich, W. Boughers, and S. Meiselwitz. 2010. Clay Calibration: Radial Dependence of Calibration Drop Depths for Body Armor Testing. Presentation to the Sixteenth Army Conference on Applied Statistics, October 20-22, Cary, N.C.

Fackler, M., and J. Malinowski. 1985. The wound profile: A visual method of quantifying gunshot wound components. Journal of Trauma 25(6):522-529.

Fung A.Y.C., and P. Tong. 2001. Advanced Series in Engineering Science—Volume 1: Classical and Computational Solid Mechanics. Hackensack, N.J.: World Scientific Publishing Co., Inc.

Harvey, E., J. McMillen, and E. Butler. 1962. Mechanism of wounding. Pp. 143–235 in Wound Ballistics. J. Coates, editor. Rockville, Md.: Office of the Surgeon General, Department of the Army.

Hosseini, S., A. Fazlali, E. Ghasemi, H. Moghaddam, and M. Salehi. 2010. Rheological properties of a γ-Fe_2O_3 paraffin-based ferrofluid. Journal of Magnetism and Magnetic Materials 322(23):3792-3796.

Huang, Z., M. Lucas, and M. Adams. 2002. A numerical and experimental study of the indentation mechanics of plasticine. The Journal of Strain Analysis for Engineering Design 37(2):141-150.

Juliano, T., A. Forster, P. Drzal, T. Weerasooriya, P. Moy, and M. VanLandingham. 2006. Multiscale Mechanical Characterization of Biomimetic Physically Associating Gels. ARL-RP-134. Aberdeen Proving Ground, Md.: Army Research Laboratory.

Lenk, R., and A. Ph. Krivoshchepov. 2000. Effect of surface-active substances on the rheological properties of silicon carbide suspensions in paraffin. Journal of the American Ceramic Society 83(2):273-276.

Metker, L., N. Prather, and E. Johnson. 1975. A Method for Determining Backface Signatures of Soft Body Armors. EB-TR-75029. Aberdeen Proving Ground, Md.: U.S. Army Armament Research and Development Command.

Moy, P., T. Weerasooriya, T. Juliano, M. VanLandingham, and W. Chen. 2006. Dynamic Response of an Alternative Tissue Simulant, Physically Associating Gels (PAG). ARL-RP-136. Aberdeen Proving Ground, Md.: Army Research Laboratory.

NIST (National Institute of Standards and Technology). 1994. Memorandum: Rheology of Clays. December 12, 1994. Gaithersburg, Md.: National Institute for Standards and Technology.

NRC. (National Research Council). 2009. Phase I Report on Review of the Testing of Body Armor Materials for Use by the U.S. Army: Letter Report. Washington, D.C.: National Academies Press.

NRC. 2010. Phase II Report on Review of the Testing of Body Armor Materials for Use by the U.S. Army. Washington, D.C.: National Academies Press.

Peltsman, I., and M. Peltsman. 1984. Low Pressure Moulding of Ceramic Materials. International Ceramic Review 33(56).

Pena, L., B. Lee, and J. Stearns. 1994. Structural rheology of a model ointment. Pharmaceutical Research 11(6):875-881.

Prather, R., C. Swann, and C. Hawkins. 1977. Backface Signatures of Soft Body Armors and the Associated Trauma Effects. ARCSL-TR-77055. Aberdeen Proving Ground, Md.: U.S. Army Armament Research and Development Command Technology Center.

Raftenberg, M. 2006. Modeling thoracic blunt trauma: Towards a finite-element-based design methodology for body armor. Pp. 219–226 in Selected Topics in Electronics and Systems, Volume 42: Transformational Science and Technology for the Current and Future Force, Proceedings of the 24th US Army Science Conference. J.A. Parmentola, A.M. Rajendran, W. Bryzik, B.J. Walker, J.W. McCauley, J. Reifman, and N.M. Nasrabadi, editors. Hackensack, N.J.: World Scientific Publishing Company, Incorporated.

Reed, J.S. 1988. Introduction to the Principles of Ceramic Processing. Hoboken, N.J.: John Wiley & Sons, Incorporated.

Roberson, C., W. Perciballi, D. Rogers, and D. Fleming. 2010. Consistently Depressed? Presentation to Personal Armour Systems Symposium 2010, September 13-17, Quebec City, Canada.

Rothon, R. 1999. Filler manufacture and characterization. Pp. 67–107 in Mineral Fillers in Thermoplastics: Advances in Polymer Science 139. J. Jancar, editor. New York, N.Y.: Springer.

Salisbury, C., and D. Cronin. 2009. Mechanical properties of ballistic gelatin at high deformation rates. Experimental Mechanics 49(6):829-840.

Suwardie, H., R. Yazici, D. Kalyon, S. Kovenklioglu. 1998. Capillary flow behaviour of microcrystalline wax and silicon carbide suspension. Journal of Materials Science 33(20): 5059-5067.

Topolnicki, M. 2004. In situ soil mixing. Pp. 331–428 in Ground Improvement. M. Moseley and K. Kirsch, editors. Boca Raton, Fla.: CRC Press Taylor and Francis Group.

Uzar, A., M. Dakak, T. Özer, G. Ögünc, T. Yigit, C. Kayahan, K. Öner, and D. Sen. 2003. A new ballistic simulant "transparent gel candle" (experimental study). Turkish Journal of Trauma and Emergency Surgery 9(2):104-106.

Weber, D.J. 2000. An Indirect Method of Measuring Impact Velocity for Non-Lethal Weapons Blunt Trauma Studies. Aberdeen Proving Ground, Md.: U.S. Army Edgewood Chemical and Biological Center.

Zorzi, J.E., C.A Perottoni, and J.A.H. Da Jornada. 2003. Wax-based binder for low-pressure injection molding and the robust production of ceramic parts. Industrial Ceramics 23(1):47-49.

5

Instrumentation and Procedures for Measuring an Indent in a Clay Backing Material

The committee was tasked to provide findings on the best instrumentation and procedures for measuring backface deformation (BFD) in clay. Accordingly, this chapter discusses relevant criteria for test instrumentation and procedures, including fixed and variable costs, precision and accuracy, and human operator considerations.

CONCEPTUAL STEPS TOWARD IMPROVEMENTS IN THE MEASUREMENT OF BFD

It is informative to review past events to learn how the instrumentation and measurements of BFD relate to the overall methodology of the original animal tests, clay selection, selection of performance specifications, and instrumentation measurements. These conceptual steps give some direction for improvements of testing procedures. It may be noted that in most experimental studies and scientific measurements, there are several conceptual and scientific stages that need to be considered and followed: (1) conceptualization of which phenomena or parameters must be measured, (2) the validity and completeness of the design of the experiment or measurement protocol to ensure a complete and accurate data set that eliminates outside variables, and (3) statistics associated with the measurements, including instrumentation accuracy vs. required accuracy. The six conceptual steps that follow trace the development of BFD measurement specifications.

Step One

Animal experiments with goats indicated that indents of 40 to 50-mm were the maximum acceptable value. Specifically, from Chapter 3 in this report, 44-mm deformation in the goats was correlated with ~10 percent probability of death from the impactor, similar to the initial program requirement of less than or equal to 10 percent lethality. This deformation level in the goats was below the lowest value for which any of the goats died in the much lower velocity impactor tests. As such, this depth was selected as the injury reference value in the clay.

This selection implies that a 44-mm goat impactor deformation is similar to a 56-mm deformation in clay or gelatin. Conversely, a 44-mm deformation in the clay is similar to 34-mm deformation in the goat for this hard impactor, as shown in Figure 3-5. (In Figure 3.5, the fatality risk at 34 mm of deformation in goats, or 44 mm of deformation in clay, is approximately 4 percent.) It was reasonable to assume that use of the blunt impactor tests to model the injury behavior from behind-armor body trauma was "likely" conservative for this impact velocity range. However, outside the velocity range typical of handgun rounds (i.e., 240 m/sec) into soft body armor, the relationship between injury and deformation response in the clay is much less certain. More recent measurements in sheep using velocities of about 800 m/sec showed significant lethality even with 34-mm deep indents (cf. Chapter 8 and Gryth et al., 2007).

Step Two

Clay BFDs were tailored to mimic such an indent. From Chapter 4 in this report, and as introduced in Chapters 2 and 3, the Roma Plastilina #1 (RP #1) modeling clay backing material used in armor testing has two important purposes. The first is to simulate the tissue response beneath the point of impact so that ballistic data generated in laboratory tests can be correlated to effects seen on the human body. The second purpose is to denote the extent of BFD during ballistic testing (Prather et al., 1977).

Multiple materials are available to simulate a body; in fact, at the time it was introduced, modeling clay was recognized to only approximate tissue response, and empirical correlations were needed to develop a probability for lethality or injury. The chief advantage of modeling clay over other materials available at the time was that it better served the function of recording BFDs; that is, when impacted, modeling clay deforms plastically, and a permanent cavity (also termed "indent," "impression," or "crater") is developed under the point of impact. Correlations were developed between the geometry of the cavity and the probability of lethal injury.

Step Three

The U.S. Army Aberdeen Test Center (ATC) has set the maximum acceptable BFD value at 44 mm for body armor plates tested using clay. This value appeared reasonable based upon the past measurements. As noted in Chapter 3, the Army does not have the medical outcomes to know whether 44 mm is a conservative value.

Step Four

Measurement instruments were used to verify the test results, as directed by the procurement specifications. Digital calipers and then laser-based instruments were used to better measure the BFD under nonideal conditions (i.e., offset and side/edge indents) (Walton et al., 2008). However, different instruments may give different BFD readings due to each instrument having different measurement precision and different measurement accuracy that is dependent upon the measurement scenario.

Steps Five and Six

Two steps have taken place since the National Research Council (NRC) was tasked to study BFD measurement techniques:

> Step Five: The office of the Director, Operational Test & Evaluation (DOT&E) developed a statistically based protocol and test processes, including measurement techniques, for first article and lot acceptance testing of hard body armor (DOT&E, 2010). (See Chapter 6.)
> Step Six: The NRC committee examined data related to the precision and accuracy of different measurement instruments, under different measurement scenarios, to gain insights into which instruments or usage might meet or exceed the precision and accuracy required to measure BFD. The different instruments and different measurement scenarios are covered in this chapter, and statistical considerations are presented in Appendixes G and M.

It is informative to keep in mind that the above conceptual steps were made to address four overall questions:

> (1) How well do the testing procedure and measurements of the BFD quantify the probability of lethality being measured?
> (2) Is the measurement a complete and scientifically valid measurement set that eliminates outside variables—e.g., the design of experiment or measurement procedure?
> (3) How well must the BFD be measured?
> (4) What are the statistical accuracy and precision of the measurements?

INSTRUMENTATION PERFORMANCE BASED ON STATISTICAL ANALYSIS

When discussing body armor testing and, particularly, the equipment required in the conduct of adequate tests, the question arises: How well must the BFD be measured? Put another way, what are the limits of acceptable error in BFD measurement? To answer this question, it is first necessary to define a number of terms.

The Phase I report (NRC, 2009) discussed the difference between the accuracy and the precision of a measuring device. Although the two terms are often used interchangeably and considered synonymous in colloquial use, they have quite different technical meanings. Accuracy is the closeness of a measured quantity to its actual (true) value. Precision is the closeness of agreement between measured values obtained by replicate measurements under specified conditions (NRC, 2009).

A measurement device is valid if it is both accurate and precise. However, a device can be precise but not accurate, accurate but not precise, or neither accurate nor precise. If a measurement device has a systematic error, then increasing the number of samples (the sample size) increases precision but does not improve accuracy. On the other hand, eliminating the systematic error improves accuracy but does not change precision.

We often quantify precision in terms of either the standard deviation or expanded uncertainty (twice the standard deviation of the repeated values) of the measurement device. Accuracy or bias is estimated as the difference between the average of a large number of repeated values under a specific set of conditions and the true value. Beginning with bias, the ideal measurement device for BFD should have no bias. That said, a biased measuring device may be acceptable in armor testing if it is consistently biased across all possible plate and test configurations. Under these conditions, the BFD test standard can simply be increased or decreased to account for the bias. However, determining that a device is consistently biased is likely to be difficult at best and unlikely to be true in practice.

Appendix G demonstrates that there are diminishing returns (and probably increasing costs) in the pursuit of ever more precise measuring devices. This result follows from the fact that the necessary level of measurement precision is a function of the overall variation in the testing process, where, for example, highly precise measurements add little value to a testing process that is itself inherently highly variable. Conversely, in any testing process, there should be a precision threshold that any measurement device must meet—again based on the overall variation of the testing process—to ensure that the measurement process itself does not add to the variability arising from the test.

For the current clay-based test methodology, the results from the Appendix G analysis suggest that a precision threshold of 0.5 mm (i.e., a standard deviation) is necessary to ensure that the measurement device does not add any appreciable variation to the body armor testing process. This value is consistent with subject matter experts who expressed to the committee their intuition that measurement precision on the order of 0.5 mm is sufficient for the current testing process. It is also consistent with injury "effect size" calculations done by the committee. It is somewhat larger than the heuristic suggested in the Phase I report (NRC, 2009) that the measurement system variance required for a test should better by a factor of 10 or more than the total measured variation (McNeese and Klein, 1991).

The original procurement specifications for body armor plates state that the BFD shall be measured with an instrument that has an accuracy of ±0.1 mm. However, the detailed analysis presented in Appendix G indicates that this value (and the accompanying wording) was probably too stringent (and somewhat ambiguous in its use of the term "accuracy"), and that a more reasonable value for an instrumentation precision of about 0.5 mm would be sufficient to correctly test and detect statistically significant effects.

Finding: Given the current clay variation, a measurement precision (standard deviation) of 0.5 mm is sufficient; instruments featuring greater precision add little practical value to the testing process. Future improvements in the inherent variability of the backing material will require instruments that are correspondingly more precise.

OVERVIEW OF CURRENT INSTRUMENTATION AND MEASUREMENTS

This section covers instrumentation that has been and is being used for BFD measurements. In particular, three instruments have been used: the Co-ordinate Measuring Machine (CMM), used as a reference instrument; the digital caliper; and a Faro laser scanning probe system. A CMM costs about $500,000; the Faro system, about $150,000; and the digital caliper, about $200. The three systems were used in extensive measurements and tests as reported in the Walton et al. (2008) report.

Coordinate Measuring Machine

The CMM is a Wenzel X-orbit Bridge type with both digital point probes and LDI laser scanner (Model SLP250); this system is has a precision of ±0.35 mm (0.00035 mm) and an accuracy of 10 mm (0.01 mm). It was considered by ATC to be highly precise and accurate and yields measurement results that can serve as a "true" value (Walton et al., 2008). The two systems that have been used to measure BFD indents made in clay during body armor testing are the digital caliper and the Faro laser.

Digital Caliper

The digital caliper (referred to as the "caliper") was used for many years as a low cost, point-to-point instrument to measure BFD depths. This was accomplished by using a manually operated depth probe integrated with an electronic digital display and paired with a bridge gauge to provide a stable base for measurement. The operator affixes a bridge gauge to the side of the box that holds the body armor and underlying clay recording medium. A baseline preshot measurement is made with a digital caliper to the point of aim where the test bullet is expected to strike the armor. The bridge gauge and caliper are removed. The test firing is conducted. The bridge gauge is replaced on the box and a measurement of the postshot BFD is made with the digital caliper. The deepest point in the BFD crater is visually located by the operator, and one reading with the caliper is taken at that point; in this case, an operator with experience and judgment is required for an accurate and consistent measurement. The two types of caliper instruments used are shown in Figure 5-1.

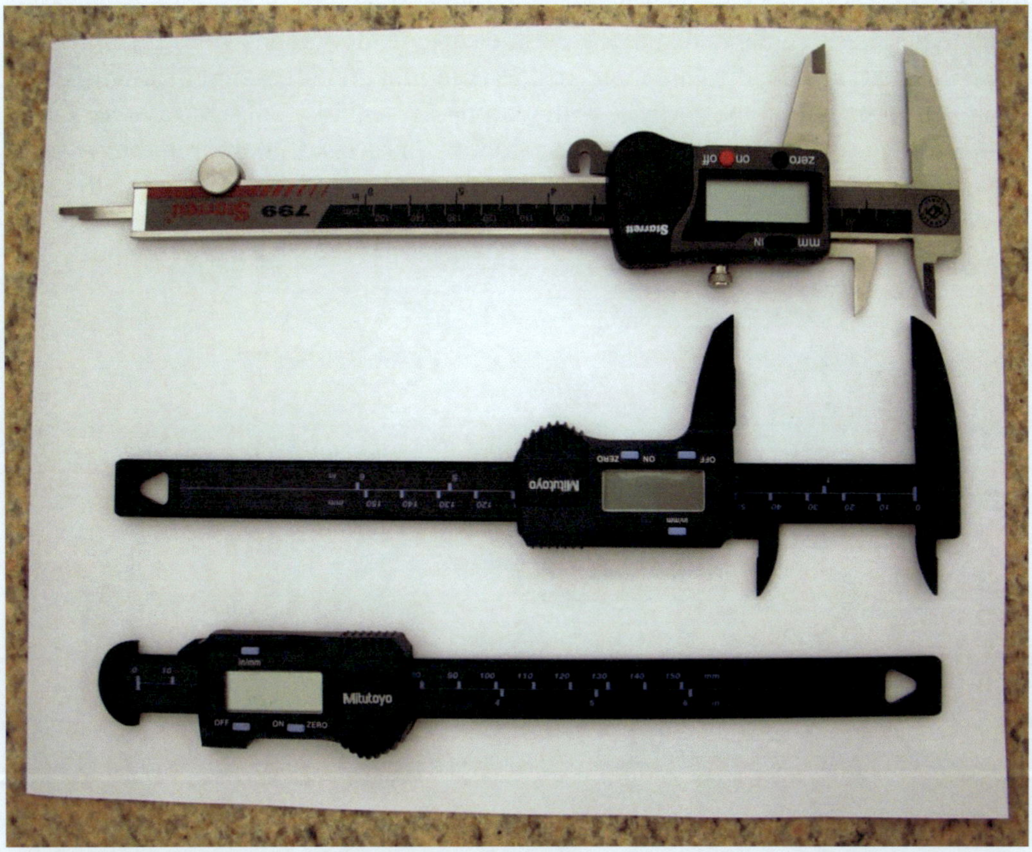

FIGURE 5-1 Digital calipers used in armor testing. The ATC standard caliper with the small end (3 mm) is shown at the top. The caliper used by commercial testers (H.P. White Laboratory and Chesapeake Testing) with a large 19-mm tip is shown at bottom. The dimension of the wide tip (19 mm) was measured by Chesapeake Testing at the request of the committee. (The center caliper is not used.) SOURCE: Courtesy H.P. White Laboratories.

Faro System

The Office of the Director, Operational Test and Evaluation (DOT&E) has designated the Faro Quantum Laser Scan Arm and Geomagic Qualify software for hard and soft body armor (referred to as "the Faro") as the device to measure BFDs (DOT&E, 2010). Laser profilometry, as used by the Faro scanning laser instrument, employs the commonly used principle of optical triangulation. A laser generates a collimated beam, which is then focused and projected onto a target surface. A lens reimages the laser spot formed on the surface of the target onto a charge-coupled device, which generates a signal that is indicative of the spot's position on the detector. As the height of the target surface changes, the image of the laser spot shifts owing to parallax. To generate a three-dimensional image of the specimen's surface, the sensor scans in two dimensions, generating a set of noncontact measurements that represent the surface topography of the specimen under inspection. The data are then used to compute the three-dimensional geometrical profile of the surface, with readings essentially continuous over the scanned region. Thus, the laser scanner produces a series of measurements over the whole surface of the clay, as opposed to the single reading obtained with the digital caliper. In addition, a laser scanning system has the ability to acquire substantial quantities of inspection data. Figure 5-2 shows the Faro Quantum Laser Scan Arm.

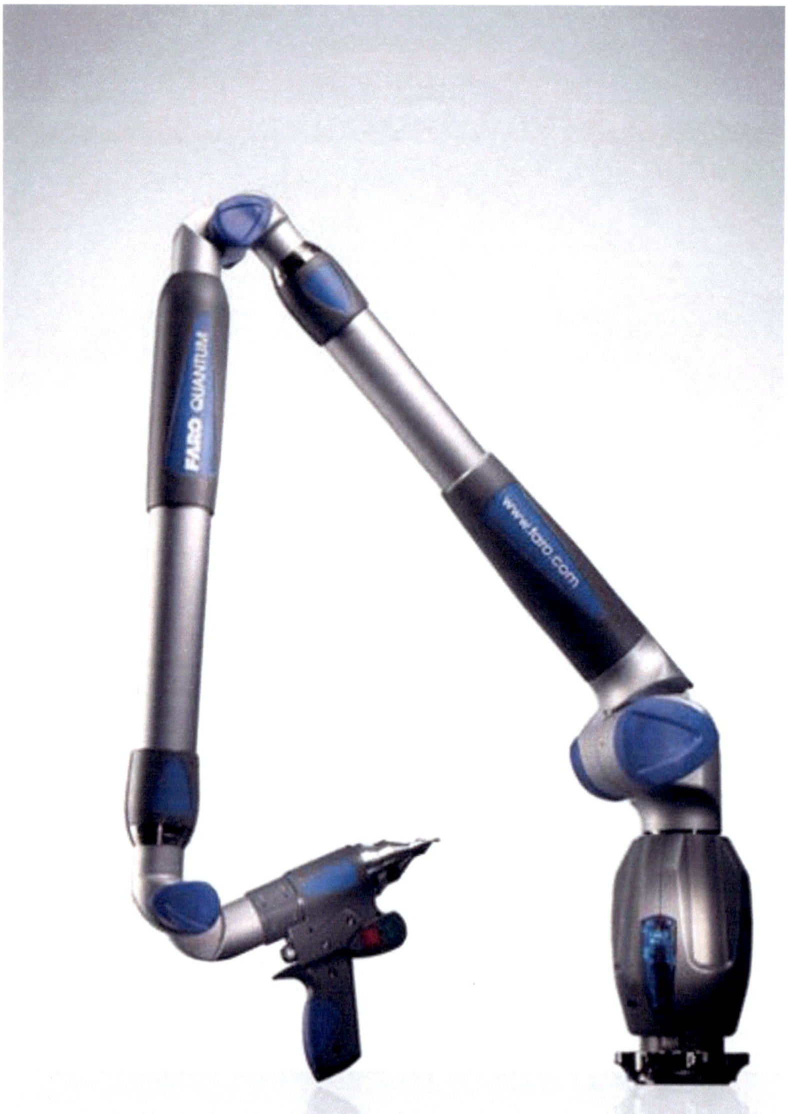

FIGURE 5-2 Faro Quantum laser scan arm. SOURCE: © 2012, FARO. http://www.faro.com/FaroArm/Home.htm. Accessed on March 8, 2011.

BFD MEASURING PROCEDURES

The precision and accuracy of instruments depend to a great degree on the associated operating procedures and on the skill of human operators. This section describes the "art of measurement" in testing procedures as observed by the committee.

Human Operator Considerations

A number of practical human operator considerations have an impact on the measurement differences and variations associated with all measuring systems. These include subjective differences in human handling, process transparency, and the selection and settings of software.

Operator-to-Operator (Human Handling) Variability

The operator-to-operator (human handling) variability associated with both measuring devices appears to be generally different. However, Walton et al. (2008) suggest that operator variability for the Faro is 0.041 mm and operator/caliper variability for the digital caliper is an order of magnitude greater, at 0.471 mm (Walton et al., 2008).

Members of the committee interviewed operators at ATC and at two commercial testing companies.[26] They learned a couple of things. The caliper end makes contact with the clay. The operator must use judgment to determine the deepest point in what may be a complex BFD and then carefully and manually push the caliper arm down so it just touches the surface of the clay but does not dent the clay. Operators state that making precise measurements is an art. Variation among operators can be 0.1-0.3 mm when measuring the same BFD in the center of an armor plate. This variation was actually observed at ATC when three different operators measured the same BFD using a digital caliper.

The Faro is a noncontact device. The operator must use judgment as to where and how fast to "paint" the armor on a prefire event (to digitally capture the surface of the armor) and similarly how fast to paint the BFD area in a postfire event. The computer program compares the pre- and postfire digits and calculates the maximum BFD. According to experienced operators, variation due to these judgmental factors can result in measurement differences among Faro users of 0.1-0.2 mm for the same BFD. Similar differences in results were observed during a demonstration to the committee when three different operators measured the same BFD with the Faro.[27]

---

[26]Site visits to the ATC, H.P. White Laboratory, and Chesapeake Testing by members of the committee on August 30 and 31, 2010.

[27]Variations were reported by Faro operators at the ATC and commercial testing sites. Variations were then observed by the committee during demonstration at the ATC, August 31, 2010.

Testing protocols should anticipate that anomalous data can occur for any number of reasons and should include procedures to ensure data quality. These protocol procedures can provide a means for operators to quickly confirm that a measurement outside a predetermined upper and lower bound is not due to a major equipment or software malfunction. The committee notes that great caution is warranted if this idea is implemented because it has the potential to lead test operators to focus on measurement differences that are the result of noise and not actual differences.

Software Variability

There is a software variability associated with the Faro resulting from a software selection that allows for smoothing the raw digital data captured by the Faro. For example, the committee was shown that a change from one level of smoothing to another resulted in a 1-mm difference in the BFD measurement for the same cavity.[28] Two software settings are in use, one for 0.7 mm and one for 1.5 mm spatial resolution. An ATC manager stated that ATC testers tended to use the most conservative setting (i.e., higher spatial resolution of 0.7 mm), which will result in the largest BFD measurement to ultimately protect soldiers.[29] Manufacturers, on the other hand, feel that their armor may be unfairly penalized due to judgment decisions that depend on the smoothing setting the operator chooses.[30]

In statistical and testing terms, the choice of the smoothing setting directly affects the accuracy of the Faro. That is, the choice of smoothing settings can introduce a systematic bias into the measurements, a bias that can make the test either harder or easier to pass depending on whether the bias results in systematically larger or smaller BFDs. As discussed in Appendix G (Key Point 4), an overly conservative setting on the Faro laser resulting in high spatial resolution may result in a design penalty that is roughly five times larger than the design space improvement achieved via better measurement precision. Thus, unless care is taken to understand the effect of the software smoothing algorithms on the indent measurement, any gains in precision achieved by using the Faro could be more than offset by a systematic bias. This result might not only make the test harder for manufacturers to pass but also might result in heavier armor if manufacturers are driven to make plates heavier to compensate for a measurement bias.

The committee considers the National Institute of Standards and Technology (NIST) to be an excellent third-party source of expertise on measurement instruments and standards, because the NIST has provided significant support to both DOT&E and the Army Program Executive Officer Soldier on both body armor testing and body armor design issues in the past.

---

[28]As observed by committee members during their site visit to the ATC, August 31, 2010.

[29]Discussion with Irene Johnson, ATC, during site visit, August 31, 2010.

[30]David Reed, President, North American Operations, Ceradyne, Inc., “Pragmatics of Body Armor Testing—Manufacturer’s View,” presentation to the committee, August 9, 2010.

Recommendation 5-1: An organization such as the National Institute of Standards and Technology should conduct a controlled study to determine the most reasonable and consistent Faro smoothing settings to be used while measuring backface deformations (BFDs) in body armor testing. Similarly, any other software selections that could cause relevant changes to BFD measurements should be studied. Corresponding values for the precision and accuracy of each software setting will need to be quantified.

Compensating for Offset between the Point of Aim and the Deepest Indent

Sometimes the deepest penetration in the clay and the initial bullet aim point are offset by a small distance. This affects both accuracy and precision of the instrumentation measurements. Operators of the caliper calibrate their instrument on the aim point but move the caliper to measure the deepest point of impact when the aim point and deepest penetration do not align, which happens frequently. The caliper measurement procedure disregards the curvature of the plate and tends to overestimate the depth of the BFD. This correction and offset value can be large and is the result of having only one preshot reading for the plate surface. As a mathematical correction for the offset, an operator referred the committee to an equation for offset contained on an ATC test instruction sheet.[31] Government and commercial operators alike felt that the equation was imprecise and would likely lead to an underestimation of BFD. Also, the equation does not make provision for the offset being changed to a positive number if the deepest point is on the upside of the aim point on an edge shot.

In comparison to the caliper, the Faro takes into account the curvature of the plate, calculates the geometry, and reduces offset errors. This computational capability allows the Faro to measure and calculate an offset value with high precision and leads to a more accurate measurement of the maximum indent.

Variability (Noise) in the Overall Testing Process

As described in Chapter 4, there is significant variability in the RP #1 modeling clay that has been used for decades as the backing material in the testing process.

- The RP #1 modeling clay was and continues to be designed for artists and not for the ballistics testing community. As a result of requests from artists for a certain feel and other characteristics the formulation has changed over time. From the standpoint of the ballistics testing community, the clay over time has been allowing less deformation than the original RP #1.

---

[31]The equation used is Offset = $-0.25 \times$ BFD. The test instruction sheet was shown to members of the committee during site visit to the ATC, August 31, 2010.

- To compensate for the change in formulation, the testing community has had to warm the clay in ovens to achieve the calibration numbers required by National Institute of Justice (NIJ) standards. Heat introduces significant variation.
- The amount of manual working that is performed using mallets to initially pound the clay bricks into the testing box or recondition the clay after a test shot introduces additional variability in clay deformation. As described in Chapter 4, some studies indicate that this human dynamic of working might introduce even more variability in the deformation of clay measurements than changes in temperature.
- It has been observed over the years of body armor testing that clay in a box used for testing has a limited useful life. Since old clay may result in unreliable deformations during testing, both government and commercial testers routinely discard clay before it is a year old. Although variability due to aging appears to be less than variability introduced by temperature and working, it is one more indicator of the significant variability that is inherent in RP #1.

Owing to the above and other considerations, the NIJ standard allows for significant modeling clay variation. Specifically, to determine if the modeling clay is ready for testing it must calibrate to a specification of 25 mm ± 3 mm. In other words, 6 mm of overall clay variability is accepted as, and perhaps understates, the noise in the testing process related to clay.

A great deal of variation is introduced into the measurement system by RP#1. As discussed in Chapter 4, there is much merit in reducing the variability in the recording medium, because with less variability in the recording medium testers can more fairly state that the differences in test results are related to plate behavior. One battlefield payoff will be greater confidence that the plates will function successfully in combat. Another is the possibility that lighter weight plates will be able to pass the tests, ultimately reducing the weight burden on soldiers.

Some additional variation occurs as a result of the ammunition that is used during testing. If, for example, a tester was to replicate the real-world threats that a soldier might face, that tester would use ammunition procured from third-world countries. Such ammunition may not have been manufactured to specifications that result in consistent velocity and bullet mass from round to round. Variation in velocity will cause some variation among BFDs that are created during testing, because the energy is proportional to the velocity squared. Bullet velocity measurements are part of ATC testing procedures and should be part of all live-fire tests. Within one batch, manufactured small arms have a velocity variation leading to a 12 percent difference in deposited energy.

Variability can also result from human handling or subjective software selection within the measurement systems. As discussed previously, such variability can result in different measurements for the same BFD.

NEED FOR A STAND-ALONE BFD ARTIFACT OR STANDARD MODEL FOR INTERORGANIZATION VERIFICATION

An important issue that should be addressed is the importance of having a measurement standard to determine the ability of any given device (caliper, Faro, etc.) to precisely measure a representative BFD regardless of the organization, measurement instrument, software version, operator, and so forth.

In the development of methods to measure BFDs, virtually no inter-laboratory testing has been carried out to date to determine sources of inter-laboratory errors as a consequence of test procedure differences or differences in the setup of the test equipment or of differences in the operation of equipment. Interlaboratory errors are often systematic, resulting in a constant statistical difference in BFD measurement from one measurement laboratory to the next. These measurement errors can lead to undue acceptance or rejection of lots of ceramic armor, which is undesirable. Interlaboratory errors can be rooted out by having several laboratories run the same test with the same or equivalent instrumentation. The source of the error can be identified and eliminated by a change in experimental procedure or equipment.

A physical artifact would replicate a standard BFD cavity and perform a gauge block function for noncontact instruments. That said, the BFD standard artifact should be more than just a gauge block. Rather it should represent a physical model of the complete BFD measurement process. It should allow operators to follow a four-step process:

1. Measure a representative preshot surface;
2. Measure a representative postshot BFD crater;
3. Subtract the two numbers; and
4. Compare the number from the previous step with the artifact's standard depth.

The result would quickly determine if the device as used was sufficiently accurate for this application. For example, a complete artifact system could be made that mimics the preshot surface with a flap that covers a replicate BFD crater. Such a model could be made of hard plastic, or, a softer coating could be applied. While the thickness of the flap would affect the absolute readings, the relative readings between organizations and operators would not be affected. A single artifact system, upon acceptance by NIST and the testing community, would become the one national standard for quickly confirming a device's precision and accuracy for measuring a BFD.

Previous work by NIST has established the usefulness of such a standard (NIST, 2010). The committee supports turning this idea into a practical solution for the entire body armor testing community. In addition to the test standard just described, evaluation of interlaboratory test variation is important for establishing test reliability. Hence, interlaboratory tests should be run in order to establish the accuracy of a test as well as its precision.

Recommendation 5-2: An organization such as the National Institute of Standards and Technology should develop a standard backface deformation artifact system and procedures to allow operators to ensure that different measurement devices at different locations are able to meet specified levels of accuracy and precision.

CHARACTERISTICS OF A "BEST UTILITY" MEASURING INSTRUMENT

Based on the preceding discussion of the instruments and procedures for measuring BFD, the committee developed criteria for a measuring device that would provide the "best utility" for the body armor testing applications. A best utility measurement device must meet the following criteria:

- Meet or exceed precision and accuracy requirements for measuring BFD;
- Achieve the lowest practical fixed and variable costs; and
- Minimize human judgment and error.[32]

In addition, it would be advantageous if the instrument could also

- Be versatile enough to measure indents behind both plates and helmets[33] and
- Be widely available and supportable here and abroad.

An example of an instrument that will have potential in the future is being tested at the Army Research Laboratory.[34] The Microscribe Model G2LX is a digital arm/mechanical scribe instrument that is being used by the Army Research Laboratory in research on the BFD cavities formed in the head forms used to test helmets.

The G2LX, which costs approximately $8,500,[35] has an advertised precision of 0.012 in. (0.3 mm). The system is connected to a computer that can capture measurements made by the operator based on a three-dimensional *x, y,* and *z* coordinate system. It also has an automated database that captures

---

[32]Capturing measurement readings in an automated database is helpful. Expert testing operators who spoke with the committee agreed that manually capturing readings can lead to transposition and other errors. There are commercially available automated database interfaces for both contact and noncontact instruments.

[33]See Chapter 7 for a description of the helmet testing process. The differences between armor plate testing and helmet testing are considerable, and all operators interviewed agreed that a laser-based measuring tool was generally preferred for helmet testing due to the complex curves of the head form, on which the helmet BFD measurements must be made.

[34]Committee site visit to the ATC, August 31, 2010; Rob Kinzler, Army Research Laboratory, "Improvements in Helmet Measurement," presentation, to the committee, October 13, 2010.

[35]Source: http://www.3d-microscribe.com.

measurements made by the user. The user can activate a finger or foot switch to notify the computer to enter the current measurement into the database.

During a demonstration to the committee, the time required to measure a clay indent appeared to be less than that required by a caliper since there is no need to set up a bridge gauge. The G2LX system is a basic one-point contact measurement system, which means it suffers from the same inability to compensate for offset as does the caliper.[36] The system combines a fairly inexpensive robotic arm capability, similar to that of the Faro system, with an inexpensive hard-mount caliper scribe end.

The MicroScribe system could be used for testing both body armor plate and helmet BFD measurements (although a finer grid pattern is needed for helmet testing) and could significantly reduce the offset error currently seen with the caliper, which uses one preshot measurement.

MicroScribe offers a more sophisticated arm advertised to achieve a precision of 0.003 in. (0.0762 mm) on the upgraded MLX model; it costs approximately $22,000. The robotic arm can be outfitted with a laser scanner similar to that used on the Faro for an additional $15,000 or so. The MicroScribe system is just one example of instruments that are available in the commercial sector. The committee believes there are also others that have "best utility" characteristics and are readily available.

Finding: The data available to the committee were not obtained through a formal gauge or "artifact standard" repeatability and reproducibility study by an independent agency. Thus, the committee can draw no quantitatively reliable conclusions about the precision and accuracy (potential biases) of the measurement systems it examined.

Late in the course of the committee's final deliberations as it prepared this report, it received additional test results data that had not been available to it earlier in the effort (see Appendix M). Considering all available data, the committee recognized (1) an insufficiency of sample sizes for all the data examined; (2) inconsistencies in the direction and magnitude of biases; and (3) presumed large differences in offset magnitudes between the data in Hosto and Miser (2008) and the more recent live-fire test experiences.

The committee determined from its analysis of the available data that remedial procedures for properly designed evaluations are needed to determine the magnitudes of accuracy and precision of current or proposed instruments in the measurement of body armor BFD before definitive conclusions can be drawn regarding best utility.

Recommendation 5-3: In anticipation of future test measurement requirements, the Office of the Director, Operational Test and Evaluation and/or the Army

[36]The offset measurement problem could be overcome by having the operator enter several point measurements on the surface of the clay near the aim point prior to the test round being fired. The extent of the grid pattern (e.g., 3 × 3 vs. 4 × 4 grid) would depend on the accuracy of the BFD measurement that was needed.

should charter an organization such as the National Institute of Standards and Technology to conduct an analysis of available candidate commercial instruments with inputs from vest users, manufacturers, testers, policy makers, and others. The goal is to identify one or more devices meeting the characteristics of "best utility" measuring instruments as defined in this study to the government, industry, and private testing labs.

The list of best utility instruments should be shared with NIJ, international allies, and others, as appropriate, to promote measuring instrument standardization for body armor testing nationally and internationally. A formal gauge repeatability and reproducibility study is required to quantify accuracy and precision as inputs to the best utility analysis.

REFERENCES

DOT&E. 2010. Memo on Standardization of Hard Body Armor Testing, 27 April 2010, Director, Operational Test and Evaluation.

Gryth, D., D. Rocksen, J. Persson, U. Arborelius, D. Drobin, J. Bursell, L. Olsson, and T. Kjellstrom. 2007. Severe Lung Contusion and Death After High-Velocity Behind-Armor Blunt Trauma: Relation to Protection Level. Military Medicine 172(10): 1110-1116.

Hosto, J. and C. Miser. 2008. Quantum FARO Arm Laser Scanning Body Armor Back Face Deformation. Report No. 08-MS-25. Aberdeen, Md.: U.S. Army Aberdeen Test Center Warfighter Directorate, Applied Science Test Division, Materials and Standards Testing Team.

McNeese, W., and R. Klein. 1991. Measurement systems, sampling, and process capability. Quality Engineering 4(1): 21-39.

NIST (National Institutes of Standards and Technology). 2010. Dimensional Metrology Issues of Army Body Armor Testing. Gaithersburg, Md.: NIST.

NRC (National Research Council). 2009. Phase I Report on Review of the Testing of Body Armor Materials for Use by the U.S. Army: Letter Report. Washington, D.C.: National Academies Press.

Prather, R., C. Swann, and C. Hawkins. 1977. Backface Signatures of Soft Body Armors and the Associated Trauma Effects. ARCSL-TR-77055. Aberdeen Proving Ground, Md.: U.S. Army Armament Research and Development Command Technology Center.

Walton, S., A. Fournier, B. Gillich, J. Hosto, W. Boughers, C. Andres, C. Miser, J. Huber, and M. Swearingen. 2008. Summary Report of Laser Scanning Method Certification Study for Body Armor Backface Deformation Measurements. Aberdeen Proving Ground, Md.: Aberdeen Test Center.

6

Statistical Considerations in Body Armor Testing

During Phases I and II of the study, the committee was requested to consider the use of statistics to permit a more scientific determination of sample sizes to be used in body armor testing. Specifically, the committee was requested to review a statistically based protocol that had been developed by the Office of the Director of Operational Test and Evaluation (DOT&E) with assistance from Army statisticians and testers. The Phase II report provided initial insights on statistics-related issues. This chapter consolidates those insights and provides more detail.

In this chapter, the committee presents its findings on statistical aspects of body armor testing with a focus on body armor plate testing beginning with general discussions of (1) what it means to conduct statistically principled testing, (2) how uncertainty and variation can influence overdesign and overmanufacture, and (3) important considerations in test protocol design.[37] The chapter then proceeds to describe the Army and the U.S. Special Operations Command (USSOCOM) historical test protocols and discusses the new first article testing (FAT) protocol and the proposed lot acceptance testing (LAT) of the DOT&E, including a discussion of the assumptions underlying the statistical methods and protocol design trade-offs.

INTRODUCTION

This introduction discusses the concepts of statistically principled testing, how uncertainty and variation drive overdesign, and key test protocol design requirements and considerations.

Statistically Principled Testing

The testing of body armor and helmets is destructive, meaning that the tested items are damaged as a result of the test and thus are no longer fit for use upon completion of the test. For this (and other) reasons, only a fraction of the items produced are tested. All such testing is inherently statistical since we use the information (i.e., the data and the resulting statistics) from a sample of tested items to learn something about the quality, acceptability, and/or fitness for use of the larger (untested) population of items. In statistical terminology this is referred to as

[37]Statistical considerations for helmet test protocols are not discussed.

"inference," where the goal is to infer something about the unobserved population based on the data obtained from the observed sample. Thus it is correct to say that all such tests, including the Army's original body armor test protocol, are statistically based. However, it is critical to note that not all statistically based tests are statistically principled.

A "statistically principled" test uses appropriate statistical methods to properly make formal inferences about the population from the sample. Formal inference means that the desired characteristic or characteristics in the population are estimated from the sample data in such a way that uncertainty inherent in the inference from sample to population is appropriately and explicitly accounted for by the statistical methods. In the case of testing, this generally means a particular sample size is specified (as well as other sampling and estimation details) to minimize the uncertainty to some acceptable level. Thus, the use of statistically principled test procedures and test methods allow decision makers, test organizations, and manufacturers to all have confidence that the test performance of the sample appropriately characterizes the performance of the population.

Uncertainty and Variation Drive Overdesign

Larger and/or thicker body armor insert plates provide additional survivability but at the cost of more weight. Heavier body armor can contribute to fatigue, may inhibit mobility and effectiveness, and, at its worst, may result in personnel choosing not to wear the body armor, completely defeating its purpose (OTA, 1992).

Body armor is designed to protect against a particular level of threat. To the extent that the armor exceeds this level, it can be thought of as overdesigned or overmanufactured, in the sense that lighter plates could have been produced to achieve the desired level of protection.

Uncertainty and variation in the manufacture, testing, and employment of body armor, as well as the natural concern for protecting personnel, tend to result in conservative decision making, which in turn can result in body armor overdesign and/or overmanufacture. For example,

- Variation in body armor manufacturing processes can drive suppliers to produce plates that are generally heavier than required to lower the risk of producing nonconforming plates.
- Variation in FAT and LAT can further drive suppliers to produce heavier-than-necessary body armor to ensure their product successfully meets the FAT and LAT test standards.
- Uncertainty about the particular threat that personnel may face can result in tighter specifications and/or testing to a higher possible threat and sometimes to threats beyond what personnel would actually experience in order to ensure that the threats are clearly met.

Furthermore, with statistically principled test protocols, variation in both the manufacturing and testing processes requires testing greater quantities of body armor to achieve a given level of certainty about performance. To the extent that variation in the manufacturing and testing processes is reduced, higher certainty about body armor performance can be achieved within a given testing protocol or, alternatively, fewer tests can be conducted, with attendant savings in cost and effort, to achieve an equivalent level of certainty.

In sum, uncertainty and variation at each step of design, manufacture, and test are frequently accounted for with safety margins, the cumulative effect of which can be overdesign. To the extent that uncertainty and variation in manufacturing and testing are minimized, body armor with the desired level of performance could be achieved with greater certainty and perhaps lighter weight.

Key Test Protocol Design Requirements

The most fundamental requirements for the new protocols are that they are (1) statistically principled and (2) implemented across the Department of Defense (DoD). As previously described, having a statistically principled test protocol ensures that acceptance decisions are based on defensible, scientifically sound principles and methodology. DoD-wide implementation of the protocols ensures that all body armor in DoD meets a common, minimum standard of performance. Both of these requirements are reflected in the DoD Inspector General (IG) report (DoD, 2009) and in a DOT&E memorandum (DOT&E, 2010a).

A corollary is that the standards in the proposed protocol, and any subsequent modifications to them, should be based on empirical evidence. There are two aspects to this:

- Body armor procured under the Army's original (statistically based but not statistically principled) test protocols have performed well in theater. As discussed in Chapter 2, there are no known deaths attributed to failure of existing body armor against the threats for which it is designed.[38] Thus, the new protocol standards should not

[38]"There have been no known soldier deaths due to small arms that were attributable to a failure of the issued ceramic body armor." (PEO-S, 2010). Likewise, as discussed in personal communication between James Zheng, Chief Scientist, Office of the Program Executive Officer, Soldier, and Larry Lehowicz, Chair, December 29, 2009, in no case has it been determined that an issued enhanced small arms protective insert (ESAPI) or enhanced side ballistic insert (ESBI) armor plate failed to prevent an armor piercing by small arms projectiles of 7.62 mm × 63 mm or less.

The committee notes that the statement in PEO-S (2010) is carefully qualified. It is possible that soldiers wearing body armor may suffer casualties when their ceramic armor is defeated by rounds of caliber larger than 7.62 mm × 63 mm, when projectiles or shrapnel strike a eliminate currently acceptable body armor designs from continued production, nor should they negatively impact the design or inappropriately incentivize changes to the design solely because of the new standards.

- On the other hand, changes to the proposed new protocol, and any changes to future protocol requirements, should be based on empirical evidence and solid statistical analyses. This is not meant to suggest that expert judgment should be ignored; such judgment is often crucial for insight and understanding. However, given that the test protocol design is intended to be a scientifically defensible, statistically principled protocol, changes to the protocol should be based on the same criteria. Under these conditions, proposed changes must be based on empirical evidence, not anecdote and opinion.

Finally, a design consideration is that any protocol should (1) be flexible enough to accommodate mission-specific needs (or lack thereof) and (2) as necessary, allow the standards to vary by threat. As for flexibility, and as previously described, it is critical that the protocol specifies requirements that ensure a scientifically sound, statistically principled test that achieves a minimum standard of body armor performance DoD-wide. However, there are likely benefits to a protocol that is not unnecessarily overly specific. As for the latter threat consideration, since the performance of the body armor varies by threat, it may be useful to have threat-specific standards. In particular, and perhaps more to the point, having a single common set of protocol standards for all threats could result in body armor that is overdesigned for the actual or most likely threat.

---

portion of the body not protected by body armor, when the blast comes from improvised explosive devices (IEDs) or other explosives, and so forth. In addition, it is also possible that casualties have occurred but were not attributed to failure in the body armor, for example, when a casualty becomes separated from issued body armor and it is not be possible to track the armor back to the original casualty.

According to Lt. Col. Edward L. Mazuchowski, Deputy Medical Examiner, Armed Forces Institute of Pathology, in a presentation to the committee entitled "Body Armor and Blunt Force Injury: A Medical Examiner's Perspective," August 11, 2010, there has been no evidence of a failure of body armor against the threats for which it is designed based on forensic analysis of the casualties. This report must be qualified by the fact that not all casualties are returned with their body armor. However, it is not unreasonable to conclude that body armor failures (against threats for which the armor is designed) must at most be quite rare since, if such failures were more common or frequent, it is likely that at least one failure would have been observed over the years that body armor has been deployed in combat.

Finally, Lt. Col. Peter Greany, USSOCOM, stated in discussion with the committee statistics working group on October 12, 2010, that there had been "zero USSOCOM fatalities due to body armor failure."

BACKGROUND

This section discusses the historical FAT protocols, the DOT&E protocol for body armor FAT, test protocol assumptions, and LAT.

Historical First Article Testing Protocols

FAT is used to ensure that body armor (and helmets) conform to all contract requirements for acceptance, including specific inspections and tests as well as drawings or other specifications. As described in the DoD IG report *DoD Testing Requirements for Body Armor*, the U.S. Army and the USSOCOM originally conduct FATs using the same measures (probability of penetration and backface deformation [BFD]) but to separate standards (DoD, 2009).

Under their original protocols, both the Army and USSOCOM assess ballistic performance using penetration and BFD under various threats and environmental conditions. They both assess V_{50}, the highest velocity of a threat at which the probability of complete penetration is 50 percent, by measuring plate penetration over a range of velocities. In addition, USSOCOM tests plate shatter gap, which occurs when a bullet penetrates body armor at a lower velocity than the body armor was designed to defeat (DoD, 2009).

As described in the IG's report, the original Army protocol for body armor testing is statistically based but not statistically principled. It is statistically based because a sample of plates is tested with the intention of inferring the properties of a larger but unobserved population of plates. However, it is not statistically principled, because small sample sizes and an ad hoc scoring rule do not support formal statistical inference of the population. In particular, for enhanced small arms protective inserts, the Army requires testing of the following:

- Three plates, each against a defined matching threat in ambient conditions,
- Three plates against a defined overmatching threat in ambient conditions, and
- One plate for each of nine environmental conditions.

In addition, the original Army protocol uses 12 plates for V_{50} testing, so that in total 27 plates are tested (Dunn, 2010). FAT failure occurs with (1) one or more catastrophic penetrations or BFD failures on

V_0 tests[39]; (2) accumulation of a limited number of failure points; or (3) failure to meet minimum V_{50}.[40]

When testing a plate, the first shot must be within ¾ in. to 1¼ in. or 1 in. to 1½ in. (depending on the threat) of an edge. The second shot (assuming the plate passes the first shot) is targeted either 3 in. to 6 in. or 4 in. to 5 in. (again, depending on the threat) away from the impact location of the first shot, at least 1.5 in. from any edge, and at the ballistically weakest point of the plate (RDECOM, 2009).

The original USSOCOM protocol is statistically principled with sample sizes that can vary from a minimum of 146 plates tested to a maximum of 480 plates tested. At the minimum, USSOCOM requires the following:

- Sixteen plates each against four defined threats, including one overmatching threat under ambient conditions, and
- Six plates for each of eight environmental conditions.

When testing a plate, the first shot must be within 0.75 in. to 1.25 in. of an edge and then the second shot (assuming the plate passes the first shot) is targeted 4 in. toward the center of the plate from the impact of the first shot. As shown in Figure 6-1, subsequent plates are tested by proceeding clockwise.

In addition, the original USSOCOM protocol uses 6 plates for V_{50} testing and another 28 for shatter gap testing. Should a plate fail in any category, the USSOCOM protocol requires additional testing in that category. Successful completion of the USSOCOM FAT is based on achieving the following:

- A 90 percent probability the plate will stop the first shot and not exceed BFD requirements (44 mm), with 80 percent confidence for all four defined threats,
- A 90 percent probability the plate will stop the second shot with 80 percent confidence for the three matching threats, and
- A 60 percent probability the plate will stop the second shot for the overmatching threat with 80 percent confidence (USSCOM, 2010).

---

[39]V_0 testing is "resistance to penetration" testing and occurs at velocities where there should be no plate penetrations.

[40]LTC Jon Rickey, Product Manager, Soldier Protective Equipment, "Historical XSAPI Results APR 09 - JUN 10," presentation to the committee's statistics working group, October 12, 2010.

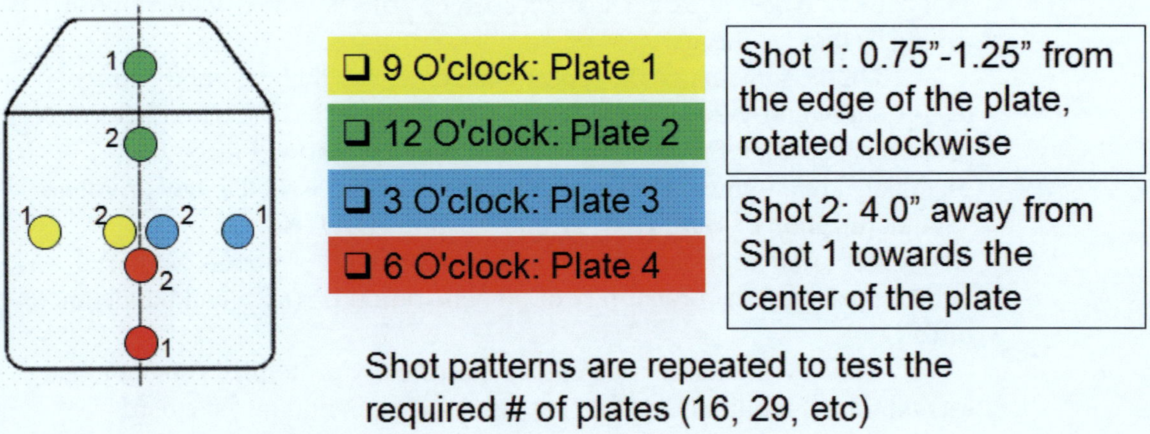

FIGURE 6-1 USSOCOM FAT shot pattern. SOURCE: P. Greany, discussion with the committee's statistics working group, October 12, 2010.

In its report, the DoD IG analyzed Army and USSOCOM protocols and, based on a first shot comparison for the overmatching threat under ambient conditions, found that ". . . the USSOCOM sampling plan provided a 27 percent better chance that defective plates are detected during first article testing. . . ." (DoD, 2009, pp. 30-31). The DoD IG attributed the 27 percent improvement in defective plate detection "primarily to the total number of plates tested" by USSOCOM (DoD, 2010).

Finding. Because of their differences, and as demonstrated in the Department of Defense (DoD) Inspector General calculations, neither the historical Army protocols nor the United States Special Operations Command protocols met the key protocol design requirement as a common standard DoD-wide. In addition, the historical Army protocol did not meet the key design requirement as a statistically principled test.

DOT&E Protocol for Body Armor FAT

In *DoD Testing Requirements for Body Armor*, the IG recommended that "the Director, Operational Test and Evaluation (DOT&E) develop a test operations procedure for body armor ballistic inserts and involve the Services and USSOCOM to verify the procedure is implemented DoD-wide" (DoD, 2009, p. i). It also stated that "standardization of body armor testing and acceptance will ensure that Service members receive body armor that has been rigorously tested and will provide uniform protection in the battlefield" and proposed that "the test procedure should include, at a minimum, requirements for sample size, shot pattern, types of testing, and acceptance criteria to verify the rigor of testing." (Ibid., p. 32)

On the same page, the report went on to say that ". . . body armor testing should provide a certain level of confidence that the manufacturing process is capable of producing an armor product that will meet the established requirements." In response, the DOT&E issued a new protocol, "Standardization of Hard Body Armor Testing" (DOT&E, 2010a). Assessment standards for both old and new DOT&E protocols are based on two measures: the probability of no penetration [Pr(*nP*)] and the depth of BFD.

The new standard establishes a statistically principled protocol that sets minimum requirements for first article tests, including "standard testing references, protocols, procedures, and analytical processes for hard body armor testing." A key component of the protocol is a 60-plate design matrix that specifies the number and sizes of plates to be tested in each of nine environments and under ambient conditions and by shot order (Table 6-1). The 60-plate design matrix is replicated for each threat. The proposed standard does not specify the threats for testing.

An important consideration when evaluating this design matrix is to recall that it is designed for acceptance (i.e., contractual) testing as opposed to operational testing. An acceptance test is intended to evaluate the hard body armor against contractual requirements—in this case, against a requirement for the probability of penetration and BFD under a variety of environmental conditions. In contrast, an operational test assesses performance under realistic operational conditions and, as such, might lead to different choices about the allocation of tests. For example, an operational test might allocate additional plates to ambient conditions if that was determined to be the most likely environment in which the plates would actually be used.

The committee notes that the design is reasonably balanced, with every size plate appearing in each environment and an equal number of tests for the two shot orders. Based on analytical results of past test data conducted by Army statisticians, the inclusion of shot order (first shot edge/second shot crown vs. second shot crown/second shot edge) in the 60-plate design matrix is relevant since plate performance varies by shot order. The committee also notes that the current design matrix requires USSOCOM to test under one ambient condition not currently tested and to test extra small plates, which USSOCOM does not use.[41]

Finding. Because the protocol requires the same 60-plate protocol for all tests, some user communities are required to test for environmental conditions and plate sizes that are not necessarily relevant to those communities.

---

[41]Lt. Col. Peter Greany, USSOCOM, discussion with the committee's statistics working group, October 12, 2010.

TABLE 6-1 60-Plate Protocol

Environment	First Shot Edge/ Second Shot Crown	First Shot Crown/ Second Shot Edge
Ambient (unconditioned)	1 extra small plate 1 large plate 1 extra large plate	1 small plate 1 medium plate 1 extra large plate
Temperature cycling	1 medium plate 1 large plate 1 extra large plate	1 extra small plate 1 small plate 1 medium plate
JP-8 soak	1 extra small plate 1 small plate 1 medium plate	1 medium plate 1 large plate 1 extra large plate
Oil soak	1 small plate 1 medium plate 1 large plate	1 extra small plate 1 small plate 1 extra large plate
Salt water	1 extra small plate 1 medium plate 1 extra large plate	1 extra small plate 1 small plate 1 large plate
Weathered	1 small plate 1 medium plate 1 extra large plate	1 extra small plate 1 large plate 1 extra large plate
High temperature	1 small plate 1 large plate 1 extra large plate	1 extra small plate 1 medium plate 1 large plate
Low temperature	1 extra small plate 1 small plate 1 extra large plate	1 small plate 1 medium plate 1 large plate
Altitude	1 extra small plate 1 medium plate 1 large plate	1 small plate 1 large plate 1 extra large plate
Total	27	27
Impacted[a]	2 extra small plates, 1 small plate, 1 medium plate, 1 large plate, 1 extra large plate	
Total plates tested	60	

[a] Shot order is not relevant for impacted plates since the first shot is taken at the most severely damaged part of the plate as identified by X-ray. In the absence of a visible crack, the first shot is taken at the crown and the second shot 5-6 in. away from the first shot impact location and not closer than 1.5 in. to an edge.

SOURCE: DOT&E, 2010a.

In addition, the committee notes that 54 out of the 60 plates are tested under other than ambient conditions. To the extent that these conditions are not often experienced during operational use, the testing may skewed toward assessing plate performance under less common conditions. On the other hand, as previously discussed, this is an acceptance test and not an operational test, so if performance in these environments and under these conditions is contractually required, then testing six plates per environment is certainly appropriate. However, it is worth noting that the resulting statistical inference is to a population of plates that experiences environmental conditions in proportion to the fraction of plates tested in each condition in the design matrix. To the extent that some of the environmental conditions are stressing, this could result in a conservatively biased test, in that the resulting estimates for probability of penetration and/or BFD may be greater than that experienced under actual operational conditions.

The committee understood that the choice of a 60-plate sample size resulted from the necessity to balance statistical precision against the real-world constraints of test range capacity as well as the cost and length of the tests, as is the case with all statistical test designs. Specifically, in the absence of constraints, more testing will provide better estimates of plate performance. However, test range capacity is not unconstrained, nor are budgets, and the time it takes to conduct a FAT must be reasonable so that production is not unduly delayed. In the case of body armor, it was determined that testing 60 plates per threat is executable and provides sufficient statistical precision to assess the aggregate performance of a manufacturer's plates for that threat. A consequence of that choice (and the design of the test matrix) is that the effects of plate size, shot location, and environment can all be estimated, as can the size by location and the location by environment two-way interactions; size and environment, however, are confounded.

Finding. The 60-plate protocol makes a reasonable (and necessary) trade-off between the precision of the statistical tests and real-world constraints, such as test range capacity and test costs.

The assessment standards for the DOT&E protocol are based on two measures, the probability of no penetration and BFD. Table 6-2 provides an overview of the statistical basis for the proposed FAT protocol. For any threat, the following is required to successfully pass the FAT:

- For the first shot, the 90 percent lower confidence bound for the probability of no complete penetration must be greater than or equal to .9.

- For the second shot, the 90 percent lower confidence bound for the probability of no complete penetration must be greater than or equal to .7.
- For the first shot, with 90 percent confidence the 90 percent upper tolerance limit for the BFD must be less than 44.0 mm.
- For the second shot, with 90 percent confidence the 80 percent upper tolerance limit for the BFD must be less than 44.0 mm.

TABLE 6-2 Proposed FAT Standards

Measure	First Shot	Second Shot
Probability of no penetration [Pr(nP)]	90 percent lower confidence bound for Pr(nP) ≥ .9	90 percent lower confidence bound for Pr(nP) ≥ .7
BFD	With 90 percent confidence, 90 percent upper tolerance limit for BFD < 44 mm	With 90 percent confidence, 80 percent upper tolerance limit for BFD < 44 mm

SOURCE: Chris Moosman, DOT&E, "ATEC Review of FAT and LAT Procedures in Army PDs and the DOT&E Protocol for NAS Statisticians," presentation to the committee's statistics working group, October 12, 2010.

A confidence interval is an interval that covers a population parameter, in this case the probability of no complete system penetration for the population of plates, with a stated level of confidence. As discussed in the chapter Introduction, this (or any other) population parameter cannot be observed without testing (and thus destroying) all the plates. The higher the level of confidence the more likely the interval includes the unobserved population parameter. In particular, for the DOT&E protocol, achieving a 90 percent lower confidence bound that is greater than .9 means that the probability of no penetration for the entire population of plates is very likely to be greater than .9. Furthermore, as described earlier, manufacturers will need to achieve probabilities of no penetration significantly higher than 0.9 to have a reasonable chance of successfully passing this protocol standard.

In contrast to a confidence interval, a tolerance interval is an interval that contains a fixed proportion of the population with a stated confidence. In this case, the protocol specifies a tolerance interval standard for back-face deformation. For the DOT&E protocol, achieving a 90 percent upper tolerance bound that the BFD is less than 44 mm at 90 percent confidence means that 90 percent of the entire population of plates is very likely to have BFDs of less than 44 mm. Note that this does not mean specifically that BFDs of greater than 44 mm cannot occur, and, in fact, it is possible to observe BFDs greater than 44 mm in some of the tested plates and still successfully pass the FAT. (For a more detailed discussion of the assumptions underlying the standard confidence and tolerance interval calculations, see the next section.)

The committee notes that the original draft DOT&E protocol specified a 90 percent lower confidence bound for the probability of complete system penetration greater than 0.8 for the second shot, and this information was subsequently reflected in the committee's Phase II letter report (NRC, 2010). However, as a result of DOT&E discussions with the Army and USSOCOM, the standard was subsequently modified to a 90 percent lower confidence bound for the probability that no complete system penetration is greater than 0.7 in the protocol promulgated by DOT&E.[42]

Finding. During the course of the committee's research and deliberations, the Office of the Director, Operational Test and Evaluation, the Army, and the United States Special Operations Command have endeavored to establish statistically principled test standards that are realistically achieveable with the current body armor designs. The committee found these collaborative efforts to be commendable.

The combination of 60 plates tested per threat, combined with the requirement that the 90 percent lower confidence bound for the probability of no penetration be greater than or equal to 0.9 on the first shot, means that out of the 60 plates tested, two can fail (i.e., have a complete penetration) and the manufacturer will still pass the FAT. While some have stated that there are no existing body armor protocols that allow a penetration on the first shot, in fact USSOCOM's historical protocol allows one or more plates to be penetrated (depending on the number of plates tested) on the first shot.[43] Furthermore, while previous Army test protocols with smaller sample sizes permitted no first shot penetrations, it does not follow that there will be no first shot penetrations for the entire population of plates eventually procured. The only way to positively ensure that the population of plates will have no penetrations is to test every plate, a physical impossibility with destructive testing.

That said, the committee recognizes that there may be a perception issue with a test protocol for which one or two penetrations can still result in passing the FAT. There are zero-failure protocol alternatives: For example, with a total sample size of 22 plates a standard of zero failures

---

[42]DOT&E (Director, Operational Test and Evaluation). 2010a. Standardization of Hard Body Armor Testing. Memorandum dated April 27, 2010.

[43]Lt. Col. Peter Greany, discussion with the committee's statistics working group, October 12, 2010.

(penetrations) results in a 90 percent lower confidence bound greater than 0.9. However, limiting the total sample size to 22 plates results in much more limited information about plate performance, particularly in terms of environmental testing, which would be reduced to testing only two plates per condition. Furthermore, a 22-plate zero failure protocol would increase the risk that a manufacturer might fail the FAT even though it manufactures plates that meet the performance standards. On the other hand, maintaining the 60-plate protocol but not allowing any first shot penetrations would substantially increase the risk that manufacturers could fail the FAT even with acceptable plates that have a very high probability of no penetration. (See section on protocol design trade-offs for a discussion about how protocol design affects the risks the government and the manufacturer face during testing.)

One solution to this dilemma discussed during committee briefings is to maintain the 60-plate protocol but not allow any perforations under ambient conditions. Instead, the two allowable penetrations can occur only under the environmental conditions, some of which, like the impacted condition, are inherently stressing on the plates, and then if two failures occur they cannot occur under the same environmental condition. That is, the FAT is failed if (1) on the first shot one or more penetrations occur under ambient conditions, (2) two or more penetrations occur for the same environmental condition, or (3) three or more first shot penetrations occur for any of the 60 plates tested.

The standard also establishes fair-hit/no-hit criteria, where data from any shot with a velocity that is too high are excluded from analysis regardless of outcome, while data from shots with velocities that are too low are included only if they completely penetrate (both plate and system) or have a BFD of greater than 44.0 mm. This biases the test results toward soldier safety, as would be expected, but it may also bias toward overdesign of the hard armor. This trade-off should be explicitly recognized. On the other hand, the DOT&E protocol specifies a shot pattern similar to the Army's historical test protocol in the sense that the first and second shots on a plate must be 5-6 in. apart. To the extent that shots closer together are more stressing on the plates, this protocol may be less stressing than the USSOCOMs historical protocol that required the second shot to be 4 in. away from the first shot.

Finding. The new Office of the Director, Operational Test and Evaluation protocol meets both key protocol design requirements; it is statistically principled and it provides a minimum Department of Defense-wide body armor test standard.

Test Protocol Assumptions

The DOT&E protocol states that "the DoD BFD requirement is a BFD (based on the calculated upper tolerance limit for the data set) that does not exceed 44.0 mm" (DOT&E, 2010a). In general, a "tolerance interval" is a statistical interval, calculated from the data, in which a particular proportion of the population will be contained with a specified level of confidence. The end points of a tolerance interval are called "tolerance limits."

As described in the NIST Engineering Statistics Handbook (2010), three types of questions can be addressed by tolerance intervals:

> 1. What interval will contain *p* percent of the population measurements?
> 2. What interval guarantees that *p* percent of population measurements will not fall below a lower limit?
> 3. What interval guarantees that *p* percent of population measurements will not exceed an upper limit? (NIST, 2010, Section 7.2.6.3)

Question 1 leads to a two-sided interval; questions 2 and 3 lead to one-sided intervals, called "tolerance bounds." For body armor testing, the relevant question is 3, which requires the calculation of an upper tolerance bound.

The correct calculation of any tolerance interval, bound, or limit depends on the underlying distribution of measurements in the population. For this reason, no single formula can be applied in all situations. The most common formula assumes the underlying population is normal; formulas have been derived, however, for many other underlying distributions (see Appendix H).

The DOT&E protocol states that an upper tolerance interval will be calculated for BFD "as a continuous normal random variable" (DOT&E, 2010a, p. 8). In so doing, the protocol is explicitly assuming that BFD is normally distributed. However, if the BFD distribution is not normal, then the resulting tolerance intervals based on this assumption will not contain the intended *p* percent of the population. While many of the empirical BFD distributions observed by the committee were bell shaped, that does not necessarily mean that the BFD distribution is normal. Further, some of empirical BFD distributions for certain threats and vendors or threats and designs looked either skewed or had one or more truncated tails, suggesting the BFD distribution in some cases is not normal.[44] The violation of the normality assumption will affect the appropriate or inappropriate acceptance or rejection of body armor differently depending on the underlying population distribution. Appendix H provides additional information on tolerance-bound calculations.

Finding. The distribution of backface deformation populations has not been proven to be normally distributed for all combinations of vendor, threat, and design; therefore, the tolerance-bound calculation specified by the protocol may not be appropriate in all cases.

---

[44]Chris Moosman, DOT&E, "ATEC Review of FAT and LAT Procedures in Army PDs and the DOT&E Protocol for NAS Statisticians," presentation to the committee's statistics working group, October 12, 2010.

The protocol specifies the requirement for probability of penetration in terms of a lower confidence limit for the probability of no penetration. A confidence interval contains a population parameter (here, the proportion of plates not penetrated) with the stated confidence level. This means that if repeated samples of the same size are taken, the confidence interval would contain the parameter the specified proportion of the time. For example, a 50 percent confidence interval contains the population parameter, on average, for 50 percent of the samples taken. Since any test considers only a sample of the plates in the population, even requiring zero failures during the test cannot guarantee (and does not mean) that there will be no failures in the larger population of plates.

The DOT&E protocol specifies that the lower confidence limit "is calculated using the Clopper-Pearson method," which is based on the cumulative probabilities of the binomial distribution (DOT&E, 2010a). Because of the discreteness of the binomial distribution, the Clopper-Pearson method results in a conservative estimate in the sense that it is guaranteed to achieve at least the specified confidence level and may exceed it. As Agresti and Coull (1998, p. 119) state, "For any fixed parameter value, the actual coverage probability can be much larger than the nominal confidence level unless n is quite large." (See Brown et al. (2001) and Agresti and Coull (1998) for additional discussion.)

Figure 6-2 shows how the Clopper-Pearson confidence lower bound behaves in terms of coverage behavior for $n = 60$ and $0.8 \leq \Pr(nP) \leq 0.999$. In particular, it shows that interval coverage, while bounded below by the specified level $1 - \alpha = .9$, oscillates dramatically and can often be substantially greater than .9.

Because the specified confidence level for the Clopper-Pearson confidence interval is conservative, in the sense that the resulting interval achieves a level of confidence of at least .90, the DOT&E protocol can achieve confidence levels well above the specified 90 percent. However, since the actual confidence level changes quite dramatically for small changes in $\Pr(nP)$—at a level of precision that is inestimable with the test sample sizes—it will be impossible to determine the achieved confidence level for any particular test. For example, at $\Pr(nP) = .913$, the actual confidence level is 0.903, while at $\Pr(nP) = .914$, the actual confidence level is .97. As Figure 6-2 shows (and as do similar figures in Brown et al., 2001, and Agresti and Coull, 1998), this type of dramatic change occurs frequently.

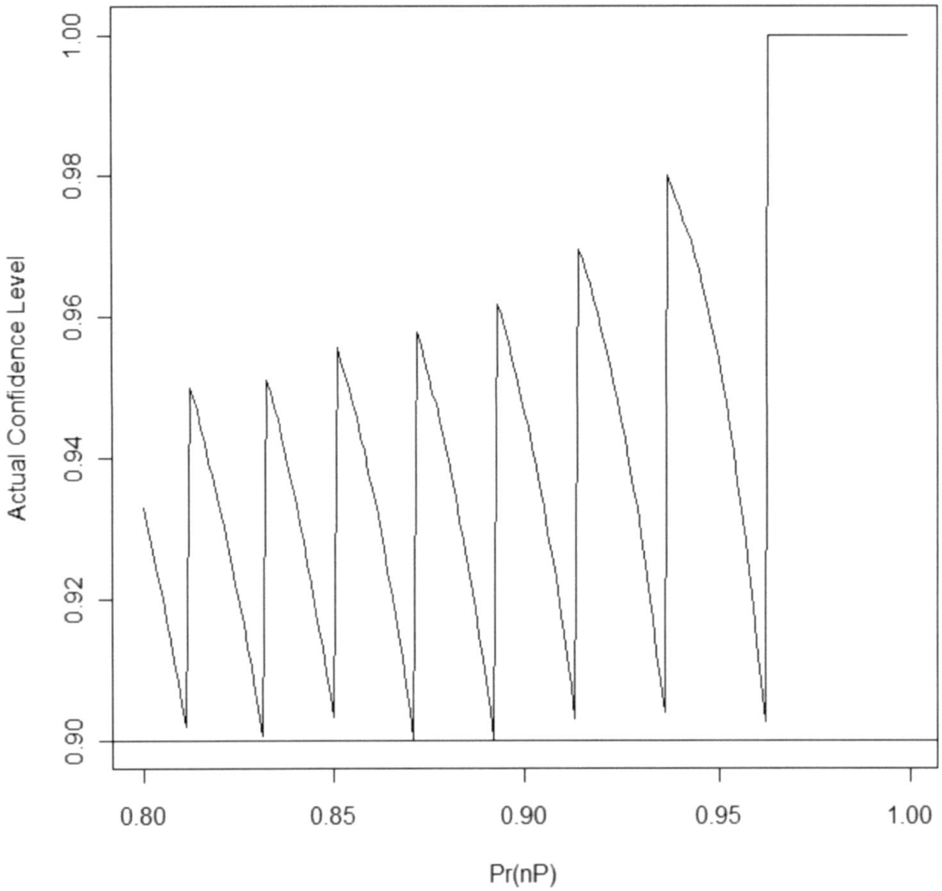

FIGURE 6-2 Plot of the actual coverage level achieved by a lower confidence bound calculated according to the Clopper-Pearson method for $n = 60$ and various Pr(nP). The horizontal line at 0.90 is the confidence level $(1 - \alpha)$ specified in the DOT&E FAT protocol.

Finding. Use of the Clopper-Pearson method for calculating the lower confidence limit is conservative, resulting in actual confidence levels that are at least as great as, and often greater than, the confidence level specified in the standard. The actual confidence level is a function of the Pr(nP) of the plates, it varies substantially depending on the particular Pr(nP), and it can be quite different for small changes in Pr(nP).

The protocol was designed (and analyses are performed) assuming that the data are independent and identically distributed. Given the current test procedures, this assumption is likely not met. In particular, biases may be introduced by the test procedures for the clay box. For example, the committee strongly suspects that if it tested a group of plates all on boxes that were 10 minutes out of the ovens and then could repeat the experiment with exactly the same plates and clay boxes except that the boxes were 40 minutes out of the oven, it would see systematically different BFD results. At present, the committee does not know how much of an issue this causes with the analyses, but it is another reason to develop clay that can be used at ambient temperature in the test.

From a statistical perspective, there are many things that can be traded off in a protocol, including sample size, confidence level, and requirements. Essentially the trade-off is in terms of risk to the DoD (of purchasing a plate design that does not meet requirements) or to the manufacturer (of having a plate design that does meet requirements fail the FAT.) A larger sample size provides more information to characterize a population and will generally lead to narrower confidence and tolerance intervals. More samples also allow for testing more combinations of factors and conditions. However, larger sample sizes usually come with higher costs. Lower confidence levels also generally lead to narrower confidence and tolerance intervals, but at the cost of less confidence that the interval contains the quantity of interest. To have a statistically principled protocol, it is critical that a high confidence level is maintained. Of course changing requirements directly impacts DoD and manufacturer risk.

Lot Acceptance Testing

Once a manufacturer has passed FAT and begins production, LAT is used to ensure that body armor continues to conform to contract requirements. Owing to the critical nature of safety when it comes to body armor, continued LAT testing is both desirable and necessary, but the committee also recognizes that modern quality control calls for manufacturing processes to be improved to eliminate as much variation as possible. As described in MIL-STD-1916, "sampling inspection alone does not control or improve quality" (DoD, 1996, p. 8). Elimination of variation can provide a number of benefits, including a more consistently performing product as well as a reduction of risks to both the manufacturer and the DoD. In addition, to the extent that such reductions in variation lead to more predictability in plate performance and testing outcomes, these reductions might lead to innovations in plate design that allow reductions in plate weight while maintaining ballistic protection.

Table 6-3 provides an overview of the statistical basis for a proposed LAT protocol. Currently, results from the protocol are being used only for government reference. The protocol is being used to provide data for evaluation of the protocol's effectiveness as a standard for lot acceptance, and, based on this evaluation, DOT&E will revise the protocol as necessary before promulgating a final, mandatory standard for use in future contracts (DOT&E, 2010b).

TABLE 6-3 Proposed LAT Standards

Measure	First Shot	Second Shot
Pr(*nP*)	4 percent AQL	15 percent AQL
BFD	With 90 percent confidence, 80 percent upper tolerance limit for BFD < 44 mm	With 90 percent confidence, 70 percent upper tolerance limit for BFD < 44 mm

SOURCE: DOT&E, 2010b.

The proposed LAT protocol is similar in many respects to the FAT protocol, including range setup, the use of clay as a backing material and its calibration, fair-hit/no-test criteria, and the definitions of complete and partial penetrations. However, there are some important differences. Most notably, LAT sample sizes are smaller than FAT sample sizes, and they vary by size of the lot. For a normal inspection schedule, the protocol requires at one extreme a sample size of 8 plates for a lot of between 91 and 150 plates and at the other extreme a sample size of 32 plates for a lot of between 1,201 and 3,200 plates.[45] Product managers have the option to implement switching procedures, and the requisite sample sizes are listed in Tables 4 thru 6 of the proposed protocol. Other differences include these:

- Because all plates are tested under ambient conditions, neither Table 6-1 nor any other such design matrix applies to the LAT.
- As shown in Table 6-3, while the BFD standard is the same in LAT as in FAT, the probability of the no penetration lower confidence bound FAT metric has been replaced with an "acceptable quality level" metric (sometimes abbreviated AQL)[46] in LAT.

[45]Sample sizes are based on special inspection level S-4 of ANSI/ASQ Z1.4-2008 (American Society for Quality, 2008). Tables 3-6 in the proposed protocol are derived directly from Tables I, II-A, II-B, and II-C of ANSI/ASQ Z1.4-2008.

[46]ANSI/ASQ Z1.4-2008 defines AQL as the "Acceptance Quality Limit." It explicitly states that "the use of the abbreviation AQL to mean Acceptable Quality Level is no longer recommended" (American Society for Quality, 2008, p. 8).

According to ANSI/ASQ Z1.4-2008, "The AQL is the quality level that is the worst tolerable process average when a continuing series of lots is submitted for acceptance sampling" (American Society for Quality, 2008, p. 2). It goes on to say, "The purpose of this standard is, through the economic and psychological pressure of lot non-acceptance, to induce a supplier to maintain a process average at least as good as the specified AQL while at the same time providing an upper limit on the consideration of the consumer's risk of accepting occasional poor lots. The standard is not intended as a procedure for estimating lot quality or for segregating lots" (American Society for Quality, 2008, p. 3).

In the course of the committee's deliberations, some have suggested that special inspection level S-3 would be preferable to inspection level S-4. However, the committee notes that this could lead to an undesirable lot rejection rate in some situations. As the ANSI/ASQ standard states, "The sampling plans in this standard are so arranged that the probability of lot acceptance at the designated AQL depends upon sample size, being generally higher for large samples than for small samples for a given AQL" (American Society for Quality, 2008, p. 2).

Figure 6-3 plots the probability that a lot of body armor passes LAT first shot requirements for the S-3 and S-4 inspection levels for various lot sizes and an AQL of 4 percent. Figure 6-3 shows that the S-4 inspection scheme does in general result in a higher probability that a lot passes LAT first shot requirements for $.9 \leq \Pr(nP) \leq 1.0$. The exception is lot sizes between 151 and 500. However, it also shows for the S-4 inspection scheme that lots with $\Pr(nP) > \sim .98$ have a very high chance of passing LAT regardless of lot size: It is at or above 99 percent with the exception that for lot sizes between 151 and 500 the probability that a lot passes is 97.3 percent. In comparison, for the S-3 inspection scheme with $\Pr(nP) = .98$ the probability of passing LAT can be as low 90 percent for lot sizes between 91 and 150. For lot sizes between 151 and 500, the probability of passing is greatest at 99 percent and for the other two it is 97.3 percent.

Figure 6-4 shows that the inspection level S-4 gives a higher probability a lot passes LAT on the second shot for all sample sizes for $\Pr(nP) > 0.827$. In addition, Figures 6-3 and 6-4 show that in general larger lot sizes tend to have higher probabilities of passing for equivalent $\Pr(nP)$ values.

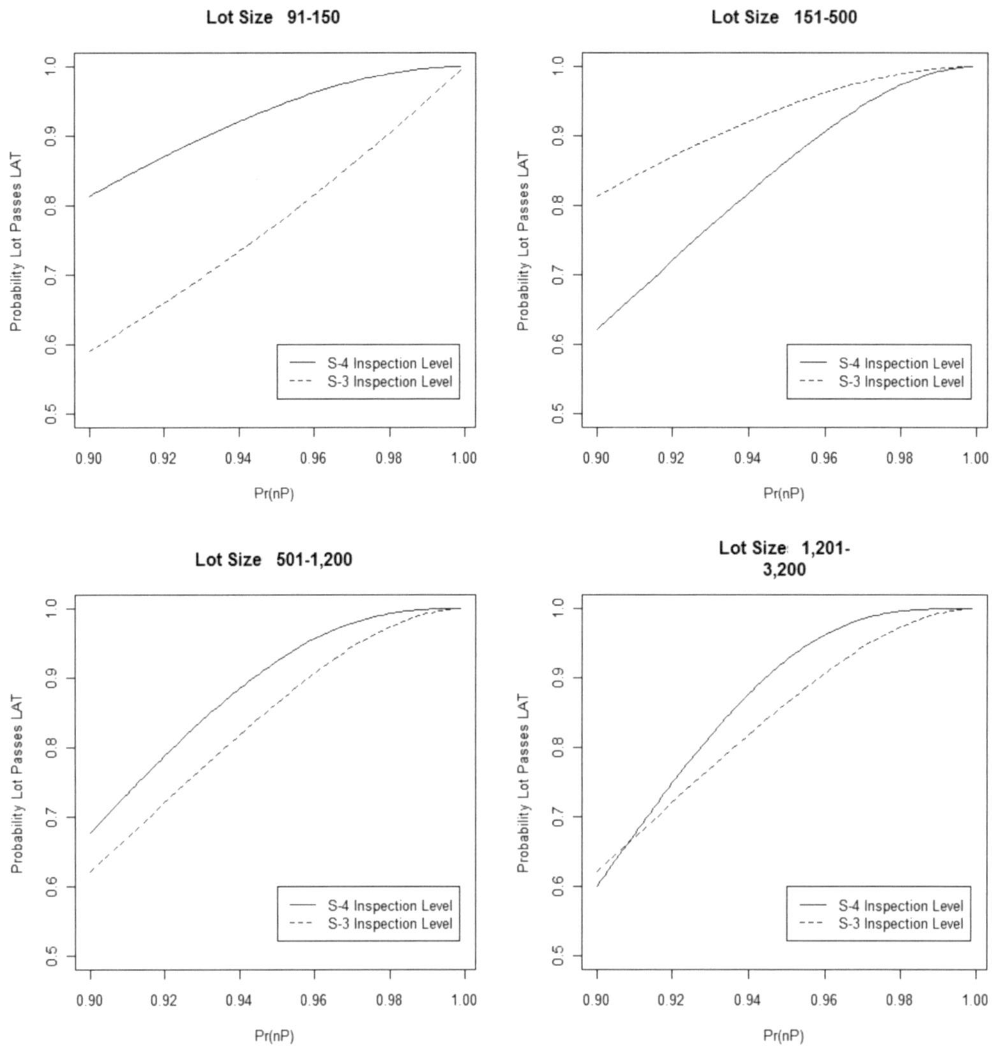

FIGURE 6-3 Probability a lot passes LAT first shot requirements for Pr(*nP*) for the S-4 and S-3 inspection levels for various lot sizes and an AQL of 4 percent.

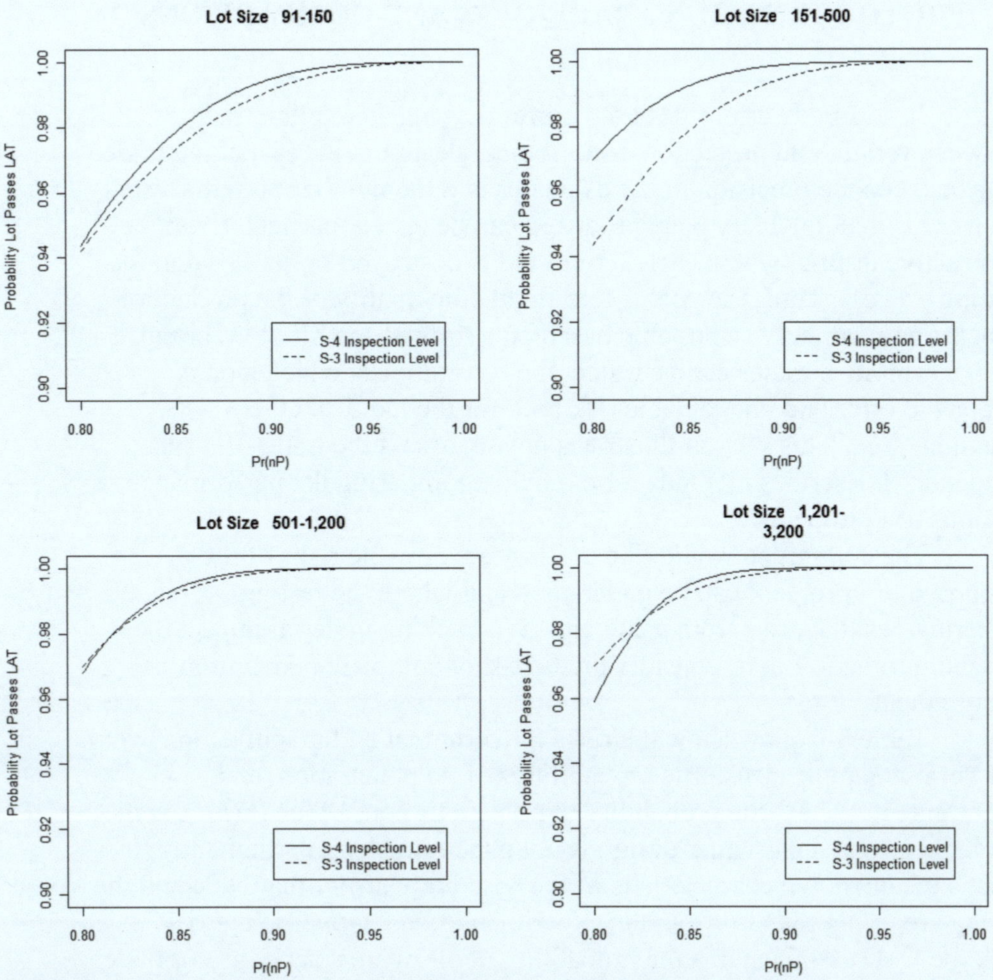

FIGURE 6-4 Probability a lot passes LAT second shot requirements for Pr(*nP*) for the S-4 and S-3 inspection levels for various lot sizes and an AQL of 4 percent.

Passing the LAT required meeting the AQL standards as well as the BFD standards. Given that the sample sizes that will be used in the BFD lower tolerance limit follow from the AQL sample sizes derived from ANSI/ASQ Z1.4-2008, the proposed protocol has necessarily lowered the LAT lower tolerance limits from 0.9 in the FAT to 0.8, and from 0.8 in the FAT to 0.7, for first and second shots, respectively (American Society for Quality, 2008).

Finding. For most lot sizes, and over the higher levels of Pr(*nP*), the S-4 inspection level results in a greater probability that lots will pass lot acceptance testing.

PROTOCOL DESIGN TRADE-OFFS AND COMPARISONS

Just as body armor design requires making an explicit trade-off between weight and protection, test protocol design requires making trade-offs between the precision of the estimates and the number of items tested. At issue is that not every plate produced can be tested, particularly in destructive testing, where each item tested is destroyed or damaged in the testing process. Thus, the goal is to estimate the quality of the production process as accurately as possible based on a limited sample. Yet, because only a sample of plates can be tested, the resulting test conclusion is subject to error and unavoidable risk both for the DoD and the manufacturer. This section illustrates how to assess the trade-offs and Appendix I describes methods for explicitly comparing the performance of various test protocols.

The committee would like to illustrate how the risks of the proposed test protocol can be understood and where the testing uncertainties that arise from using clay as a backing material impact the 60-shot protocol. Let us consider the first-shot complete penetration requirement.

Table 6-4 shows how the risks (government and manufacturer) vary for various sample sizes, true probabilities of no penetration, and requirements. The "true probability of no penetration," True Pr(*nP*), is the probability that a particular design is not penetrated by a particular threat—this is the unknown characteristic of the hard body armor that DoD and the Army are trying to learn from the experimentation. "Government risk" is a risk the DoD assumes; it is the probability of allowing a set of armor plates that just meets the "no penetration" requirement to pass the test. "Manufacturer risk" is the probability that a set of armor plates that meets or exceeds the no-penetration requirement will fail. These risks are a function of the sample size required in the sampling plan and the manufacturer's quality.

TABLE 6-4 Risk Comparisons for Probability of Complete Penetration

Sample Size	Allowable Failures	True Pr(*nP*)	Requirement	Government Risk	Manufacturer Risk
15	0	.98	.86	.104	.261
22	0	.98	.90	.098	.359
40	1	.98	.90	.080	.190
60	2	.98	.90	.053	.119
60	2	.99	.90	.053	.022
60	2	.92	.90	.053	.868
300	9	.98	.95	.000	.082
6,000	134	.98	.975	.000	.092

For example, the fourth line in Table 6-4 is interpreted as follows: A test requirement that the 90 percent lower confidence limit must exceed 90 percent means that a successful test of 60 plates can have no more than two failures. Under these conditions, a manufacturer's plates, each of which has a probability of passing the test (i.e., of no penetration) of .98 stand an 11.9 percent chance that at least 3 or more of the 60 plates will fail the test, so that manufacturer will fail the test. Conversely, under these test conditions, the government stands a 5.3 percent chance that a manufacturer's marginally performing plates that have a no-penetration probability of .90 will pass the test.

The first three lines of the Table 6-4 demonstrate that reducing the sample size from 60 shifts the risk to the manufacturer. For a sample size of 15 it is not possible to pass the test because the sample size is too small to demonstrate a 90 percent requirement with high (90 percent) confidence. The last two lines of Table 6-4 show the sharp increases in required sample size when the requirement is increased beyond .9 and the risks are held roughly constant.

Table 6-4 also shows that, for a sample size of 60, a manufacturer must produce hard body armor that has a true probability of no penetration substantially higher than .9 to have a reasonable chance of passing the test. Figure 6-5 plots the manufacturer's risk for various Pr(*nP*), where it is clear that to have a reasonably high probability of passing the protocol's first-shot, no-penetration standard, the manufacturer's plates must have a Pr(*nP*) substantially higher than .9. From a soldier safety perspective, this is appropriate.

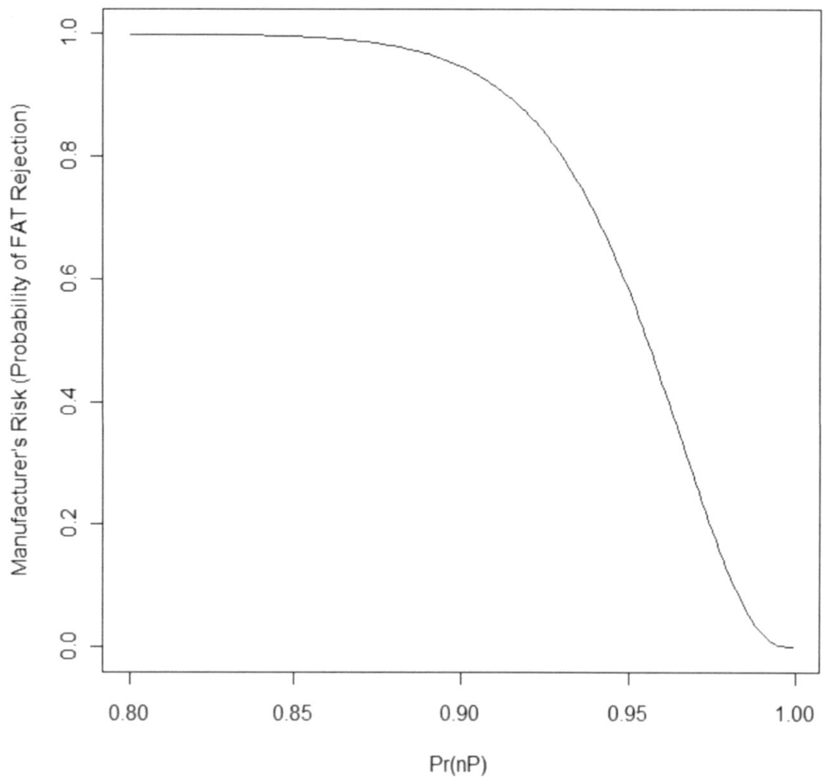

FIGURE 6-5 Plot of the manufacturer's risk for various Pr(*nP*) under the DOT&E protocol. To have a reasonably high probability of passing the protocol's first-shot no-penetration standard (out of 60 plates tested, no more than two penetrations are allowed), the manufacturer's plates must have a high Pr(*nP*).

Because of the issues discussed in earlier sections of this report, it is difficult to tell if the observed variation in BFD for hard body armor is attributable mainly to the variation in plates, to the variation in the test process, or to both. As a result, all observed variation is being attributed to the plates. While this is clearly incorrect, without a better understanding of the specific sources of variation, it is impossible to do otherwise. This probably results in overdesign and/or overmanufacture of the plate to ensure a high probability of passing FAT and LAT.

Figure 6-6 illustrates the potential impact on manufacturers by simulating the effects of the BFD test on the probability of a manufacturer failing FAT under various conditions. In Figure 6-6 (a), the assumption is that the plates resulting from a manufacturer's process have a mean BFD of 38 mm. The solid line (100 percent variance) shows the results when all observed variation is attributed to the plates. The amount of variation is shown on the x-axis in terms of standard deviations, and the probability of failing to meet the BFD criterion is shown on the y-axis. The plot shows that the probability of failure ranges from zero for standard deviations just above 3 to nearly one for standard deviations just less than 5. The dashed curves show the impact of attributing less of the observed variation to the plates. Notice that the percentage attributed to the plates decreases as the probability of passing the test increases. Figure 6-6 (b) shows a similar result for a mean BFD of 40 mm.

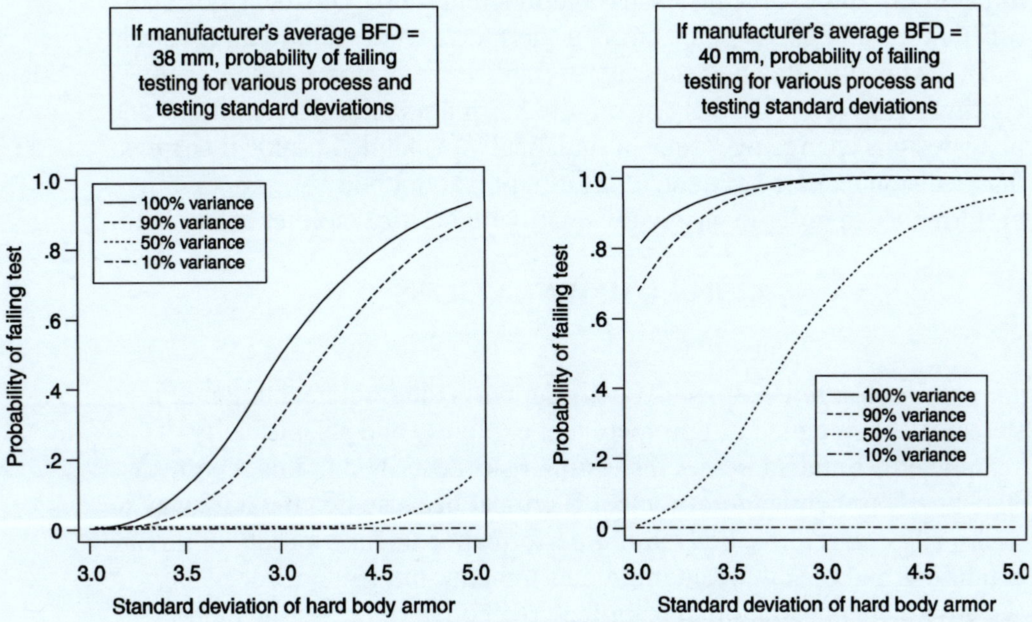

FIGURE 6-6 Risk comparisons for BFD assuming in the left plot that the manufacturer's true mean BFD is 38 mm and in the right plot is 40 mm; the associated fraction of variation is shown on the *x*-axis. The plots show that decreasing variability in BFD, either via more consistent manufacturing processes or as a result of more repeatable testing measures, lowers the manufacturer's chances of failing testing (given that the manufacturer's plates do meet standards and holding everything else constant).

The plots show that decreasing variability in BFD by means of more consistent manufacturing processes or more repeatable testing measures lowers the manufacturer's chance of failing testing (given that the manufacturer's plates do meet standards and that all other factors are constant). At issue is the current impossibility of estimating what fraction of the variation in BFD is attributable to variation in the plates and what fraction is attributable to the testing methods. The experiments recommended in Chapter 4 should provide a much better estimate of the test process variation. As discussed in earlier sections, there are known but not well-quantified issues that relate to variations in the thermal and stress properties of the clay medium itself, variations caused by different individuals handworking the clay as it is prepared for testing, variations in calibration, and other factors. Information on how the existing process performs will facilitate improving the process (minimizing excess variation, should it exist.)

Finding. Using a statistically principled protocol enables decision makers to explicitly address the necessary and inherently unavoidable risk trade-offs that must be faced in testing. Furthermore, while additional research and coordination may be necessary to finalize the protocol design, and continuing review will likely be required as manufacturing conditions and plate designs change over time, a statistically principled protocol ensures that decision makers have sound information about body armor performance in order to ensure the quality of a critical soldier safety item.

RECOMMENDATIONS

The committee unequivocally supports the implementation of a statistically principled test protocol that explicitly and scientifically acknowledges and addresses the testing risks described in this report. A statistically principled test protocol is critical because it is the only way to rigorously characterize body armor performance under a variety of threat conditions and operating environments to better inform DoD decisions. Because there is variation in manufactured body armor, testing alone cannot ensure that body armor is 100 percent effective. One can, however, develop higher confidence in the effectiveness of the body armor by using a statistically principled and rigorous assessment with sufficient sample size. The committee commends DOT&E for its leadership in establishing such statistically principled protocols for body armor first article testing and lot acceptance testing.

Any test protocol involves some risk that bad body armor will pass the test and good body armor will fail. In setting the standards within the protocols, the DoD has a responsibility to be explicit about these risks and to design a test protocol that balances cost, performance, ability to execute, fairness to the manufacturer, and risk to the soldier. Trade-offs can be made to result in statistically principled protocols that are both scientifically rigorous and practical in application. This conceptual approach is supported by the current DOT&E protocol.

Due diligence and deliberate caution are warranted during the change from the old test protocols to the new protocols. In particular, because manufacturers have strong incentives to build armor that has a high chance of passing FAT and LAT, there is some chance that the change in test protocol could have unintended impacts on body armor design and/or performance. Given the success of the current body armor in the field, changes in testing protocols should be made with deliberate caution to ensure that plate performance is maintained (or improved) while also ensuring that the best science is brought to bear on testing body armor.

The committee commends DOT&E for its ongoing discussions with the Army, USSOCOM, and other stakeholders and its willingness to reconsider and revise the confidence bounds and tolerance interval levels of the proposed protocols as appropriate and necessary. Within these discussions the committee recommends that the following three considerations should continue to be explicitly addressed.

- First, it is important to reach consensus on what constitutes a BFD failure and how such failure relates to soldier injury or death. Accordingly, Chapter 8 highlights the need for research to quantify the medical results of blunt force trauma on tissue and to use those results to underpin a scientifically based BFD standard.
- Second, the current clay-based test methodology is probably introducing extra variation into the test results. In particular, as described in the Chapter 4 section "Roadmap for Improving the Testing Process," replacing Roma Plastilina #1 with a backing material that can be calibrated at room temperature has the potential to eliminate substantial variation. Thus, Recommendation 4-1, to expedite development of a standard replacement that can be used at room temperature, is critical for improving both the testing process and the statistical assessment of body armor performance.
- Third, it is important that the proposed statistically principled protocol be seen not just as another in a long line of standards but as an improvement that incorporates input from all of the stakeholders and that embodies the best science. In so doing, it is particularly important to develop broad-based support for the statistically principled protocol and to ensure that its adoption will neither undo many years of successful body armor engineering nor result in other undesirable outcomes.

In terms of the DOT&E FAT and LAT protocols for body armor, the committee has four specific recommendations.

Recommendation 6-1: The Office of the Director, Operational Test and Evaluation (DOT&E) should continue to conduct due diligence to carefully and completely assess the effects, large and small, of its statistical protocol as it is adopted across the body armor testing community. In particular, DOT&E should continue to

- Collaborate with the Army and the United States Special Operations Command (USSOCOM) to revise the test protocol as necessary, based on the results of Army and USSOCOM "for government reference" first article testing test results and

other empirical evidence, to ensure that currently acceptable plate designs are not eliminated under the new protocol; and
- Regularly assess the impact or impacts of the new protocol on plate design, particularly plate weight, to ensure the test protocol results in body armor that achieves the requisite soldier safety while not negatively, inappropriately, or inadvertently affecting plate design.

Recommendation 6-2: The Office of the Director, Operational Test and Evaluation, should consider modifying the first article testing protocol to

- Generalize the description of the backface deformation (BFD) upper tolerance interval calculation to allow for nonnormal BFD distributions;
- Specify a confidence interval calculation methodology that has better coverage properties, such as the Agresti-Coull interval recommended by Brown et al. (2001) and described in detail in Agresti and Coull (1998); and
- Specify guidelines that will accommodate deviations in environmental conditions and/or plate size from the current 60-plate design matrix.

For example, DOT&E could revise the current protocol to specify that if a procurement contract does not require testing under one or more of the environmental conditions listed in the design matrix, the plates listed under that condition would then be tested under ambient conditions.

Recommendation 6-3: The Office of the Director, Operational Test and Evaluation, and the Army should continue to consult and engage statisticians throughout the process of assessing and revising protocols, comparing the performance of the new and old protocols, assessing the effects of the new protocols, and considering possible changes.

Testers and statisticians should continue to work together as a team (1) to quantify in a statistically rigorous manner the portion of variation in BFD attributable to the testing process and that attributable to the plates and (2) to ensure these results are appropriately reflected in an updated protocol. In particular, the statisticians involved with developing and implementing the statistically principled protocol should be involved with the experimentation recommended in Chapter 4.

Over the course of the committee's research and deliberations, the DOT&E, Army, and USSOCOM have endeavored to establish statistically principled test standards that are realistically achievable with the current body armor designs.

Recommendation 6-4: The Office of the Director, Operational Test and Evaluation, the Army, and the United States Special Operations Command should work together to arrive at an acceptable set of test standards for lot acceptance testing that is both statistically principled and realistically achievable with current body armor designs.

REFERENCES

Agresti, A., and B. Coull. 1998. Approximate is better than "exact" for interval estimation of binomial proportions. The American Statistician 52(2):119-126.

American Society for Quality. 2008. American National Standard Sampling Procedures and Tables for Inspection by Attributes. ANSI/ASQ Z1.4-2008. Milwaukee, W.I.: American Society for Quality.

Brown, L., T. Cai, A. DasGupta. 2001. Interval Estimation for Binomial Proportions. Statistical Science 16(2):101-117.

DoD (U.S. Department of Defense). 1996. DoD Preferred Methods for Acceptance of Product. MIL-STD-1916. Arlington, Va.: DoD.

DoD. 2009. DoD Testing Requirements for Body Armor. Report Number D-2009-047. Arlington, Va.: DoD Inspector General.

DoD. 2010. Memorandum for National Research Committee. Arlington, Va.: Department of Defense Inspector General.

DOT&E (Director of Operational Test and Evaluation). 2010a. Standardization of hard body armor testing. Memorandum dated April 27, 2010.

DOT&E. 2010b. Memorandum: Standard for lot acceptance ballistic testing of hard body armor. July 2, 2010.

Dunn, N. 2010. ATEC proposed Army protocol and background for National Academy of Science Statisticians. Aberdeen Proving Ground, Md.: Army Testing and Evaluation Center.

NIST (National Institutes of Standards and Technology). 2010. Engineering Statistics Handbook. Available online at http://www.itl.nist.gov/div898/handbook/. Last accessed March 15, 2011.

NRC (National Research Council). 2010. Phase II Report on Review of the Testing of Body Armor Materials for Use by the U.S. Army. Washington, D.C.: National Academies Press.

OTA (Office of Technology Assessment). 1992. Police Body Armor Standards and Testing, Volume II: Appendices. OTA-ISC-535. Washington, D.C.: Office of Technology Assessment.

PEO-S (Program Executive Officer, Soldier). 2010. Interceptor body armor (IBA). Available online at <https://peosoldier.army.mil/factsheets/SEQ_SSV_IBA.pdf.> Accessed January 11, 2011.

RDECOM (U.S. Army Research, Development and Engineering Command). 2009. Amendment of solicitation/modification of contract W91CRB-09-D-0001/P00004. Aberdeen Proving Ground, Md.: RDECOM.

USSOCOM (U.S. Special Operations Command). 2010. USSOCOM first article testing protocol. Aberdeen Proving Ground, Md.: U.S. Special Operations Command.

7

Helmet Testing

This chapter discusses helmet ballistic testing methodologies. It describes helmet design and suspension systems as well as current and proposed clay head forms used in the ballistic testing of helmets.

BALLISTIC HELMET TEST METHODOLGIES

The development of modern military helmets based on aramid fiber composites has been an outstanding success (e.g., Carey et al., 1998). Numerous soldiers and civilian police have been saved from threats that would have defeated earlier metallic helmets (Carey et al., 2000). However, though current protective levels have proved to be well matched to the threats they are designed to protect against, increasing threats on the battlefield, especially from high velocity rifle rounds, will likely require new or modified helmet test methodologies assess the risk of injuries while using improved ballistic protective helmets.

On the battlefield, mobility is often the key to survival. The development of robust test methodologies is crucial to comparing the effect of potential trauma to ergonomic and other trade-offs required for personal protection. The mass of the protection is particularly important, as it may impede mobility.

The standoff between the head and the backface of the helmet in the current helmet systems was designed to be 1.3 cm (0.5 in.) or greater (McManus, 1976). Substantial research has been performed on traumatic brain injury (TBI), but much of the work is not applicable to military threats. For example, TBI may occur from blunt impact during vehicle crashes, falls, and sports impacts. However, there are important physical differences between these lower velocity events and impacts from the backface of military helmets.

The difference in incoming momentum between several representative rounds ranging from 9 mm to 0.50 cal and a typical American football head contact is shown in Figure 7-1. The football impact typically has a much larger transfer of momentum, implying much greater overall head motion and more global internal brain deformation. This difference and the much more localized contact from a helmet backface deformation (BFD) impact raises questions about the applicability of existing head injury criteria.

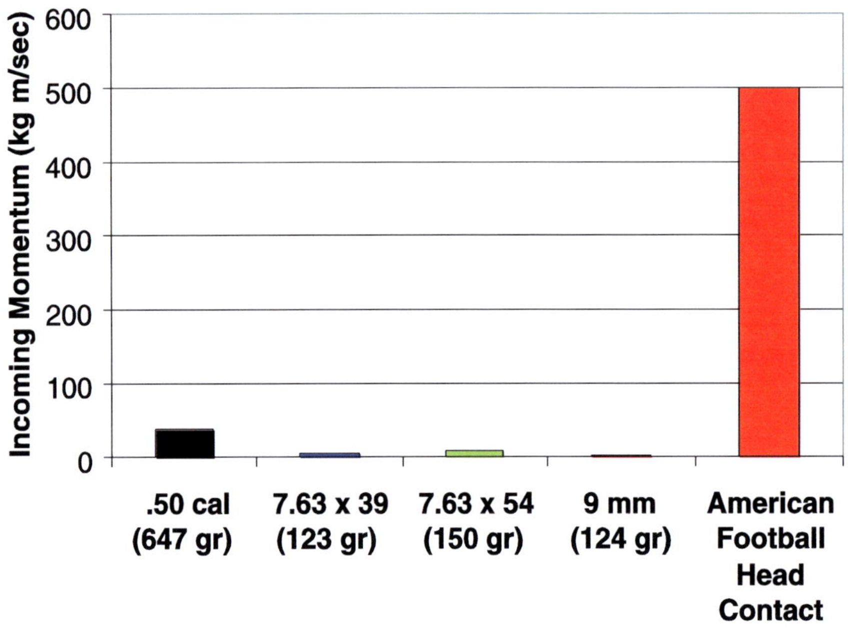

FIGURE 7-1 Effective momentum of high-rate ballistic impacts at muzzle velocity and low-rate football impact. SOURCE: Cameron Bass, Duke University.

This difference between conventional blunt trauma and ballistic blunt trauma is emphasized by considering typical timelines for ballistic impact. The need to decelerate an incoming round from hundreds of meters per second to zero over a span of centimeters implies a relatively rapid interaction between the head and the deforming helmet. A typical interaction timeline is shown in Figure 7-2.

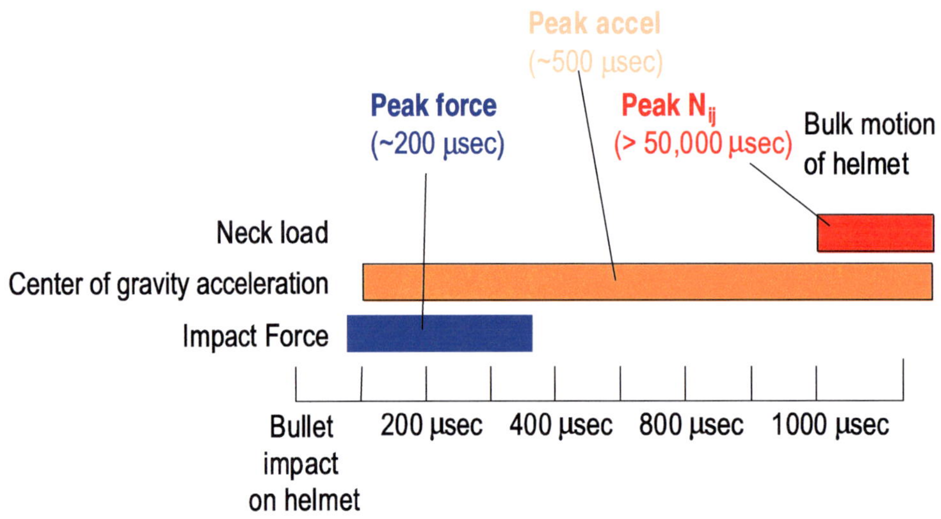

FIGURE 7-2 Ballistic impact injury timescales. SOURCE: Bass et al., 2003.

The peak impact force occurs approximately 100 µsec after helmet/head interaction. In contrast, automobile impacts and falls typically occur at time scales of 1 msec or greater, often 5-15 msec, which longer by a factor of 10-150. These momentum time scale and rate effects may play a large role in the causation of head trauma.

Finding: The existing helmet test methodologies, including the current Army test methodology, do not relate directly enough to human injury to confidently assess injury risk from back-face trauma to the head. Improving the link between test methodology and human injury is an urgent matter in light of the newer helmet systems with lower areal densities and increased threat velocities.

Injury Modes

There are two important injury modes with ballistic protective helmets and/or facial ballistic protection (Figure 7-3). First, penetrating injuries may be incurred because the helmet is defeated by the projectile. Second, impact loading injuries—also known as behind-armor blunt trauma (BABT)—may be caused by translational and/or rotational acceleration of the head. These may occur either from local contact of the deforming undefeated helmet onto the head/underlying skull or from more regional helmet/head contact, with acceleration loads transmitted through the helmet webbing or padding to the skull (Bass et al., 2003).

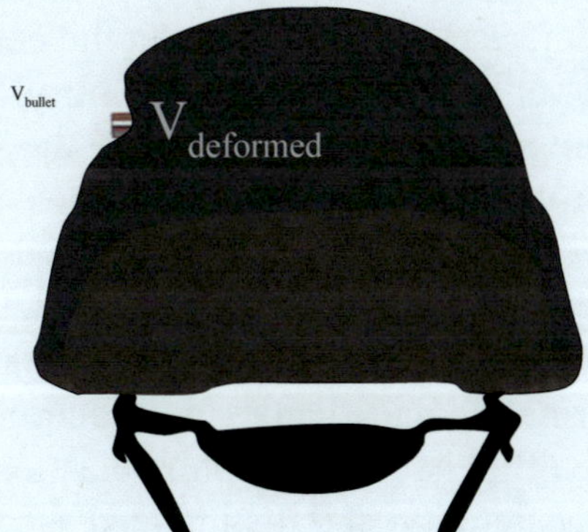

FIGURE 7-3 Likeness of a deformed personnel armor system for ground troops helmet. SOURCE: Cameron Bass, Duke University.

Generally, the function of the helmet is to prevent penetration and minimize the injurious effects of BABT. Owing to different techniques for assessing penetration and BABT, some methodologies may assess the penetration separately from the BABT.

Existing Human Injury Criteria

Substantial work has been done by the automobile, sports, and occupational biomechanics communities on head and neck injury (Pilkey et al., 2009; Mueller et al., 2008; Fuller et al., 2005). It is clear, however, that much of the work on existing injury tolerance values does not translate well to high-rate, low-momentum transfer events typical of ballistic impact events (e.g., Bass et al., 2003). Therefore, to evaluate the performance of ballistic helmets in protecting the wearer from injury, a test methodology should assess the mechanical performance of the helmet while addressing the associated risks of injury to the head/brain and neck.

Head Injury

The mechanisms that have been proposed for mechanical injury to the brain fall into six general categories: (1) direct contusion of the brain from skull deformation or fracture; (2) brain contusion from movement against rough interior surfaces of the skull; (3) reduced blood flow due to infarction or pressure; (4) indirect (countercoup) contusion of the brain opposite the side of the impact; (5) tissue stresses and strains produced by motion of the brain hemispheres relative to the skull and each other; and (6) subdural hematoma produced by rupture of bridging vessels between the brain and the dura matter (Melvin, 1993). It is hypothesized that the latter three mechanisms are involved in both impact and non-impact head injury (Ommaya, 1985). Using a variety of experimental models, researchers have been able to reproduce some of the above injury mechanisms. However, no satisfactory experimental model succeeds in producing the complete spectrum of brain injury seen clinically and yet is sufficiently well controlled and quantifiable to be a useful model for experimental studies. To improve the experimental models, it is necessary to improve our understanding of the mechanical and directional properties of brain tissue as well as intracranial deformations, relative motions, and interfaces, especially at ballistic impact rates.

Using head-impact models of primates, Gurdjian, Lissner, and associates attributed intracranial damage to skull deformation and change in intracranial pressure (Gurdjian, 1944). Holbourn, on the other hand, using photoelastic models of the head, proposed that head rotational acceleration and the induced shear strains in brain tissue are the causes of diffuse TBI (Holbourn, 1943). Gurdjian et al. later showed that injuries resulting from relative motion between the skull and the brain can be caused by head rotation (Gurdjian, 1954). Ommaya and coworkers, using acceleration models of primates, revised Holbourn's rotational theory by proposing that there is similar brain injury potential from rapid head rotation and from skull distortion owing to direct impact (Ommaya et al., 1966, Ommaya et al., 1971, Ommaya et al., 1973). Ommaya also proposed the centripetal theory, which states that the distribution of damaging diffuse strains induced by inertial loading decreases in magnitude from the surface to the center of the approximately spheroidal brain mass. Genarrelli and Thibault continued the investigation into the relative roles of translational and rotational accelerations by using more elaborate experimental models of subhuman primates (Gennarelli et al., 1982; Ommaya and Gennarelli, 1974). They concluded that diffuse injures to the brain occurred only in the presence of head rotational motion. They also investigated the effect of pulse duration on diffuse brain injury and concluded that diffuse brain injuries occurred at lower angular deceleration levels as pulse duration increased (Gennarelli and Thibault, 1982). It was shown that the incidence of unconsciousness with prolonged coma was greatest in purely lateral (i.e., coronal) plane acceleration (Gennarelli et al., 1982). Further analysis of injuries produced in the primate and human head models led Margulies et al. to conclude that in coronal plane rotational acceleration, the critical shear strain associated with the onset of diffuse axonal injury (DAI) was about 10 percent and the rotational acceleration threshold for severe DAI was about 16,000 rad/sec^2 (Margulies et al., 1990). It should be noted that in brain tissue, the threshold of functional failure is far lower than the threshold of mechanical failure (Varney and Varney, 1995). In some instances, no penetration is required to produce injury. For example, it has been shown that inertial loading alone to the head can cause DAI, an important cause of fatality due to head injury (Gennarelli and Thibault, 1989).

The early works of Lissner et al. and Gurdjian et al. emphasized skull deformations and intracranial pressure gradients as sources of brain injury. This work resulted in the Wayne State tolerance curve and was the basis for Head Injury Criteria (HIC), defined by the National Highway Traffic Administration in Federal Motor Vehicle Safety Standards (FMVSS) 208 in 1972 (Gurdjian et al., 1964; Lissner et al., 1960). HIC is based on the time history of the resultant translational acceleration of the center of gravity of the head and is currently used to assess head injury potential in automobile crash test dummies (Melvin, 1993). The HIC has been criticized by many investigators as a measure of head injury mainly because it does not distinguish between different types of head injury, nor does it address brain damage due to rotational accelerations. More recent tolerance frameworks for head injury have incorporated rigid body rotations with accelerations (e.g., Newman et al., 2000), but none are widely accepted, and their usefulness with ballistic impact is questionable. Finally, recent mild/moderate brain injury has been characterized based on sports data (Funk et al., 2007); timescale and momentum differences between sports impacts and behind armor ballistic impacts make extrapolations uncertain.

Currently, although some measures based on internal stresses and/or strains have been proposed as the injury criteria for the brain, no improved injury criteria have been developed that are universally acknowledged (Goldsmith, 1981). The accuracy of the proposed measures relies on accurate modeling of the geometry, material properties, and interfaces of the head-brain complex. By using advanced medical imaging techniques, the geometry of a head-brain model can be significantly improved. However, owing to their highly inhomogeneous, anisotropic, and nonlinear nature, no universally acknowledged method of modeling head and neck constitutive properties and interfaces has been developed, and correlates of traditional impact TBI research with ballistic impact research are uncertain.

Head Injury from Ballistic Impact

Limited research exists on head injury at ballistic rates (e.g., Viano et al., 2004; Bass et al., 2003). In 2003, Bass and coworkers performed a study to develop injury criteria for skull fracture in human heads during ballistic loading of a protective helmet. The effort further assessed experimental measures to quantify the risk of brain injury under impact loading (Figure 7-4). Two series of ballistic impact tests were performed, including tests with various initial test round velocities; these included nine cadaver tests with helmets and 9-mm test rounds and four cadaver tests with various compliant direct impacts. These ballistic impact tests were used to assess the risk of skull fracture and other head injuries from nonpenetrating BABT for military helmets. Skull fractures were produced in five of nine tests, with a single artifactual fracture at a preexisting unhealed craniotomy. Injuries ranged from simple linear fractures to complex combinations of linear fractures and a depressed fracture. Other injuries included bruising of the dura and severe local skin friction injuries.

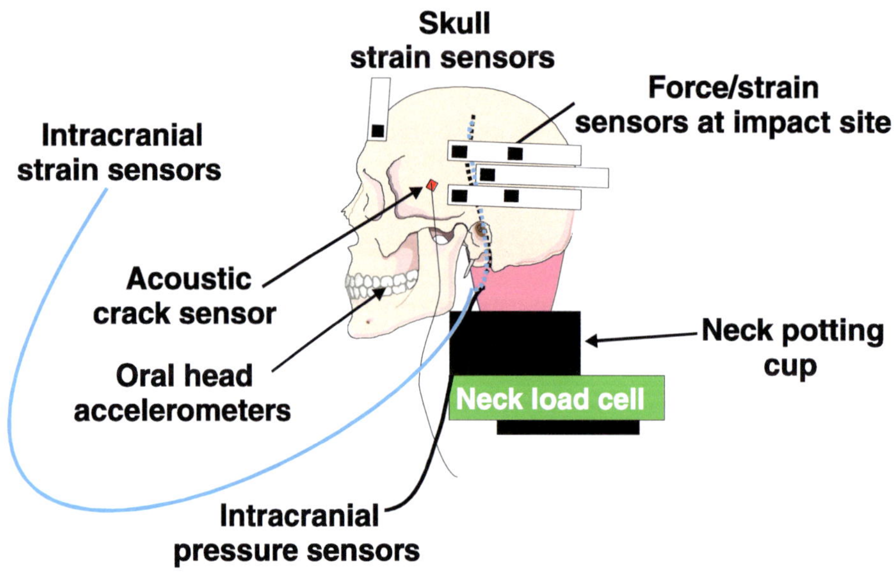

FIGURE 7-4 Cadaver instrumentation overview. SOURCE: Bass et al., 2003.

This study developed an injury criterion for both test round velocity and cadaver peak pressure. For this injury risk function, there is a 50 per cent risk of skull fracture for a peak impact pressure of 51,200 kPa as measured by the force/strain instrumentation. Using a simple velocity correlation between a dummy and a cadaver, a dummy injury risk function is developed: namely, there is a 50 per cent risk of skull fracture for dummy peak impact pressure of 15,220 kPa. This injury risk function may be used with a general helmet and the Hybrid III dummy used in previous testing.[47]

Automobile injury criteria, including the HIC, were not found to be a good predictor of cadaveric injury. Indeed, for all fracture tests, the calculated HIC was well below the injury reference value. Skull fracture from ballistic BABT is an intrinsically high-rate event. Energy is deposited locally, and local skull deformations are significant. Use of HIC requires essentially rigid body motion of the head at a much lower rate than is characteristic of ballistic events. In addition, the risk of neck injury from lateral impact was found to be small for the 9-mm projectile at up to 460 m/sec velocity, and no neck injuries were found in the cadaver tests.

Neck Injury

The principal existing neck injury criterion is promulgated by the National Highway Traffic Safety Administration for frontal impact testing. This injury reference value is termed the National Institute of Justice (NIJ) criterion and is based on human cadaver, volunteer, and animal data; it is intended for use with the Hybrid III dummy (Eppinger, 1999). The NIJ criterion is a composite of injury indicators based on a linear combination of neck loads and moments. The loads include neck axial tension and compression, and the moments include neck flexion and extension. The postulated injury levels for these combined loads have been validated using human cadaver, volunteer, and animal subjects. For this injury criterion, a NIJ value of 1.0 represents a 22 percent risk of an abbreviated injury scale value ≥ 3 neck injury (Eppinger, 1999).

The series of tests directly used in developing the neck injury criterion used live pigs and human cadavers. The data from these tests suggested that tension was the best predictor of out-of-position neck injuries; however, the tests were limited to tension-extension, which is the primary mode seen in automobile field data. Predominantly lateral impacts may result in significant lateral shear and bending modes that are not represented in the existing injury assessment criterion. An extension of this NIJ method has been proposed by Bass et al. (2000). Additional neck injury criteria have been proposed that include the effect of head supported mass such as helmets and night vision.

As motion of the neck from ballistic impact onto a helmet occurs on a time scale similar to that seen in vehicle crashes or falls, automotive injury criteria are likely applicable. For the low-momentum transfer that occurs from current helmet threats, the risk of neck injury is quite low. Direct measurements of the neck loads associated with the ballistic impact on a helmet from a 9-mm full metal jacket (FMJ) round at various bullet velocities used for human cadaver tests are shown in Figure 7-5 (Bass et al., 2003). Injury assessment values indicate

[47]In the early 1970s, the automotive industry developed the Hybrid III 50th Percentile Male anthropomorphic testing device (ATD).

very low risk of neck injuries for these scenarios, and no neck injuries were seen in the testing.

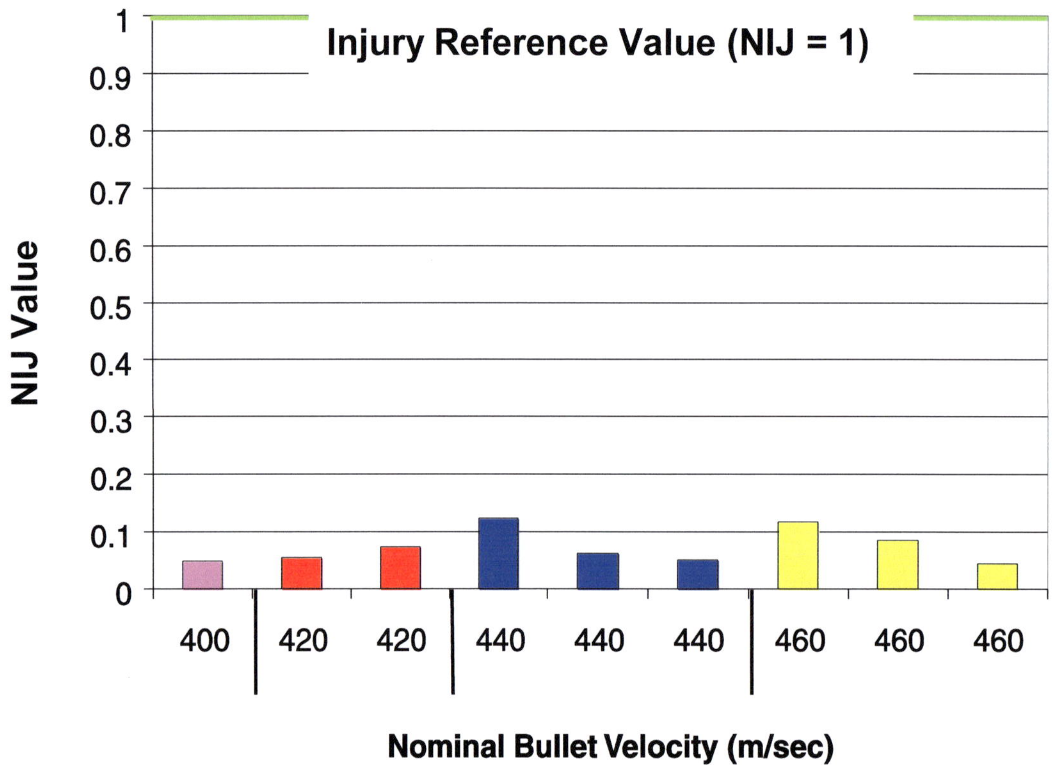

FIGURE 7-5 Neck injury assessment value for 9-mm FMJ test round at various velocities into helmeted human cadavers. SOURCE: Bass et al., 2003.

For future threats, estimates for typical incoming rounds based on momentum arguments suggest that the risk of injury is low for 7.62 × 54 mm rounds at muzzle velocity but may become substantial for .50-cal threats at substantial fractions of the muzzle velocity. The potential for neck injury is proportional to the amount of momentum and resulting head velocity. Thus, head velocity is a conservative estimate of the injury potential. An estimate of the potential for neck injury from various potential threat rounds is shown in Figure 7-6. If the average neck injury risk for the current helmet is taken as shown as in Figure 7-5, the risk of neck injuries for a 7.62 × 54 mm threat is less than 0.1 per cent, but the committee estimates it would be greater than 22 per cent for a .50-cal threat at the current areal density.

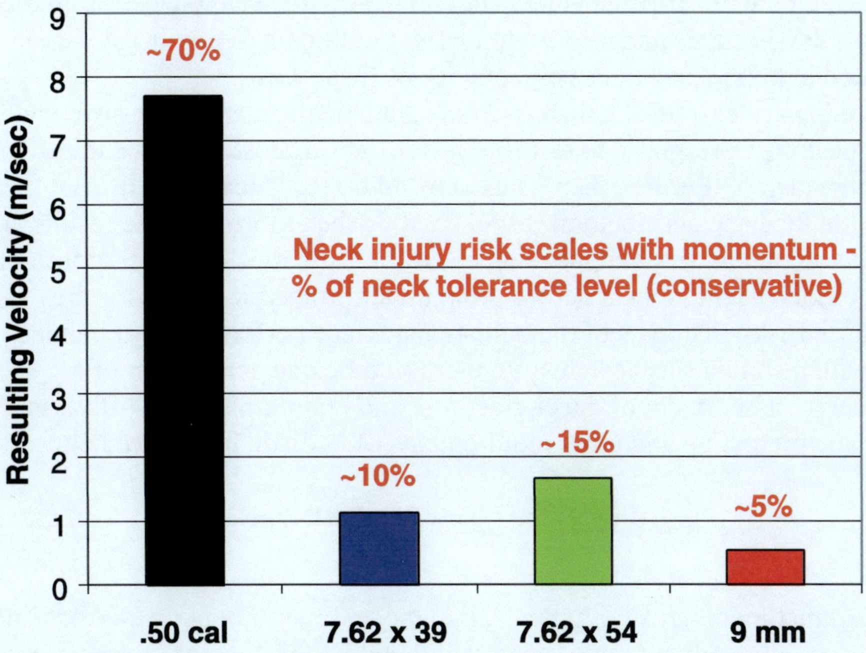

FIGURE 7-6 Residual head/neck velocity from momentum transfer to the helmet/head system. SOURCE: Cameron Bass, Duke University.

The glaring weakness of current methodologies, especially clay-based methodologies, is a link to human injury. The Army should immediately investigate human response and injury from helmet BABT to provide injury assessment values and dynamic response values to support the creation of a well-validated test methodology for helmet BABT. Some of the work done on skull fracture should be generalized and extended to other potential brain injury modes incorporating existing epidemiological, cadaveric, and animal studies.

Recommendation 7-1: The Army should perform research to define the link between human injury and the testing methodology for head behind-armor blunt trauma.

HELMET DESIGN AND SUSPENSION SYSTEMS

A potentially important aspect of ballistic protective helmet design is the suspension system that provides helmet standoff from the head, an important factor in ballistic protection. A recent study investigated potential backface trauma to the skull using two ballistic protective helmet interior systems, one a webbing-based suspension system and the second a foam-based padding system (Bass et al., 2006). Both interior systems were installed into a current military ballistic protective helmet. The back face trauma risk assessment was performed using the test methodology outlined by Bass et al. (2003). Fifteen helmet systems were tested, nine with suspension and six with padding. Two types of round were used: 9-mm FMJ rounds and steel right circular cylinders (64 grain) at velocities that resulted in helmet backface contact with the head form.

For the systems tested, there were no statistically significant differences in backface peak force response to test rounds for the suspension and padding systems considered. Further, the results showed no statistically significant difference in the head acceleration response of the head form. These results are consistent with the understood mechanisms of ballistic mechanical response of a helmet/interior system to local deformation under projectile impact for the systems tested. Assessments of the ballistic backface performance of helmet systems with installed suspension systems should be one component of a comprehensive assessment of the engineering and ergonomic trade-offs associated with ballistic protection and other requirements of ballistic protective helmets.

Existing Helmet Test Methodologies

Test methodologies for helmets, like those for assessing the performance of body armor, generally separate penetration and BFD behavior as separate assessment criteria. Although there is an extensive literature on helmet test methodologies, existing standards are largely based on requirements from motor vehicle and sports impacts. Of the 29 helmet test standards listed in the Advisory Group for Aerospace Research and Development Advisory Report on Dummies for Crash Testing (AGARD, 1996), only one, NIJ-0106.01, is intended for ballistic impact. Differences between crash test impact standards and the ballistic standards are emphasized by the effective impact energy as shown in Figure 7-7. The peak impact energy for the ballistic standard is an order of magnitude larger than that for a typical crash. Further, the impact momentum for the ballistic standard is generally far lower than that for the crash test helmet standards. These differences will increase with higher velocity threats, such as rifle threats.

Impact trauma assessment in all of the current blunt helmet standards is based on similar concepts. The principal implicit assumptions are that concussion or head injury is well correlated with skull fracture (reference) and that the head acts as a mostly rigid body so that "mean" acceleration may be associated with skull fracture (Bass et al., 2003). More recent work suggests that the correlation between skull fracture and brain injury is not good (Viano and Lau, 1985). Further, the skull does not generally act as a rigid body for ballistic deformations of the helmet backface, even for handgun rounds. Measurements taken from cadavers with and without skull fracture show no correlation with the Wayne State Tolerance Curve or similar concepts (Bass et al., 2003), as shown in Figure 7-8. Indeed, for all fracture tests, the calculated HIC was well below the injury reference value.

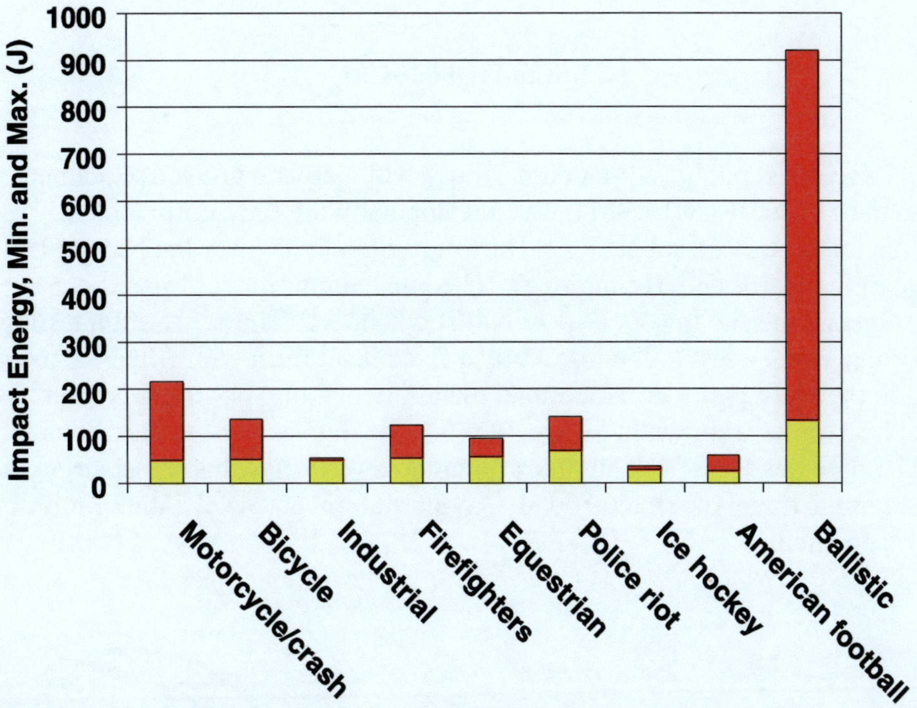

FIGURE 7-7 Impact energy for helmet standards. SOURCE: AGARD, 1996.

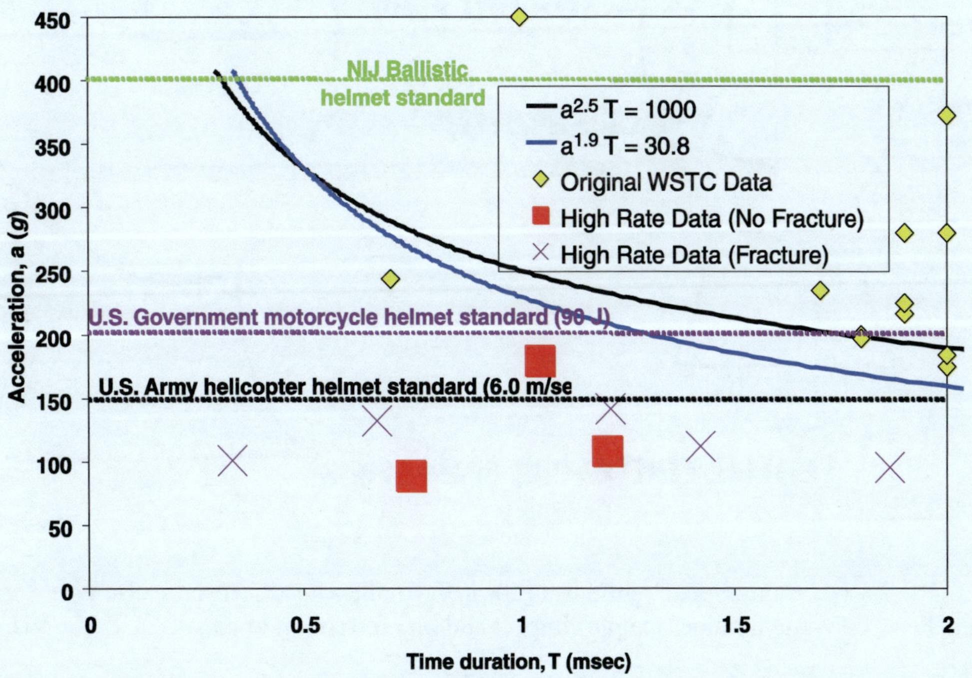

FIGURE 7-8 Ballistic (high-rate) skull fracture data vs. impact injury criteria for typical blunt injury. SOURCE: Bass et al., 2003.

NIJ Standard 0106.01

The earliest published standard for use with ballistic protective helmets, NIJ Standard 0106.01 (NIJ, 1981), was developed by the Law Enforcement Standards Laboratory of the National Bureau of Standards (now the National Institute of Standards and Technology). The NIJ helmet standard specifies penetration and inertial impact tests of ballistic helmets. The penetration testing is performed using a test round impacting a fixed head form with witness panels located in the midsaggital or midcoronal planes, depending on shot direction. The saggital head form is shown in Figure 7-9. For the impact tests, the test round impacts a rigid head and neck complex mounted on a trolley that translates in the direction of the travel of the test round. An inertial impact acceleration limit of 400 *g* is specified.

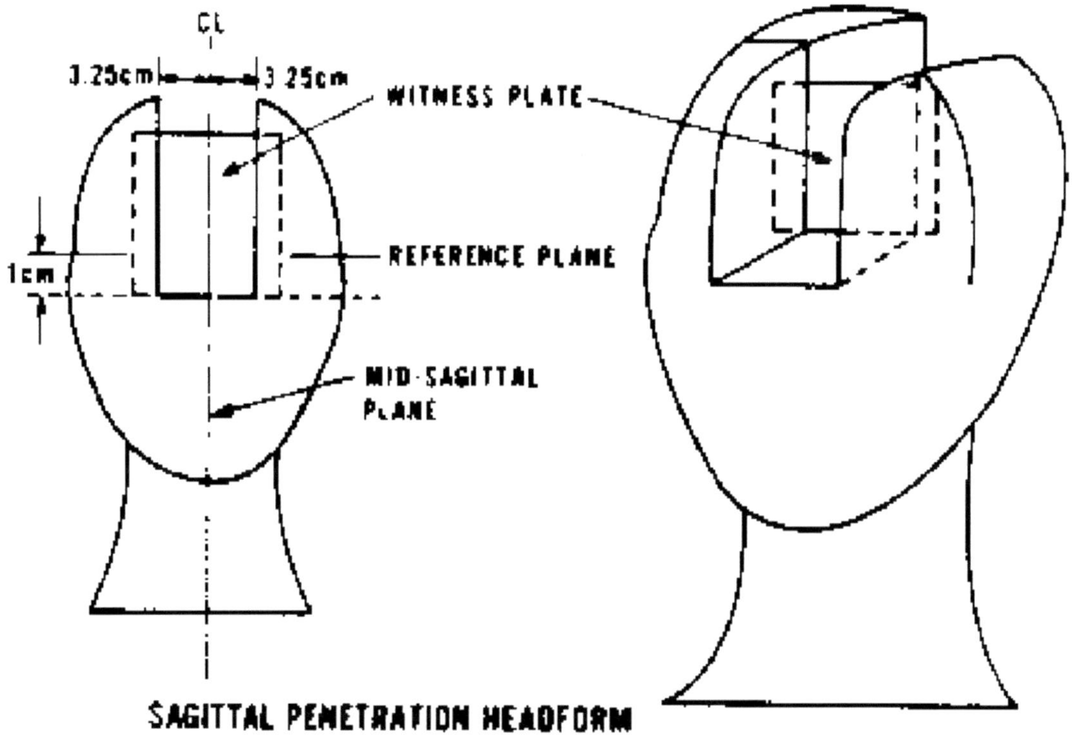

FIGURE 7-9 NIJ sagittal penetration head form. Two other head forms are also used for the helmet tests: one for side (temple) impact and one for crown impact. SOURCE: NIJ, 1981.

Threat

Characteristics of the test rounds for each helmet type specified in the standard are shown in Table 7-1. Bullet velocities for different test rounds range from 259 m/sec to 425 m/sec, with energies ranging from 133.1 J to over 900 J.

TABLE 7-1 Characteristics of Test Rounds from NIJ Standard 0106.01

Helmet Type	Test Ammunition[a]	Nominal Bullet Mass (g)	Required Bullet Velocity (m/sec)	Nominal Bullet Energy (J)
I	22 LRHV Lead	2.6	320 ± 12	133
	38 Special RN Lead	10.2	259 ± 15	342
II-A	357 Magnum JSP	10.2	381 ± 15	740
	9 mm FMJ	8.0	332 ± 15	441
II	357 Magnum JSP	10.2	425 ± 15	921
	9 mm FMJ	8.0	358 ± 15	513

[a]LRHV, long rifle high velocity; RN, round nose; and JSP, jacketed soft-point pistol.

Human Injury Criterion

A simple translational head acceleration limit of 400 *g* is used in the NIJ ballistic helmet standard. It is uncertain whether the impact attenuation test has ever been used for assessing ballistic protective helmets.[48] For impact loading injuries, this standard can obscure potentially injurious shocks in some realistic situations and is overly conservative in others (Bass, 2003). Further, typical automobile injury assessment filtering techniques require low-pass filtering to 1650 Hz, although typical timescales of back-face impact occur at similar or

[48]Personal communication between Kirk Rice, National Institute of Standards and Technology, and Dale Bass, committee member, on August 10, 2010.

higher frequencies, as discussed above. This makes it inappropriate to use the NIJ 0106.01 BFD criterion (impact attenuation) for ballistic impact.

ARMY CLAY HEAD FORM

The Army clay head form is an aluminum head form based on the penetration head form specified in NIJ 0106.01, shown in Figure 7-10. The empty spaces on the head form are filled with Roma Plastilina No. 1 (RP #1) clay.

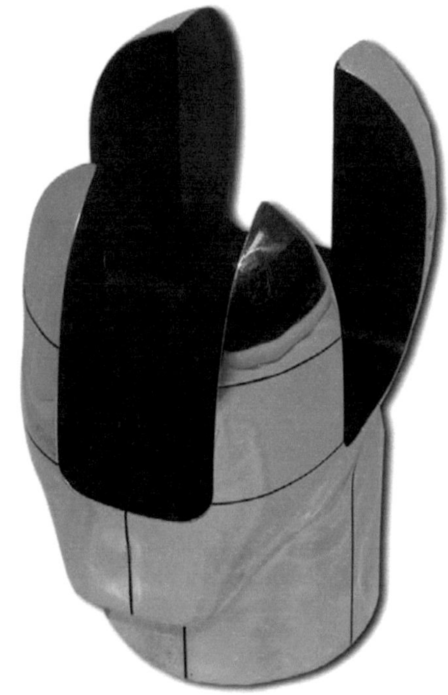

FIGURE 7-10 Army clay head form. SOURCE: Courtesy of Rob Kinsler, U.S. Army Research, Development and Engineering Command/ARL.

This is the same type of clay used to certify ballistic vests. The plastic property of the clay allows it to record BFDs caused by impact of nonpenetrating projectiles during the ballistic testing of hard body armor. Helmet testing standards and practices are derived from body armor testing standards and practices and, as in body armor testing, are based on the use of RP #1 as the test recording medium. As such, they capitalize on existing infrastructure and organizational experience with those test methods, yet they also suffer from all the attendant issues and weaknesses associated with these methods. The test standards are derived from the observed performance of a sample of helmets rather than experimental evidence relating test outcomes to likely injury.

Finding: It is uncertain how clay response correlates with human head/skull/brain response. Yet, clay response serves as the basis for current clay-based helmet methodologies.

Testing Standards

Helmets are subjected to a series of ballistic and nonballistic tests as part of both first article testing and lot acceptance testing. Nonballistic tests include the inspection and verification of various aspects of helmet construction such as edging adhesion, adhesion of the coating, and barcode labeling.[49] Ballistic tests assess the helmet's ability to (1) withstand penetration and (2) not exceed a BFD limit.

Penetration

In the testing of body armor plates, a "plate penetration" occurs when a round penetrates the soft armor behind the plate. In the testing of helmets, a "helmet perforation" occurs when the helmet is visibly penetrated by a round during the test. Both events are defined in the test procedures and both result in a test failure. Since there is no practical way to determine or measure "degree of penetration," both the plate and helmet tests must be attribute-based.

Current Aberdeen Test Center (ATC) testing practice is to assess a helmet's resistance to penetration in terms of penetration and perforation. According to Page and Humiston:[50]

> Perforation occurs when the threat defeats the sample. This is noted when the threat and/or sample fragments have entered the witness medium. . . . Penetration occurs when a threat comes in contact with the sample but does not defeat it.

The draft DOT&E Military Combat Helmet Standard for Ballistic Testing (DOT&E, 2010) redefines the penetration and perforation in terms of partial and complete penetration:

> A complete penetration shall be defined as complete perforation of the shell by the projectile or fragment of the projectile as evidenced by the presence of that projectile, projectile fragment, or spall in the clay, or by a hole which passes through the shell. …Any fair impact that is not a complete penetration shall be considered a partial penetration.

---

[49]Matthew Page and Travis Humiston, ATEC Protective Equipment Division, "Head Protection Testing: Processes, Issues," presentation to the committee, October 13, 2010.
[50]Ibid.

Resistance to penetration (or V_0 testing) is measured as a sequence of five ballistic impacts, one each to the front, rear, left, and right sides of the helmet as well as to the crown of the helmet. Current ATC protocol also tests the V_{50} ballistic limit using a series of 7 to 14 shots to the five regions of the helmet at varying velocities.[51] V_{50} testing is not part of the currently proposed DOT&E protocol.

Backface Deformation

Helmet BFD (also referred to as ballistic transient deformation) is assessed using the same sequence of five ballistic impacts, one each to the front, rear, left, and right sides of the helmet as well as to the crown of the helmet. As with current body armor testing practices, the BFD is assessed using a laser scanner and it is defined as the maximum impact depression depth in the clay, as measured from the original clay surface. Under current ATC testing protocol, the BFD can be no greater than 16 mm for crown, left, and right impacts and no greater than 25.4 mm for front and rear impacts.

The many unknowns in the use of clay as a medium make it unclear as to whether the BFD response in clay methodology is appropriate for helmet testing, especially since the mechanical backface response of the head surrogate may govern both penetration and impact tolerance portions of the test.

Recommendation 7-2: The Aberdeen Test Center should ensure the following:

1. Dynamic mechanical strain/deformation response of the head surrogate is similar for both types of loading at loading rates typical of behind-helmet response;
2. Response of the head surrogate is similar to that of the human head;
3. Required head quality control calibration is either performed on the head surrogate itself or is shown to be demonstrably represented by a surrogate for the head itself (i.e., by a sample box filled with clay) in controlled testing using a standard test procedure; and
4. Response of the clay for the low-rate calibration tests is shown to be similar or scalable to the high-rate BFD response of the surrogate in controlled testing using a standard test procedure.

Test Procedures

Test range setup is in accordance with ATC Test Operating Procedure 10-2-210 (ATC, 2008). The test is conducted in accordance with NIJ Standard 0106.01 with the following four exceptions:

---

[51] Matthew Page and Travis Humiston, ATEC Protective Equipment Division, "Head Protection Testing: Processes, Issues," presentation to the committee, October 13, 2010.

- Test items may be conditioned as needed.
- Test distances may be altered.[52]
- The test head form is modified with slots in both the coronal and midsagittal directions, with a depth of approximately 143 mm below the head form crown, with drill points along the headform pillar for use by the laser scanner.
- Striking velocities are calculated according to the U.S. Army Test and Evaluation Command International Test Operating Procedure 4-2-805 in order to determine if a shot is fair (DOT&E, 2010).

Clay Preparation, Conditioning, and Calibration

Helmet testing is based on a head form with slots in both the coronal and midsagittal directions. There is only one head form size, although there are between four and six helmet sizes depending on the type of helmet. The slots in the head form are packed with RP #1 as the recording medium for both penetration/perforation and BFD. As shown in Figure 7-11, the clay is shaped to create a smooth, uninterrupted surface with the headform.

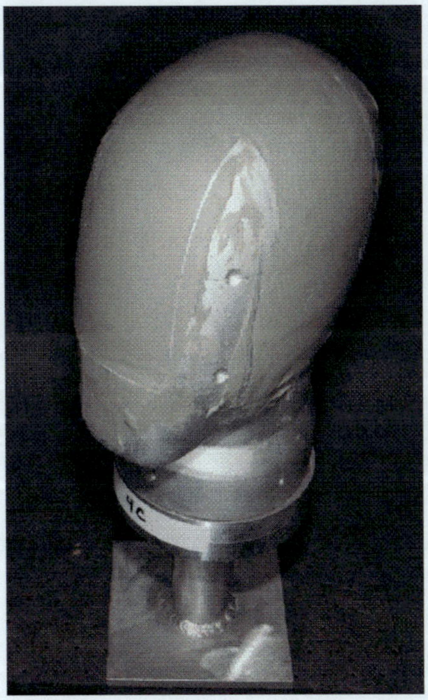

FIGURE 7-11 ATC head form with clay. SOURCE: Rob Kinsler, U.S. Army Research, Development and Engineering Command/ARL.

[52]Test distances are given in Figure 6 of NIJ Standard 0106.01 (NIJ, 1981).

As described in the ATC Internal Operating Procedures for the Head Protection Testing (ATC, 2010), the clay in the head form is calibrated in a 12 in. × 12 in. × 4 in. plywood-backed box, analogous to the box form used for RP #1 in the body armor plate testing procedure. Up to eight head forms may be conditioned with each box so long as the clay in the box and in the head forms comes from the same lot and the head forms are conditioned within 12 in. of the box (Figure 7-12). Once conditioned, calibration of the box is performed via drop test in which 2.2-lb, 1.75-in.-diameter steel cylinders are dropped from a height of 78.7 ± 0.8 in., into the clay box. The clay is considered to be within calibration if the indentations made by the steel cylinders are all within 1.0 ± 0.1 in., as measured by a digital caliper. The clay head form removed from the oven with the clay box may be used for up to 45 min after the third drop, and the remaining head forms may be used for up to 4 hours from the time of the third drop and for up to 45 min after being removed from the oven.

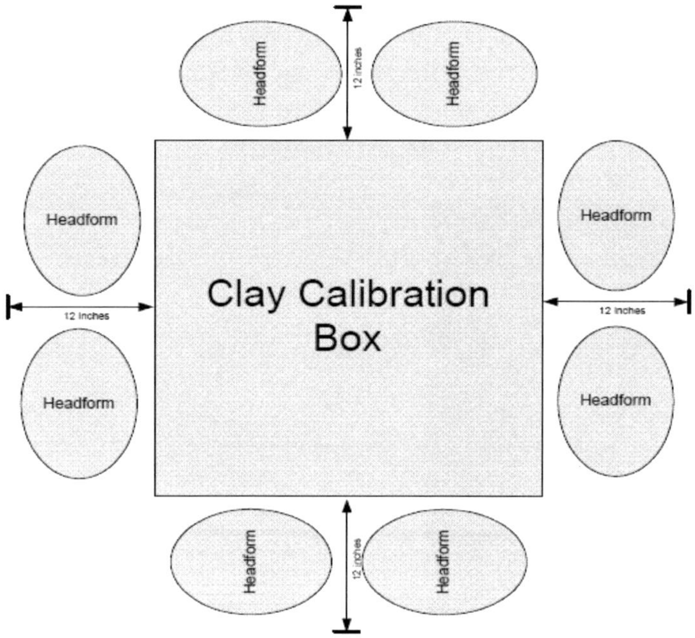

FIGURE 7-12 Head form clay conditioning by analogy. SOURCE: Kevin Reilly, Program Manager Infantry Combat Equipment, Marine Corps Systems Command, "PM ICE Helmet Testing Overview," presentation to the committee, August 10, 2010.

Threats

Helmets are tested against the following threats:

- Remington 9-mm, 124-gr FMJ projectile;
- 2-gr right circular cylinder (RCC) fragment;

- 4-gr RCC fragment;
- 16-gr RCC fragment;
- 64-gr RCC fragment; and,
- 17-gr fragment simulating projectile.[53]

Test Item Configuration and Impact Locations

The head form used in the test is mounted on a test fixture that is capable of being rigidly fixed with 6 degrees of freedom to allow for positioning the head form in all required positions. The helmet is mounted on the head form in the as-worn configuration and position, using the helmet's suspension/retention system to hold it on the head form. Prior to mounting, the helmet is marked to show the desired impact locations. As illustrated in Figure 7-13, the current test uses five preset impact locations that are shot in the following order:

- Crown at the intersection of midsaggital and coronal planes;
- Left side (as facing) on the coronal plane 50 mm above the earflap;
- Right side (as facing) on the coronal plane 50 mm above the earflap;
- Front on the midsaggital plane 85 mm above the edge and at least 1.5 in. from the night vision goggles hole; and
- Back on the midsaggital plane 75 mm above the edge (DOT&E, 2010).

In addition, at least one of the noncrown shots must be at a bolt. A high-speed camera is used to record the bolt shot (to ensure the hit is within 1/2-in. as per the standard), and the FARO laser is used to measure BFD.

[53]Matthew Page and Travis Humiston, ATEC Protective Equipment Division, "Head Protection Testing: Processes, Issues," presentation to the committee, October 13, 2010.

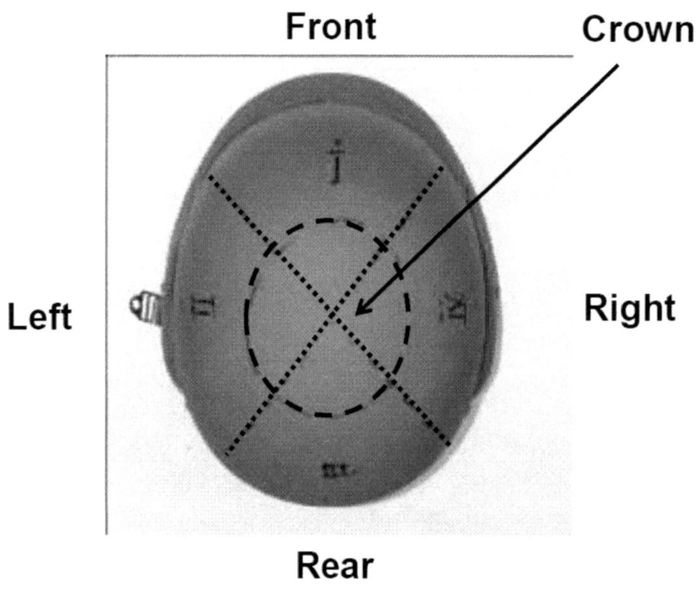

FIGURE 7-13 Test impact locations. SOURCE: Linda Moss, Survivability/Lethality Analysis Directorate, U.S. Army Research Laboratory, "Statistical Issues Related to Helmet Testing," presentation to the committee, August 10, 2010.

Test Process

The head form and helmet attached to the test fixture are mounted on the test frame shown in Figure 7-14. The helmet is aligned to ensure the target location achieves the required obliquity and, for bolt shots, the high-speed cameras are aligned. The helmet is removed from the head form and the clay surface is scanned with the laser. The helmet is reattached to the head form and the shot taken. The helmet is then removed from the head form and inspected for penetration and perforation. The clay is rescanned with the laser to calculate BFD. A fair hit is recorded if the shot location, obliquity, yaw, and shot velocity are within required limits.

FIGURE 7-14 Test frame. SOURCE: Matthew Page and Travis Humiston, ATEC Protective Equipment Division, "Head Protection Testing: Processes, Issues," presentation to the committee, October 13, 2010.

H.P. White Laboratory Test Procedure

The helmet testing procedures used by H.P. White Laboratory, a private test laboratory, were developed for testing helmets for law enforcement agencies and have been adapted to the testing of military helmets. The test procedure was developed specifically for bullet penetration or for excessive BFD of the helmet material, and does not include biomechanical shock. It is similar to the one used currently by the Army and is based on the original NIJ 106.01.

The projectile types are described in the NIJ Standard 0101 and the head form used for the deformation studies is described in NIJ Standard 0106. The test procedure evaluates the helmets for penetration or excessive deformation by bullet type, impact velocity, and bullet caliber. Six different threat levels are used in the test: I, IIA, II, IIIA, III, and IV (see Table 7-2). Helmets passing the test are certified only for these six threat levels. A helmet size equivalent to a hat size 7-1/4 is considered standard for the H.P. White test. Four helmets of each design are tested for threat levels I, IIA, II, and IIIA; two helmets are tested for levels III and IV.

TABLE 7-2 H.P. White Laboratory Test Procedure

TABLE I. BALLISTIC PROTECTION/THREAT LEVELS

Level (a)	Test Ammunition Caliber	Bullet Weight (gr)	Bullet Type	Required Bullet Velocity (fps) Minimum	Required Bullet Velocity (fps) Maximum	Required Shots Penetration (b)	Required Shots Deformation (b)
I	.22 LR	40	Lead	1050	1100	5	5
	.38 Special	158	RN Lead	850	900	5	5
IIA	9x19mm	124	FMJ	1090	1140	5	5
	.357 Magnum	158	JSP	1250	1300	5	5
II	9x19mm	124	FMJ	1175	1225	5	5
	.357 Magnum	158	JSP	1395	1445	5	5
IIIA	9x19mm	124	FMJ	1400	1450	5	5
	.44 Magnum	240	SWC-GC	1400	1450	5	5
III	7.62x51mm	150	Ball, M80	2750	2800	5	5
IV	.30-06	166	AP	2850	2900	5	5
V	Special Category(c)		-	-	-	5	5

(a) Duplicate of ballistic threats and velocities as specified by NIJ-STD-0101.03, BALLISTIC RESISTANCE OF POLICE BODY ARMOR.
(b) One (1) shot in each quadrant and helmet crown.
(c) This procedure may be used to test the resistance of other bulletted ammunition or Fragment Simulating Projectiles (FSP) conforming to MIL-P-46593A or Drawing HPW-02-010-00.

SOURCE: H.P. White.

The test procedure involves five impacts on each of two helmets with bullets of two different calibers. Helmets are impacted on five sites: front, back, left, right, and top. The helmet is held on the head form only by the chin strap and may be replaced on the head form if it is knocked loose during the test. Shots must be "fair," i.e., normal to the helmet surface, of the correct velocity, and not yawed more than 3 deg. If the shot is unfair, a second shot can be made at least 3 in. from the first. One fair shot must be made at each of the five locations, and if one penetrates the helmet or if the impact crater in the clay on the head is deeper than 25 mm, the helmet will not be certified.

Tests are made at 70°F after wetting the helmet with water. The shot is made at a distance of 16.5 ft from the helmet. The muzzle velocity (see Table 7-1) is measured at a distance of 8 ft for each shot. After each shot, the helmet is removed from the head form and examined for penetration by the bullet. All locations on the helmet must be tested even if the helmet has failed at one impact site.

BFD is tested on a second helmet. The deformation test uses a head form, Figure 7-15, filled with RP #1 modeling clay. The clay is first qualified by dropping a 2.2-lb, 1.75-in. diameter steel sphere from a height of 78.7 in. onto the clay already mounted in the head form. To be acceptable for the ballistic test, an impression of 25 mm ± 1 mm deep should be left behind. Three drop tests are carried out for each head form prior to shooting the helmet. The depression left is measured after each shot to the nearest millimeter, but the clay in the head form is qualified only prior to the first shot. As with the penetration tests, the deformation tests are carried out until all five locations on the helmet are tested. No reference to the acceptable penetration depth could be found in the H.P. White written procedure.

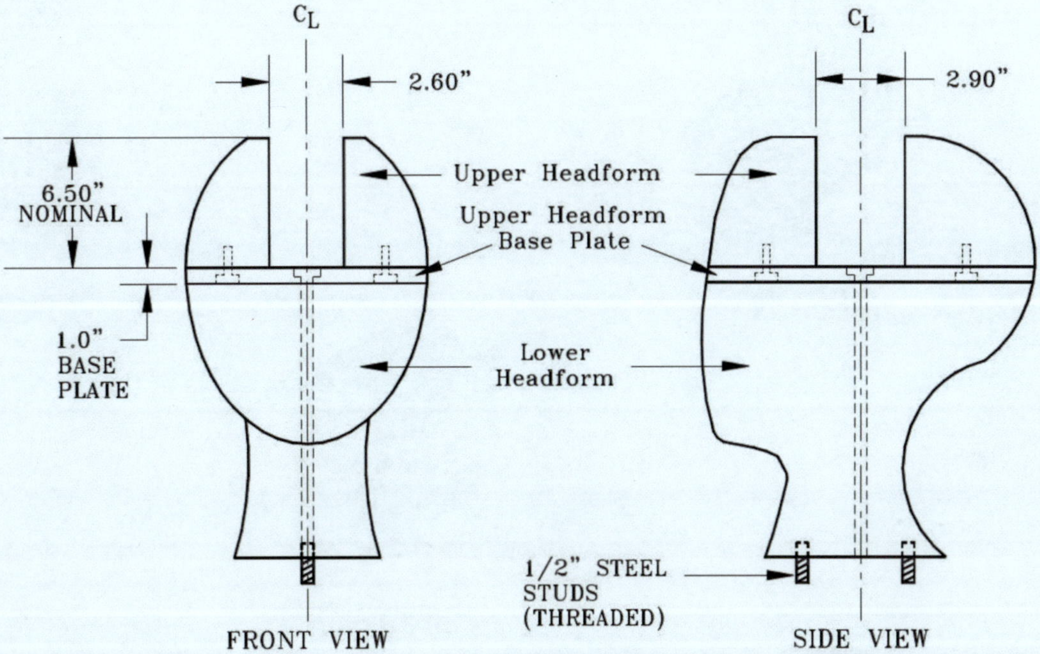

FIGURE 7-15 H.P. White head form: Only one headform is used for all impact tests. Overall dimensions comply with NIJ-STD-0106.01 (hat size 7¼). Upper portions, including the base plate, are made of 6061-T6 aluminum or equivalent. Lower head form is of USG epoxy #303. SOURCE: H.P. White Laboratory, Inc., 1995.

Hot tests, 120ºF, or cold tests, −20ºF, can also be carried out by this procedure. Only the helmet is temperature conditioned. The test environment is maintained at 70ºF and the tests are to be completed in 30 min. Wet conditioning is not carried out for helmets tested under cold conditions. Penetration of the helmet by any fair shot or an excess of deformation deny the helmet certification.

This test methodology is similar to the current Army methodology and provides a record of BFD in the clay. No displacement limits are specified in the document, and human injury has not been linked with this assessment technique.

Peepsite Head Form

A new head form, referred to as the Peepsite head form (Figure 7-16) was developed by the U.S. Army Research Laboratory to avoid potential drawbacks of the NIJ head forms.[54] A big shortcoming of the current test head forms is that the clay used to measure the BFD of the helmet upon impact is contained between two solid aluminum parts of the head form (see Figure 7-10).

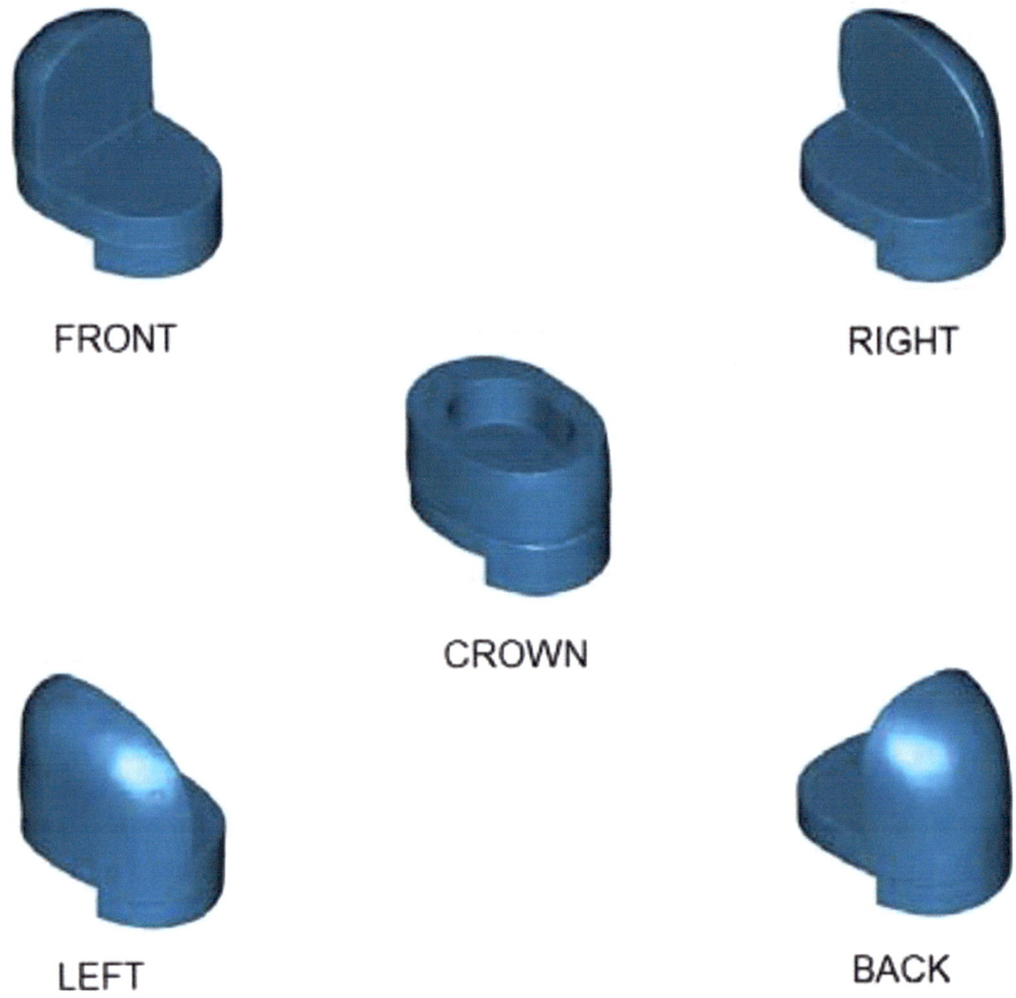

FIGURE 7-16 Peepsite head forms: different head forms for different shot directions.

[54]The new head form was demonstrated to the committee by Rob Kinsler, ARL, on October 13, 2010.

This causes three potential problems. One, the solid aluminum will constrain both the outward and inward flow of the clay during the impact, giving a smaller displacement than might actually occur during use of the helmet in combat. Since blunt trauma by the impact is the primary cause of injury to the soldier for a bullet that does not penetrate the helmet, the actual displacement of the back surface of the helmet should be measured with some accuracy. The Peepsite head form reduces this concern by eliminating the metallic petals near the backface impact.

The second problem is that the backface contact can span the aluminum petals, preventing further impact or altering the BFD response and backface signature. Again, elimination of the petals in the Peepsite headform eliminates the potential for helmet/head form interactions to alter the backface response.

The third potential problem arises from the fact that the clay and the helmet have very different temperature characteristics. In such tests the clay is normally heated above room temperature to achieve the desired rheological behavior. Testing on the Peepsite head form, however, is done at room temperature, which means that the rate of cooling of the clay and the aluminum head form will be very different from one another, resulting in thermal gradients and residual strains and stresses in the clay that may affect the impact event.

As in the tests developed by NIJ and H.P. White, five surfaces are tested for impact: left, right, front, back, and crown. Instead of only one impact in each area, three are used, each on different helmets, and the displacement is measured after each impact. The test matrix is illustrated in Table 7-3. Three replicate shots are made for each area, and five threats are tested.

TABLE 7-3 Helmet Test Matrix

Table 1. Shot Matrix

THREAT vs. TARGET	9mm FMJ 427 m/s	9mm FMJ 457 m/s	225gn RCC 284 m/s	225gn RCC 316 m/s	total shots per shot location
FRONT	3	3	3	3	12
LEFT SIDE	3	3	3	3	12
BACK	3	3	3	3	12
Total shots per threat/velocity	9	9	9	9	36 shots grand total

SOURCE: H.P. White.

Clay deformations may be measured by several methods, including manual calipers, digitization arms, digitization arms with laser scanning capability, and purpose-built laser scanners. Since the deformation of the head form is three-dimensional, laser arm and laser scanning systems can be used to maximize the information detail of the deformations.

As there is no link to human injury for this methodology, especially with the current generation of room-temperature clay, the U.S. Army Research Laboratory (ARL) provided results of comparison testing of an existing helmet system against a new candidate helmet system.[55] This time of comparison testing is essential for establishing a baseline in the absence of a link to human injury. Additional head form sets have been supplied to five commercial testing laboratories to more widely disseminate the capability for Peepsite head form testing.

Finding: The Peepsite head form reduces or eliminates several potential problems with the National Institute of Justice head form that is used in the current clay test methodology.

Since testing with the clay head forms is based on an unproven assumption that clay deformation is correlated in some way with human injury, an essential prerequisite to the development of the Peepsite head form as a viable test methodology is correlation of the current helmet system performance with deformations in room-temperature clay at desired threat levels. This would provide a benchmark for the clay methodology while human response and injury metrics are developed.

The Peepsite head form and test procedure have clear advantages, including these:

- Use of room-temperature clay;
- Inherent quality control since the drop test procedure uses the head form itself;
- BFD characteristics that are not limited by the petals at the edge of the clay-containing region but are limited only by the characteristics of the clay itself.

Recommendation 7-3: The Army should investigate use of the Peepsite head form currently in development by the Army Research Laboratory with room-temperature clay. This head form and the procedure have potential as a near-term alternative to testing using the National Institute of Justice clay head form tested at elevated clay temperatures.

---

[55]Presentation to the committee by Rob Kinsler, ARL, on October 13, 2010.

UVA Head Form

To simulate the impact response of the human, the automotive industry developed the Hybrid III 50th Percentile Male anthropometric test device (ATD) in the early 1970s. It has since become a validated tool for the evaluation of automotive impacts, and can accommodate a wide range of instrumentation and transducers. It is the required test device for automotive crash testing (DOT, 1998) and is robust enough to perform repeatedly in ballistic environments (Bass et al., 2003).

For ballistic testing, a collaborative effort between Natick labs, Defense Research and Development Canada -Valcartier, and the University of Virginia (UVA) developed a ballistic version of the Hybrid III head augmented with impact pressure sensors (Bass et al., 2003). In what is called the "UVA head form," shown on the left side of Figure 7-17, instrumentation for the Hybrid III head and neck region consisted of three linear accelerometers and angular rate sensors at the center of the ATD headform and six-axis upper and lower neck load cells. Injury metrics assessed using this headform include the HIC and the NIJ neck injury criteria. With the Hybrid III head form modified to accept the Dynasen pressure sensors, the pressure measurements at various locations were recorded, analyzed, and compared to human cadaver results (Bass et al., 2003).

An injury risk assessment was developed based on cadaver tests with force sensor gauges as shown on the right side of Figure 7-17. This head form has been evaluated by the ARL,[56] UVA (Bass et al., 2003), and by Duke University/Applied Research Associates.

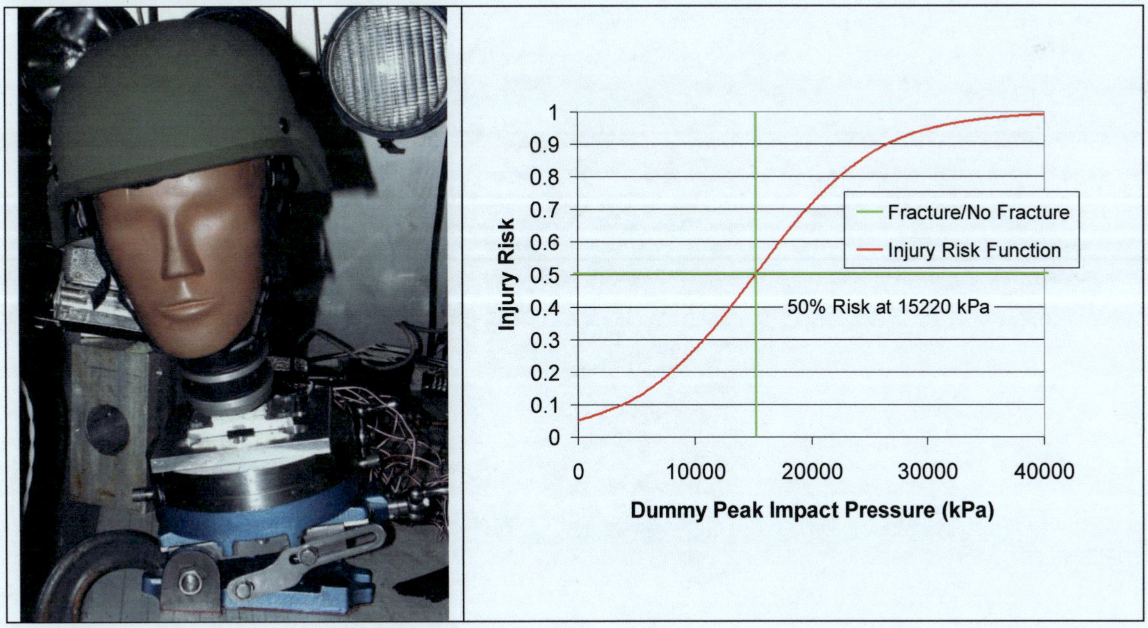

FIGURE 7-17 Left, UVA head form; right, risk assessment.

[56]Presentation to the committee by Dixie Hisley, ARL, on August 10, 2010.

BLS Head Form

The ballistic load sensing (BLS) head form (Biokinetics and Associates, LP, Ottawa, Ontario, Canada) is an evolution of the UVA head form consisting of two load-measuring head forms based on the International Organization for Standardization J size. One head form measures impact forces to the front and back of the head form, the other measures forces applied to the left and right sides of the head form. Both head forms are mounted on a Hybrid III ATD neck (Figure 7-18).

The BLS head form enables a direct measurement of the dynamic load imparted to the skull by the deformation of a ballistic helmet caused by nonpenetrating projectiles. To measure the force of impact, the BLS head form is equipped with an array of seven piezoelectric load cells residing under hexagonal aluminum pieces. These load cells were positioned directly under the impact site. The load cell arrangement is shown in Figure 7-19. The load cells are covered by a flat rubber sheet to simulate normal skull load distribution response. This head form is not able to record the global dynamic response of the head form.

Originally, the force data were correlated with the injury risk assessment of Bass et al. (2003). However, there is currently no suggested correlation between the BLS force data and injury data. ARL researchers have evaluated this head form and it is the subject of continuing evaluation.[57] Additional ballistic impact response data have been collected by TSWG for use in assessing head form response.

FIGURE 7-18 BLS head form. SOURCE: Courtesy, Biokinetics and Associates, Ltd.

---

[57]Ibid.

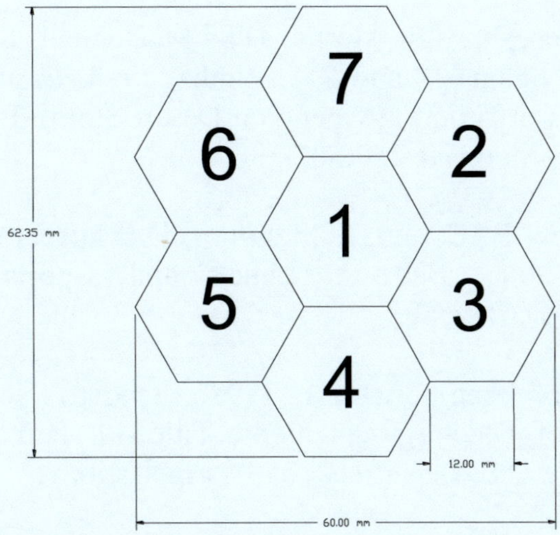

FIGURE 7-19 Arrangement and dimensions of load cells in the BLS head form.
SOURCE: Courtesy, Biokinetics and Associates, Ltd.

REFERENCES

AGARD (Advisory Group for Aerospace Research and Development). 1996. Anthropomorphic dummies for crash and escape system testing. AGARD Advisory Report 330.

ATC (U.S. Army Aberdeen Test Center). 2008. Test Operations Procedure (TOP) 10-2-210 Ballistic Testing of Hard Body Armor Using Clay Backing. Aberdeen Proving Ground, Md.: Aberdeen Test Center.

ATC. 2010. Internal Operating Procedures for the Head Protection Testing. (remainder of reference needed)

Bass, C.R., B. Boggess, M. Davis, E. Sanderson, G. Di Marco, and C. Chichester. 2000. A methodology for evaluating demining personal protective ensembles for AP landmines. UXO/Countermine Conference, New Orleans, La.

Bass, C.R., B. Boggess, B. Bush, M. Davis, R. Harris, M.R. Rountree, S. Campman, J. Ecklund, W. Monacci, G. Ling, G. Holborow, E. Sanderson, and S. Waclawik. 2003. Helmet behind armor blunt trauma. Specialist Conference on Human Factors in Medicine, Koblenz, Germany.

Bass, C., R. Salzar, S. Lucas, M. Davis, L. Donnellan, B. Folk, E. Sanderson, and S. Waclawik. 2006. Injury risk in behind armor blunt thoracic trauma. International Journal of Occupational Safety and Ergonomics 12(4):429-442.

Carey, M.E., A.S. Joseph, W.J. Morris, D.E. McDonnell, S.S. Rengachary, C. Smythies, J.P. Williams II, and F.A. Zimba. 1998. Brain wounds and their treatment in VII Corps during Operation Desert Storm, February 20 to April 15, 1991. Military Medicine 163:581-586.

Carey, M.E., M. Herz, B. Corner, J. McEntire, D. Malabarba, S. Paquette, and J.B. Sampson. 2000. Ballistic helmets and aspects of their design. Neurosurgery 47(3):678-689.

DOT (Department of Transportation). 1998. Standard No. 218: Motorcycle Helmets, Code of Federal Regulations, Title 49, Part 571, FMVSS 218. Washington, D.C.: U.S. Department of Transportation.

DOT&E (Office of the Director, Operational Test and Evaluation). 2010. Standard for Lot Acceptance Ballistic Testing of Hard Body Armor. Memorandum dated July 2, 2010. Arlington, Va.: Director, Operational Test and Evaluation.

Eppinger, R., E. Sun, F. Bandak, M. Haffner, N. Khaewpong, M. Maltese, S. Kuppa, T. Nguyen, E. Takhounts, R. Tannous, A. Zhang, and R. Saul. 1999. Development of improved injury criteria for the assessment of advanced automotive restraint systems – II. National Highway Safety Administration, US Department of Transportation.

Fuller, C.W., A. Junge, and J. Dvorak. 2005. A six year prospective study of the incidence and causes of head and neck injuries in international football. British Journal of Sports Medicine 39:i3-i9.

Funk, J.R., S.M Duma, S.J. Manoogian, and S. Rowson. 2007. Biomechanical risk estimates for mild traumatic brain injury. Annual Proceedings - Association for the Advancement of Automotive Medicine 51:343-361.

Gennarelli, T.A. and L.E. Thibault. 1982. Biomechanics of acute subdural hematoma. Journal of Trauma 22(8):680-686.

Genarrelli, T., L. Thibault, J. Adams, D. Graham, C. Thompson, and R. Marcincin. 1982. Diffuse axonal injury and traumatic coma in the primate. Annals of Neurology 12(6):564-574.

Gennarelli, T., and L. Thibault. 1989. Clinical Rationale for a Head Injury Angular Acceleration Criterion. Proceedings of Head Injury Mechanisms: The Need for an Angular Acceleration Criterion. Washington, D.C.

Goldsmith, W. 1981. Current controversies in the stipulation of head injury criteria, Letter to Editor. Journal of Biomechanics 14:883-884.

Gurdjian, E.S., and H.R. Lissner. 1944. Mechanism of head injury as studied by the cathode ray oscilloscope preliminary report. Journal of Neurosurgery 1(6):393-399.

Gurdjian, E.S., H.R. Lissner, J.E. Webster, F.R. Latimer, and B.F. Haddad. 1954. Studies on experimental concussion. Neurology 4(9):674-681.

Gurdjian, E.S., L.V. Roberts, and L.M. Thomas. 1964. Tolerance curves of acceleration and intracranial pressure and protective index in experimental head injury. Journal of Trauma 6(5):600-604.

Holbourn, A.H.S. 1943. Mechanics of head injuries. Lancet 2:438-441.

H.P. White Laboratory, Inc. 1995. Test Procedure – Bullet Resistant Helmet. HPW-TP-0401.01B. Street, Md.: H.P. White Laboratory, Inc.

Lissner, H.R., M. Lebow, and F.G. Evans. 1960. Experimental studies on the relation between acceleration and intracranial pressure change in man. Surgery, Gynecology & Obstetrics 111:329-338.

Margulies, S.S., L. Thibault, and T. Gennarelli. 1990. Physical model simulation of brain injury in the primate. Journal of Biomechanics 23(8):823-836.

McManus, L.R., P.E. Durand, and W.D. Claus. 1976. Development of a One Piece Infantry Helmet. Report 76-30-CEMEL. Natick, Mass.: U.S. Army Natick Research and Development Command.

Melvin, J. and A.M Nahum. 1993. Accidental Injury, Biomechanics and Prevention, edited by A. M. Nahum and J. W. Melvin. New York, N.Y.: Springer-Verlag.

Mueller, B.A., P. Cummings, F.P. Rivara, M.A. Brooks, and R.D. Terasaki. 2008. Injuries of the head, face, and neck in relation to ski helmet use. Epidemiology 19(2):270-276.

Newman, J.A., N. Shewchenko, and E. Welbourne. 2000. A proposed new biomechanical head injury assessment function - the maximum power index. Stapp Car Crash Journal 44(November):215-247.

NIJ (National Institute of Justice). 1981. NIJ Standard for Ballistic Helmets. Standard-0106.01. Washington, D.C.: U.S. Department of Justice.

Ommaya, A.K., A.E. Hirsch, E.S. Flamm, and R.H. Mahone. 1966. Cerebral concussion in the monkey: An experimental model. Science 153(3732):211-212.

Ommaya, A.K., R.L. Grubb, Jr., R.A. Naumann. 1971. Coup and contre-coup injury: Observations on the mechanics of visible brain injuries in the rhesus monkey. Journal of Neurosurgery 35(5):503-516.

Ommaya, A.K, P. Corrao, and F.S. Letcher. 1973. Head injury in the chimpanzee. Part 1: Biodynamics of traumatic unconsciousness. Journal of Neurosurgery, 39(2):152-166.

Ommaya, A.K., and R.A. Gennarelli. 1974. Cerebral concussion and traumatic unconsciousness: Correlation of experimental and clinical observations on blunt head injuries. Brain 97(4):633-654.

Ommaya, A.K., 1985. Biomechanics of Head Injury: Experimental Aspects. Pp. 245-269 in The Biomechanics of Trauma, edited by A.M. Nahum and J.W. Melvin. Norwalk, Conn.: Appleton-Lange.

Pilkey, W.D., D.V. Balandin, and N.N. Bolotnik. 2009. Injury Biomechanics and Control: Optimal Protection from Impact.

Varney, N. and N. Varney. 1995. Brain injury without head injury. Some physics of automobile collisions with particular reference to brain injuries occurring without physical head trauma. Applied Neuropsychology 2(2): 47-62.

Viano, D.C., and I.V. Lau. 1985. Thoracic impact: A viscous tolerance criterion. Tenth International Conference on Experimental Safety Vehicles, Oxford, England.

Viano, D.C., C. Bir, T. Walilko, and D. Sherman. 2004. Ballistic Impact to the forehead, zygoma, and mandible: Comparison of human and frangible dummy face biomechanics. Journal of Trauma-Injury Infection & Critical Care 56(6):1305-1311.

8

Medical Basis for Future Body Armor Testing

This chapter discusses thoracic ballistic test methodologies, blast injury criteria and blastlike mechanisms, injury scales, potential adverse effects of body armor in blast exposures, possible injury to body organs remote from blunt trauma, and instrumented simulators for research.

THORACIC BALLISTIC TEST METHODOLOGIES

As discussed in Chapter 3, modern body armor can defeat incoming pistol and rifle rounds, trading energy and momentum of the round for deformation of the armor. This deformation, however, has the potential for creating injuries in the thorax behind the armor, as well as injuries to remote organs, that may generally be characterized as blunt trauma.

Introduction to Behind Armor Blunt Trauma

Injuries to the thorax due to deformations of the armor are often termed behind- armor blunt trauma (BABT). Backface deformation (BFD) of the armor can cause local and distant fractures, contusions and hemorrhage in the thorax, as has been demonstrated in numerous animal studies (e.g., Prather et al., 1977; Clare et al., 1975; Cooper et al. 1982; Suneson et al., 1987; Lidén et al., 1988; Knudsen and Gøtze, 1997; Sarron et al., 2000; Gryth et al., 2007; Mayorga, 2010). These injuries are the result of physical deformation of the back face of the armor and associated stress waves that propagate through the thorax. While BABT is a known phenomenon, what is not known is the extent to which there may be significant injury to organs more distant from the point of impact, such as the brain, heart, spinal cord, and gut due to significant pressure waves transmitted through the body armor that result in pressure waves or shear waves in the body.

The injury risk for BABT will generally depend on the type and configuration of armor, the round, and the delivered energy of the round that results in an impact displacement and profile. This impact displacement and profile also depend on the physical characteristics of the person wearing the body armor. For both the body armor and the thorax, the impact location is important, and for the thorax, the rate sensitivity of the impact may be large. Many comprehensive discussions of penetrating ballistic trauma exist (e.g., Ryan et al., 1997), but there are relatively few such discussions on the topic of nonpenetrating ballistic trauma, especially those that might be relevant to injuries associated with bullets and shrapnel striking body armor. This section presents the physics, biophysics, and methods of studying nonpenetrating blunt trauma with the goal of optimizing the design and testing of manufactured body armor.

A bullet is a localized source of energy that can cause high local compression and shear forces, penetrating protective layers. The most effective bullets deposit energy, shear, and momentum rapidly in the target. One general strategy for protection is to blunt the penetration of the incoming round, picking up as much mass as possible in the body armor while decreasing the round energy and increasing the contact area. Thus, the protective effect of any ballistic protective vest is provided by increasing the area of impact, thus transferring energy and momentum to the vest. However, effective transfer of large amounts of energy and momentum from the incoming round into the body armor generally implies some deformation of the rear, or back face, of the body armor.

The BABT deformation is generally larger under soft body armor for a given incoming round. An interesting comparison of energy and momentum scales may be seen by comparing characteristics of various rounds, as shown in Table 8-1. Energy varies by a factor of over 30 between the relatively slow 9 mm handgun round and the .50 caliber (12.7 mm) rifle round due to the differences in mass and velocity.

TABLE 8-1 Muzzle Parameters for Various Types of Rounds

Device	Muzzle Velocity (m/sec)	Round Mass (g)	Energy (kJ)	Momentum (kg m/sec)
9 mm	358	8	0.5	2.86
5.56 × 45 M193 ball	991	3.6	1.7	3.57
7.62 × 51 NATO ball	838	9.6	3.4	8.13
12.7 mm 50 M2	890	42	16.6	37.4

SOURCE: Ness, 2011.

A further elucidative comparison may be made between the impact energy and momentum scales of low-rate blunt trauma events such as automobile impacts and high- rate impact events such as BABT. The energy and momentum for various potential blunt trauma situations are shown in Table 8-2 and are plotted in Figure 8-1. It is apparent from Figure 8-1 that a nonpenetrating ballistic impact involves much lower total momentum transfer than typical low-rate blunt impacts. However, the energy transfer is comparable, depending on the round and impact velocity. This implies increased localization of energy transfer and shorter interaction time and likely increased localization of injury.

TABLE 8-2 Energy/Momentum for Various Typical Thoracic Trauma Situations

Action	Velocity (m/sec)	Mass (g)	Energy (kJ)	Momentum (kg m/sec)
U.S. Football Block	5	100,000	2.5	500
Automobile Thoracic Dash Impact	5	50,000	0.6	250
Automobile Head Impact	5	5,000	0.06	25
Blast	~300	~	0.0004[a]	0.0013
Ultrasound Damage	~1,500	~	0.00001[b]	6.7×10^{-6}

[a] Based on assumed total lung volume of 3,000 mL.
[b] Based on applied lung volume of 300 mL.

SOURCE: Cameron Bass, Duke University

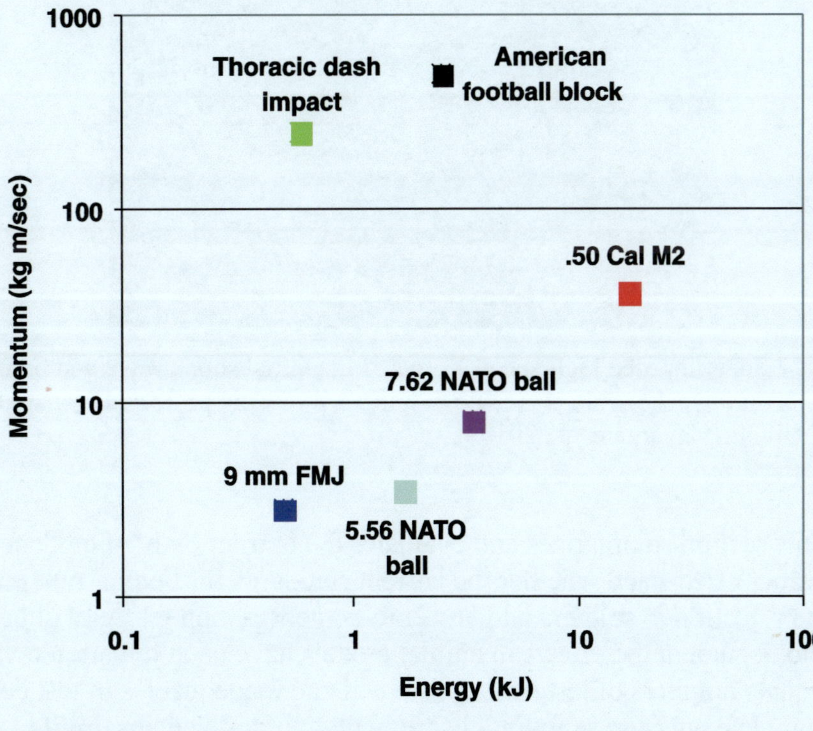

FIGURE 8-1 Initial energy and momentum for ballistics and other blunt impacts.
SOURCE: Bass, unpublished.

Blast injury can occur with very small impact momentum and energy over very short time scales. In the limit, as the duration of such impacts becomes very short, an interesting comparison may be made with damage using ultrasonic energy. At high rate (~4 MHz), less than 20 cycles of acoustic energy delivered to lung tissue with a peak pressure of approximately 1 megapascal (MPa) will cause tissue damage (see Raeman, 1996). It is uncertain if these high frequency effects occur with BABT.

Figure 8-2 shows a high speed X-ray of deformation of hard body armor after rifle round impact. The chest deformation shown here may lead to trauma to ribs, lungs, heart, liver, and other organs. Data are needed to determine the optimum vest design that provides protection to the body, potentially including organs remote from the site of impact while minimizing weight that the soldier must carry.

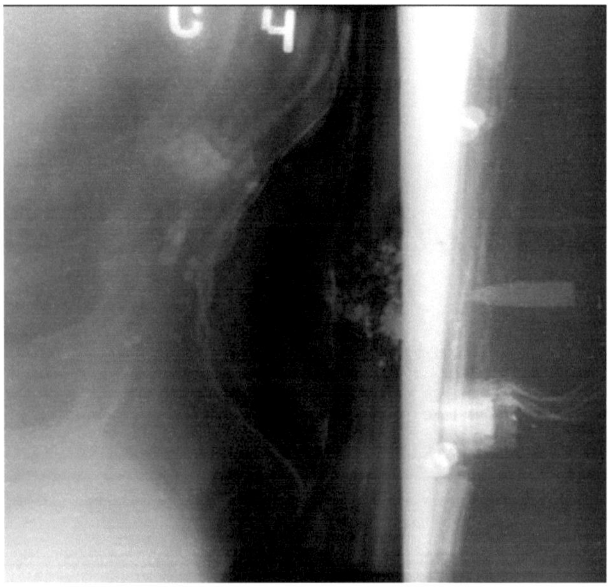

FIGURE 8-2 Superimposed high-speed X-rays of the initial shock wave and deformation of the thorax during a 7.62 mm projectile live-fire test in a pig protected by hard body armor. SOURCE: Mayorga et al., 2010.

This section summarizes and evaluates the current body of evidence for behind- armor effects and whether the current standards for body armor provide sufficient protection to soldiers and law enforcement personnel. As will be seen, attempts to document the effects in animal models have been impaired owing to the inadequate numbers of test subjects studied and inadequacies in test design (e.g., incomplete pressure sampling, instrumentation deficiencies, limited measurement methods, short duration of follow-up). Much is still unknown about the injury mechanisms of BABT. The information provided can be used so that more informed recommendations can be made about the types of further studies needed, such as studies in large animals, physical models, and computer-based simulations. It is important to conduct experiments to define the thresholds of energy transfer and the biophysical mechanisms that produce significant injury, as well as to evaluate if and how body armor can be improved to prevent remote injury and/or incapacitation.

Ergonomics of Body Armor

Until the eighteenth century, combat infantry soldiers are estimated to have carried 15-30 kg (33-66 lb), with the remaining equipment and supplies being transported in a baggage train (Knapik et al., 1996). The weight burden in recent times has increased substantially. Negative consequences of substantial load carriage include potential heat stress (Barwood et al., 2009), decrements in psychomotor performance (Bensel, 1975), and ergonomic factors that may limit mobility (e.g., Harman et al., 1999). Beekley (2007) found also that significant increases in oxygen consumption, respiration, and heart rate for loads to 70 percent of lean body mass in male U.S. Army personnel.

Current body armor basic system mass is 7.1 kg (15.7 lb), accounting for 15 percent of typical maximum load carriage, but can be as high as 15.1 kg (33.3 lb), or 31 percent of maximum load carriage. Clearly, reductions in body armor mass are a potential method of reducing total load carriage and increasing mobility on the battlefield.

Finding: Carried mass, such as that associated with body armor, may decrease a soldier's mobility and lead to fatigue.

Additional studies of the relationships between injury and the energy and momentum transferred to a body protected by armor innovations could lead to lighter weight armor that provides survivability equivalent to that of current military issue. However, the potential benefits of reducing the load carriage of body armor must be carefully weighed against the advantages of enhanced deformation that may arise from reduced areal density. The U.S. government and the North Atlantic Treaty Organization (NATO) are currently pursuing these goals.[58]

Injury Biomechanics

Beyond the development of techniques to identify injuries from BABT, it is important to develop a technique for assessing the risk of injury to humans behind body armor. One technique that has been shown to be effective in many fields of injury biomechanics is the use of an instrumented surrogate (dummy) to

---

[58]See, for example, the U.S. Army Research Institute of Environmental Medicine (USARIEM). Available online at www.usariem.army.mil/pages/download/LoadCarriagePDF.pdf and http://www.nwguardian.com/2010/12/09/9152/jblm-soldiers-provide-feedback.html.

evaluate the risk of injury from blunt trauma in automobile crashes. Elements of this technique include the following:

- *Biofidelic surrogate.* A dummy that is robust, gives a repeatable physical response, and responds in a human manner. A dummy may be physically very simple and may represent only a part of a human body. For example, an instrumented beam has been used successfully to represent an arm (Bass et al., 1997), and clay is currently used to represent the human thorax (Prather et al., 1977). However, dummies might be very complex, such as the anthropomorphically correct dummies being developed for the automobile industry. Generally, a surrogate should be as simple as possible while still representing the relevant human response.
- *Engineering measurement.* A physical parameter such as force or acceleration may be used to quantify the physical response of the dummy. Dummies may be instrumented to produce accepted or proposed injury criteria.
- *Injury risk evaluation.* A correlation between an engineering measurement and some injury model. For example, in frontal thoracic blunt impacts, an injury threshold of 60 times the force of gravity is used in the automobile industry.
- *Validation by injury model.* The injury risk evaluation is correlated to a physical model of injury. An injury risk model is without value if it has not been successfully validated using (1) epidemiology or physical reconstruction of an actual injury event, (2) an animal injury model, or (3) a human cadaveric injury model, as shown in Figure 8-3. Development of a relationship between a robust surrogate for injury and a validated injury model is crucial to the success of this approach.

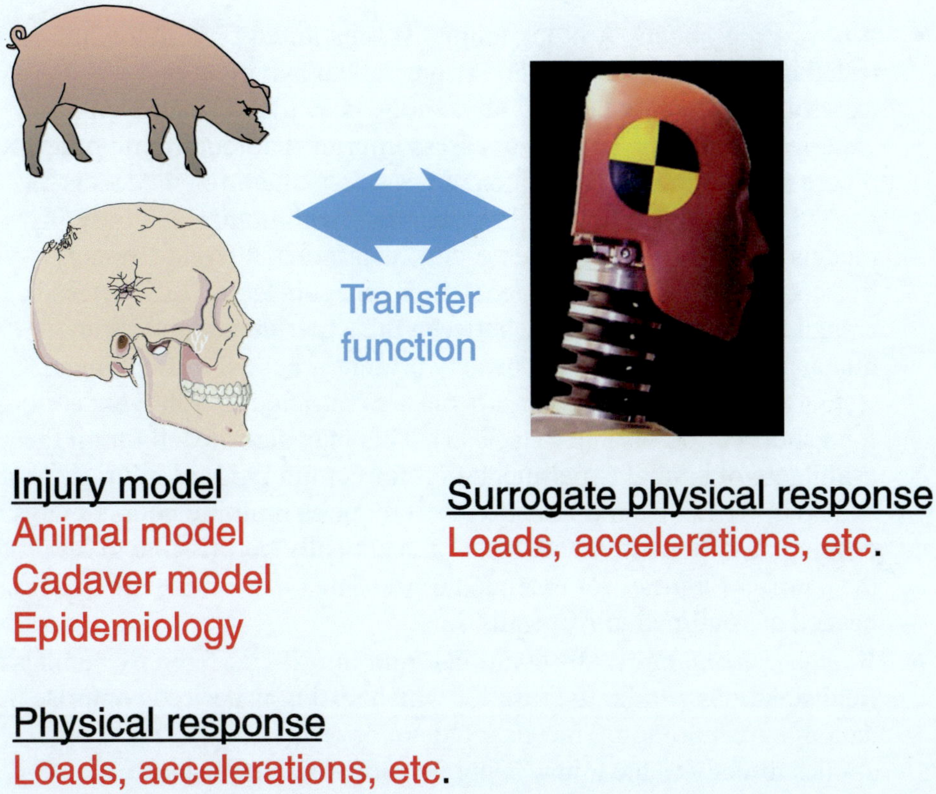

FIGURE 8-3 Development of surrogate injury model. SOURCE: Bass, 2000. Reprinted with permission of the Center for International Stabilization and Recovery.

As discussed above, there are only three ways of obtaining direct injury data. Each technique has its strengths and weaknesses as follows:

- *Cadaver experiments.* A human cadaveric specimen is substituted for a living human body and tested in a realistic manner. The strong advantage of cadaveric experimentation is that the anatomy closely matches that of a living human. Both skeletal injury and tissue damage may be assessed using a cadaveric model. In addition, body kinematics and kinetics may be accurately determined. The principal weakness of cadaveric models is the lack of human physiology. It is not possible to assess certain injuries (e.g. commotio cordis,[59] adult respiratory distress syndrome (ARDS), or diffuse axonal injury) using cadavers, as these pathologies require life processes to develop, and some pathological manifestations do not appear until weeks or years after the trauma.

[59]Commotio cordis is a disruption of heart rhythm that occurs as a result of a blow to the area directly over the heart at a critical time during the cycle of a heart beat. It frequently results in sudden death.

- *Animal experiments.* A living animal is substituted for a living human and tested in a realistic manner. The strong advantage of animal experimentation is that living physiology is available. Thus, animal experimentation can be used to assess injuries that require life processes before they manifest. The principal weakness of animal models is the limited correspondence of animal anatomy with human anatomy. Typical models in current use include porcine, caprine, and bovine models. As livestock is typically quadrupedal, there are substantial differences in cranial and thoracic anatomy between the experimental model and humans. Ethical considerations may present practical difficulties in testing with animals. Often protocols are restricted to following animals for a short period of time (e.g., 2 hr). This may significantly limit the usefulness of animal experimentation for certain types of injuries. As an example, ARDS from traumatic insult requires multiple hours or days to develop. When justified, however, it is ethically permissible to investigate the course of injuries for extended times. The types of experimentation needed are outlined in Appendix J.
- *Human epidemiology.* Observations from injuries suffered by humans in field situations similar to those for which testing is desired comprise the data of epidemiology. The advantage of epidemiology is that it often applies directly to the injury being investigated. For instance, epidemiologic data on injuries and conditions from automobile crashes are collected by the U.S. Department of Transportation for all fatalities and large numbers of nonfatal injuries each year. These data may be used to develop injury models and to focus the development of countermeasures. There are several limitations with epidemiological data. First, there is often limited information on the circumstances under which the injury occurred since the injuries do not occur in a controlled laboratory environment. Second, the data are always retrospective. Epidemiology does not have information on future systems or systems that are not in use. Third, in military environments, the collection of data may be quite difficult, and the information may be sensitive.

Volunteer models may not generally be used to obtain injury data (Figure 8-4); ethically, researchers must keep impacts in volunteer experiments below injury thresholds.

FIGURE 8-4 Volunteer experimentation. SOURCE: Stapp, 1949.

The most effective models use as many of these three means of obtaining injury data as possible or as are available for a given injury situation. The models must be appropriately biofidelic (lifelike). In other words, the models must adequately represent a human body in the situation analyzed. It is useful to employ several models (e.g., cadavers and animals) using consistent instrumentation and test conditions. Given appropriate injury modeling, the injury risk might be found to be as realistic as possible; otherwise, there is a potential for increasing the injury risk with inappropriate modeling.

Injury Criteria and Experimentation

There are a number of potential sources of injury from BABT; these include the initial contact shock, the subsequent displacement of the thoracic wall, and, in some cases, the propagation of pressure. The initial shock may occur with substantial high- frequency components and a relatively low resulting displacement. This shock pressure peak occurs because of the transmission of a pressure impulse from the rear of the body armor into the thoracic wall. Later bulk displacement may occur following significant local momentum transfer between the back of the body armor and the body. There has been extensive investigation into the relative effect of the initial shock and resulting displacements. Animal experiments at Oksbøl, discussed below, were designed to investigate this (Sarron, 2000). This issue, however, has not been completely resolved.

Pressure profiles have been measured in tissue simulants for impacts behind body armor. Pressure data from measurements in gelatin simulant material behind 6.4 mm ceramic (24 kg/m^2) with aramid composite (10 kg/m^2) and a fragment protective vest show an initial pressure pulse of 7.5 MPa less than 0.05 msec wide followed by a second pulse 0.8 msec later (van Bree, 2000). Stress wave propagation and concentration of reflected and refracted waves may enhance injury (Lui et al., 1996). This is discussed further in the Oksbøl animal section.

The experimental basis for high-rate impact is not well established. Variables that might significantly affect injury potential include delivered energy, thoracic wall displacement, contact area or contact profile, loading rate, and location of impact. It is clear that the more localized the energy, the greater the potential deformation into and through the body armor. Large local displacements may cause destructive local shearing or compression of tissue. Distributed thoracic loading has a different injury pattern than localized loading (Crandall, 1998). Indeed, relatively sharp indent profiles may be assumed by penetrations behind soft body armor (Lewis, 2001) or behind full penetrations of ceramic armor captured in the soft backing material (DeMaio, 2001). The sensitivity of human tissue response to the applied loading rate may be large, and the location of impact is important (e.g., anterior thorax versus lateral thorax). Additional factors may include gender, age, body mass, stature, and other anthropometric parameters. Experimental programs designed to develop standard injury test methodologies usually focus first on a single relevant subject population. This may imply a focus on mid-sized males as being appropriate for a military population. However, it is perhaps necessary to consider gender-related size differences for general applications.

In the next section, injury mechanisms and mechanical correlates with injury are discussed. These sections are followed by a discussion of animal, cadaveric, and epidemiological experimentation for assessment of BABT.

BABT Injury Mechanisms

Thoracic anatomy, as shown in Figure 8-5, emphasizes the importance of the thoracic region for BABT. Indeed, ballistic protective measures have been designed specifically to protect this region and portions of this region. The majority of organ systems necessary for life are located in the thorax. The mediastinal region is particularly important. Notable structures in the mediastinum in addition to the heart include major blood vessels (aorta, pulmonary artery, and vein), branch points of the lungs (trachea), and connections to the gastrointestinal system (esophagus). The thorax includes the lungs, which offer a large impact surface area, and the liver, that are not completely protected by the rib cage. On the posterior region of the thorax, the spine presents an additional impact location that is potentially debilitating or life threatening.

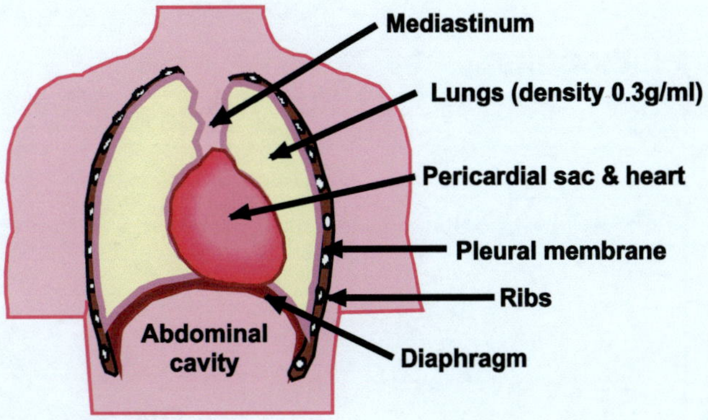

FIGURE 8-5 Human upper torso.

There are a number of potential mechanisms for injury that have been seen in animal, cadaver, or human epidemiological studies. These include physiological mechanisms such as commotio cordis (Link et al., 1999) and ARDS (Miller et al., 2002). Commotio cordis, a disruption of heart rhythm that occurs from blunt trauma directly over the heart, is thought to occur during only a small window of the cardiac cycle. So, the risk of such cardiac rhythm disruptions may be small.

Several types of injuries may be attributed to tissue displacement. These include large displacements that may cause shear or crushing injuries. These injuries may include puncture caused by bony fracture. An example of this is a fractured rib penetrating into a lung and disrupting the plural cavity, which may result in pneumothorax or hemothorax. Fung et al. (1988) suggested that lung injuries might be related to compression of the alveoli under mechanical stress. However, at high rate, for ultrasonic forcing, the influence of local cavitation has been suggested. Indeed, negative pressures have been seen recently in ballistic animal experiments (Sarron, 2001).

Pulsation mechanisms similar to those seen in ultrasound tissue damage may be relevant to BABT tissue damage. For short-duration ultrasound, stress waves and cavitation have been proposed that may be BABT injury mechanisms. Ultrasound impulses involve low displacement with relatively high frequency (50 kHz-1 MHz). The effect of exposure duration on the threshold injury pressure is important (Carstensen et al., 1990) (Figure 8-6). Thresholds for damage in lung tissue in the murine model are significantly lower than those for damage in other tissues, as shown in Figure 8-7, and have been found to be frequency dependent.

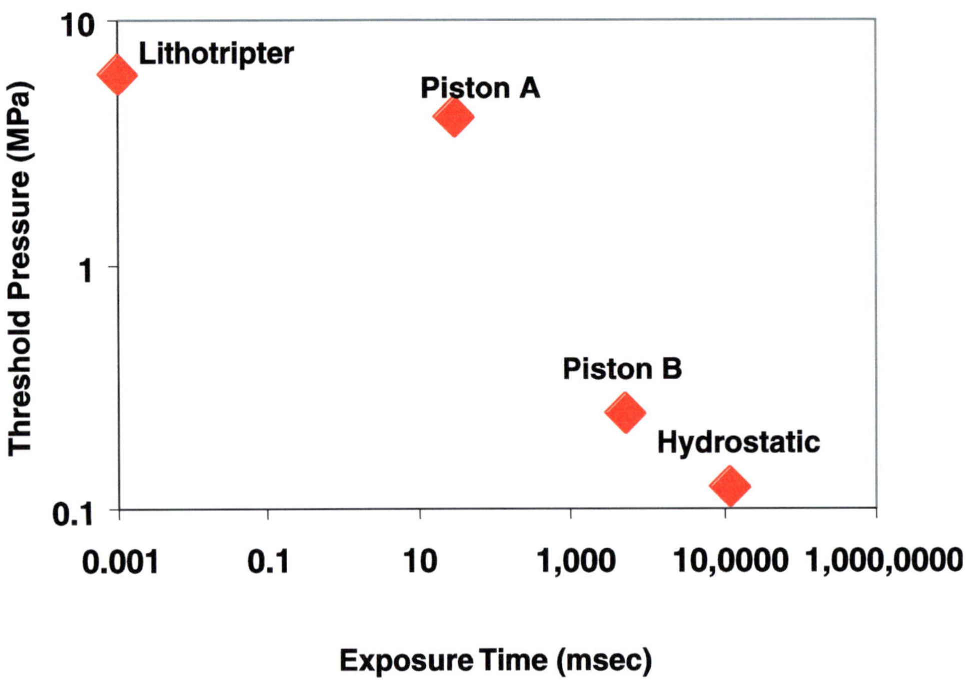

FIGURE 8-6 Threshold pressures and exposure times needed to damage drosophila larvae using various loading devices. SOURCE: Carstensen, 1990.

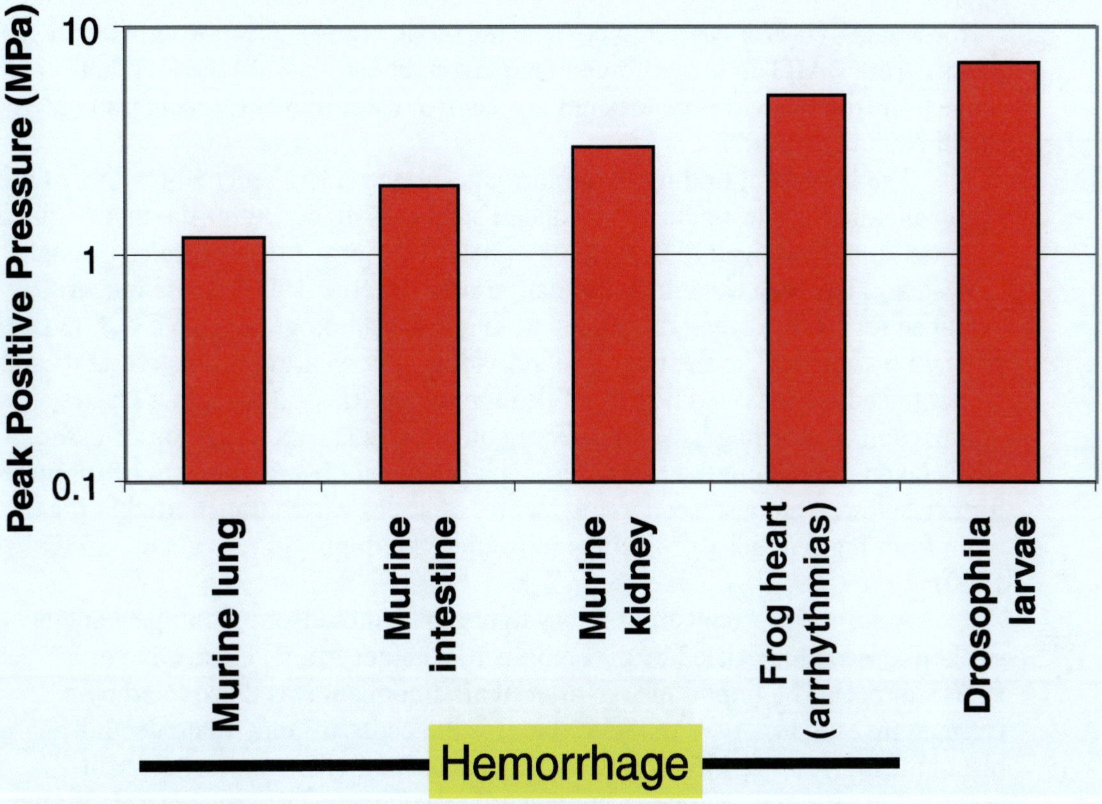

FIGURE 8-7 Threshold damage for various tissues. SOURCE: Carstensen, 2000.

Unfortunately, no single-pulse thresholds are currently known; current thresholds involve five cyclic pulses. Of the nonthermal mechanisms, spalling from an interfacial impedance mismatch may be a BABT injury mechanism. Spalling may occur at alveolar surfaces, causing local hemorrhage.

For high rate BABT, there is a potential for ablation injuries caused by the friction of the impacting surface on the body. This has been seen in epidemiological studies (Mirzeabassov et al., 2000) and in cadaver experiments (DeMaio et al., 2001). These ablation injuries may be quite severe. DeMaio reported large deep bilateral chest wounds under certain circumstances, and Bass et al. (2002) reported injury from ablation behind helmets with relatively low-velocity incoming projectiles.

A final class of BABT injuries comprises the potential for injuries remote from the direct backface contact. The earliest observations on the effects of penetrating injuries to the nervous system remote from the site of penetration were case studies from the Civil War of temporary and sometimes long term motor and sensory paralysis (Mitchell et al., 1864). Damage from the transmission of kinetic energy from the point of impact on the torso to remote body organs in humans has been observed in a number of cases (Carroll and Soderstrum, 1978;

Sperry, 1993; Akimov et al., 1993; Cannon et al., 2001; and Krajsa, 2009). These corroborate Civil War case studies (Mitchell et al., 1864). A report on human trauma from BABT in law enforcement personnel emphasized the fact that protection from penetration does not protect from significant thoracic trauma (Wilhelm and Bir, 2008).

The important finding of hemorrhages in the sclera and conjunctiva of the eye in an anterior chest gunshot-wounded subject is evidence for the theory that a ballistic impact can lead to the remote transfer of a large pressure pulse through, in this case, the vena cava and vascular circuits (Sperry, 1993). More recent evidence for remote organ damage is from a histopathological analysis of 33 cases of death by gunshots to the thorax in individuals not wearing protective vests and without head wounds or a history of head trauma (Krajsa, 2009). In all cases, microscopic hemorrhages were observed on histological examination of tissue slices taken throughout the brain. These pathological observations, to be discussed further below, were also seen in pig studies of stress waves transmitted to the brain from high-velocity projectiles impacting the thigh (Suneson et al., 1987, 1990).

A second mechanism of injury to organs remote from the impact site is strokelike ischemia caused by air embolism, whether from a blast wave or a BABT impact. The importance of arterial air embolism was discovered by a German investigator, who reported that it is the cause of immediate death from blast injuries (Rossle, 1950). Others reported air embolism as an important mechanism for organ trauma in humans (Clemedson and Hultman, 1954; Weller-Ravell et al., 1975; Mayorga, 1997). Air embolism might be expected from the modeling studies that show alveolar rupture from lung compression (Fung et al., 1988), although the Fung hypothesis might be more relevant to posttraumatic lung edema.

Mechanical Correlates with Injury

An injury criterion might be used to study and categorize blunt trauma if developed with a physical measure (e.g., acceleration) and correlated with a surrogate test device to evaluate injury. The earliest such criteria were based on global acceleration, but these may not be appropriate for injuries with localization to specific parts of the body.

Several mechanical correlates have been proposed for high-rate impingement on the thorax. Cooper and Jonsson (1997) propose a correlation between lung injury and peak acceleration of the lung for short-duration waves. The threshold injury is found at approximately 300 g (ca. $3000 m/s^2$) for the right lateral thorax of porcine test subjects and decreases as approximately the logarithm of the acceleration. Here, the damage mechanism might be based on propagating stress waves. Gross chest wall velocity is unlikely to provide the injury mechanism for high-amplitude pressure application of very short duration. It is also be an unlikely correlate for severe crushing injuries with large displacements. Between these limits, however, the chest wall velocity coupled with rate effects, might be strongly related to injury.

A mechanism for injury from blast loads that is based on chest wall velocity has been proposed (Axelsson and Yelverton, 1996). This mechanism is based on a simple lumped mass thorax model and animal experiments with sheep. They correlated the model results with the injuries in the animal experiments and defined an injury scale called the Adjusted Severity of Injury Index. This index is composed of a sum of injuries as a percentage of the maximum graded score for lung, pharynx/larynx, trachea, gastrointestinal tract, and the intraabdominal space. Results from this assessment are shown in Figure 8-8 for 177 sheep exposed to complex blast waves inside three enclosures with 11 m^3, 18 m^3, and 36 m^3 volumes. The advantage of this model is that it provides a good correlation for complex blast waves. Its drawback is that it is a simple global model that is unlikely to represent the local interactions that actually cause injury and so may be misleading in certain circumstances, such as blunt trauma from small projectiles.

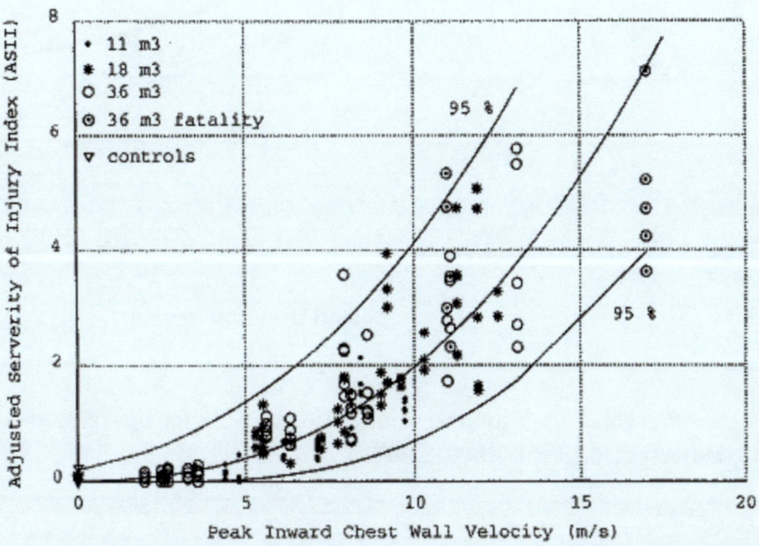

FIGURE 8-8 The Adjusted Severity of Injury Index values versus peak inward chest wall velocity. SOURCE: Axelsson and Yelverton, 1996. Copyright Williams & Wilkins, 1996.

BLAST INJURY CRITERIA AND BLASTLIKE MECHANISMS

Blast and blastlike injury criteria were extensively studied by the Lovelace Foundation supported by the Defense Atomic Support Agency and the U.S. Army Medical Research and Development Command (Martinez, 1999). These studies generally focused on free field application of pressure to the whole body (Bowen et al., 1968). Updated injury risk assessments have been recently published by Bass et al. (2008). The classic blast pressure thresholds, including the 1 percent fatal and 50 percent fatal free field curves, were developed from data on blast overpressure vs. duration compiled from numerous animal experiments (Bowen et al., 1968). These are shown in Figure 8-9. These thresholds may be a benchmark for injuries that occur from very sharp initial peak pressures owing to differences in acoustic impedance between the back of the body armor and the thorax. This is discussed further in the context of the Oksbøl animal experiments below.

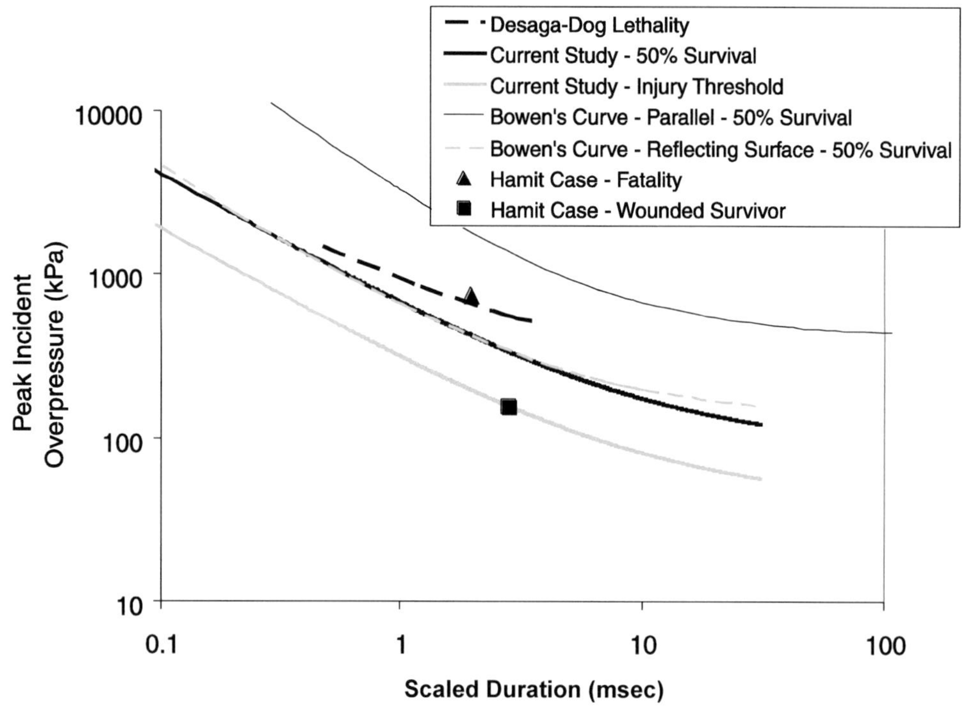

FIGURE 8-9 Overpressure/duration blast injury criteria. SOURCE: Derived from Desaga, 1950; Bowen et al., 1968; Bass, 2008.

Two common injury mechanisms may also be appropriate for blastlike BABT injury mechanisms (Maynard et al., 1997): These are

- Damage to the epithelial surfaces within the lungs owing to a stress wave passing through the parenchyma. As the wave passes, it encounters surfaces of varying density leading to impedance mismatches and local damage.
- Compression and re-expansion of alveoli owing to the passage of a shock wave.

These mechanisms may be implicated in the primary pressure wave seen in BABT impacts.

A systematic treatment of the extensive blast injury literature has been provided by Bass et al. (2008). Care must be taken in directly relating these data to blunt impact injuries, as the area of impact and rates of loading can be different.

Low-Rate Blunt Trauma Mechanisms

A large body of research has been performed on thoracic blunt trauma and thoracic injury mechanisms (Kent et al., 2006). Such blunt injury mechanisms may be relevant for trauma from larger, slower projectiles defeated by the body armor. The most widely used criteria for thoracic injury are the compression criterion and the viscous criterion for frontal impacts and the thoracic trauma index for side impacts. The compression criterion relates the relative chest deformation with respect to the chest depth to the level of injury. According to Kroell and coauthors (1971, 1974), 30 percent and 40 percent chest compressions cause abbreviated injury scale (AIS) level 2 injuries and AIS level 4 injuries, respectively. (The AIS injury scale of 0 to 6 and other injury scales are described in the next section.)

In the viscous criterion, the rate dependency of soft tissue injury is taken into account and VC_{max}, the maximum product of velocity of deformation and relative compression, is proposed as an effective predictor of injury risk. In an analysis of 39 unembalmed cadaver sternal impacts, a VC_{max} of 1.3 m/sec was associated with a 50 percent probability of AIS > 3 (Viano and Lau, 1985). Eppinger et al. (1984) analyzed a large number of side impact test results and proposed the thoracic trauma index that is proportional to age, mass, and the average of the peak values of fourth struck-side rib and twelfth thoracic vertebra accelerations.

Human Epidemiology for Battlefield BABT

There is a strong need for the investigation of epidemiology appropriate for behind-armor effects. Indeed, the sparsity of data on high-rate incidents limits evaluation of the performance of current and future ballistic protection. The value of such epidemiology is seen in many fields. Use of such studies in the field of anesthesia substantially lowered the incidence of death from adverse events over the last 20 years (Hawkins et al., 1997). Epidemiological and retrospective studies are used in aircraft accidents and automobile crashes worldwide. Such critical data gathering has been useful in designing countermeasures to injury.

There have been several efforts to gather military injury data; these include the current Army program Joint Trauma Analysis and Prevention of Injury in Combat, the Navy-Marine Corps Combat Trauma Registry, the national injury center at Fort Rucker, Alabama, and past efforts such as the Vietnam-era Wound Data Munitions Effectiveness Team database for combat injuries. There is, however, a great need for a centralized repository of data from military trauma incidents.

INJURY SCALES

An organized and robust injury scale is necessary for the evaluation of injuries using a common basis for animal experiments, cadaveric experiments, and epidemiology. There are several extant injury scores, but none is completely adequate for use in scoring BABT injury. An accepted standard for the assessment of thoracic injury is the AIS of the Association for the Advancement of Automotive Medicine. The numerical rating system ranges from 0 (no injury) to 6 (maximum, virtually not survivable). According to the 1990 revision, a flail chest with four or more rib fractures and/or bilateral lung laceration is AIS 4. The most serious thoracic injury is aortic lacerations, which is ranked from AIS 4 to 6 (Cavanaugh, 1993). As discussed below, however, this criterion is not generally sufficiently discriminative for BABT research.

The Injury Severity Score, or ISS (Baker, 1974) is used as an overall score for multiple injuries. Each body region (head, face, chest, abdomen, extremities, external) is assigned an AIS score. The highest AIS score for each region is selected, and the three highest of these are squared to produce the ISS score as:

$$ISS = AIS_{max1}^2 + AIS_{max2}^2 + AIS_{max3}^2$$

The ISS score value ranges from 0 to 75. This score has been correlated with outcome for thoracic trauma, potential chest, abdomen, and external injuries.

A scale for combat-specific BABT trauma was developed from data obtained during the Soviet conflict in Afghanistan (Mirzeabassov, 2000). The scale shown in Table 8-3 and the associated injury scale in Table 8-4 are based on levels of damage suffered from BABT.

TABLE 8-3 Description of Levels of Thoracic Trauma

Level of Trauma	Nature of Injuries
I (slight)	Scratches on the skin, ecchymoses, and restricted subcutaneous haematomas. Isolated focal subpleural haemorrhages.
II (medium-gravity)	Contused cutaneous wounds. Focal intramuscular haemorrhages. Plural focal subpleural haemorrhages. Isolated focal haemorrhages into the intestinal mesentery.
III (grievous)	Closed and open rib fractures. Lacerations of the pleura, haemorrhages into the pulmonary parenchyma. Subepi- or subendocardial haemorrhages. Subcapsular haematomas of parenchymal organs of the abdominal cavity and retroperitoneal space. Subserous haemorrhages into the intestines, ruptures of the mesentery. Restricted haemopneumothorax, haemoperitomeum. Vertebral fractures without injury to the spinal cord.
IV (extremely grievous, lethal)	Ruptures and crushing of internals. Closed trauma of the vertebral column followed by an injury to the spinal cord.

SOURCE: Mirzeabassov et al., 2000.

TABLE 8-4 Combat Effectiveness vs. Level of Thoracic Trauma

Gravity Level of the Trauma	Characteristic of the Loss of Fighting Efficiency	Probability of Rehabilitation (percent)	Class of Losses
I (light)	Loss of fighting efficiency for 1 to 3 min. Limited fighting efficiency during 15 min. Complete restoration within 24 hr.	99	Left in action
II (medium)	Loss of fighting efficiency for 3 to 5 min. Limited fighting efficiency up to 10 days. Complete restoration within 15 to 20 days.	85	Combat sanitary (recoverable) losses
III (high)	Complete loss of fighting efficiency, limited fighting efficiency within 15 to 20 days, complete restoration within 30 to 60 days. Possible fatal outcome.	25	Combat sanitary (recoverable) losses
IV (extremely high)	Immediate death. Death caused by complications. Invalidism and complete loss of fighting efficiency in surviving persons.	0	Unrecoverable losses

SOURCE: Mirzeabassov et al., 2000.

Mirzeabassov et al. (2000) report the most extensive epidemiology available. It includes data from 17 subjects hit in the thoracic region wearing body armor with either 1.25 mm titanium (6B2) or 6.5 mm titanium (6B3TM) plates. The data were acquired during the Soviet experience in Afghanistan. Data include location of impact, injury description, long-term consequences of the impact, and age of the patient. The ballistics data include the type of weapon fired (either 7.71 mm Enfield or 7.62 AKM), firing distance, and impact kinetic energy.

Bullet kinetic energy was plotted against injury severity for both human epidemiology and animal experiments, as shown in Figure 8-10. The lighter body armor (6B2) had a significantly lower threshold for the onset of severe injuries. The most serious injury reported was hemopneumothora (i.e., accumulation of blood and air in the pleural space) in two patients that progressed to abscesses. In addition, an impact in the left rib cage was reported to have resulted in a large ecchymoses[60] that extended from the groin to the knee.

While coverage of the plates is not reported, substantial injuries occurred to the back from relatively low-energy impacts, implying that the impacts were occurring in an area without plates. The researchers developed an injury level

[60]Ecchymoses is more commonly known as a "bruise."

scale that relates the initial bullet kinetic energy with the severity of injury in both the human epidemiology and animal experiments, as shown in Figure 8-10.

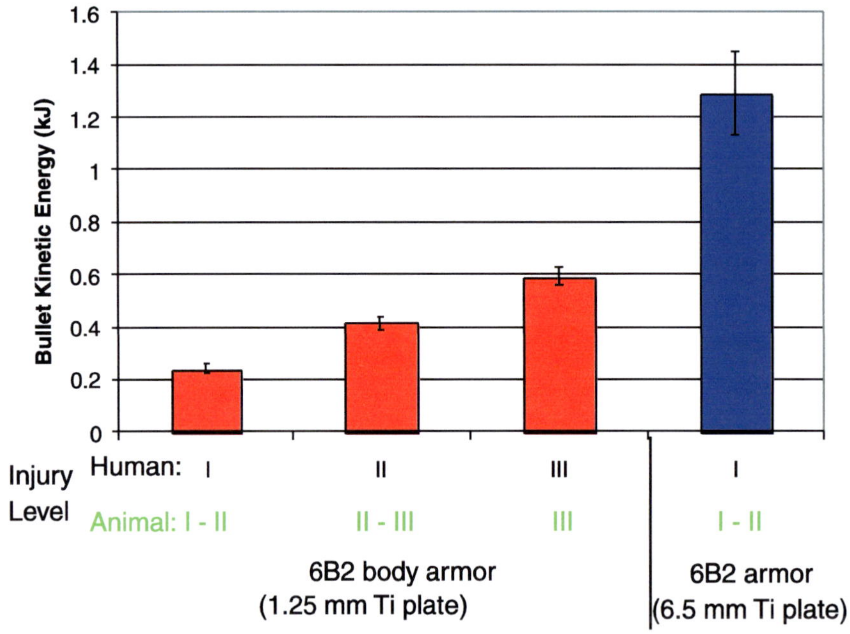

FIGURE 8-10 Kinetic energy vs. injury severity. SOURCE: Mirzeabasov et al., 2000.

To understand the Russian epidemiology data, it is important to consider details of the Soviet military medical service. Extensive data analysis is available for mine trauma victims from the Soviet experience in Afghanistan (Nechaev et al., 1995). Over 90 percent of the patients were evacuated by air, but only 4 percent were delivered to the central military hospital (CMH) within 6 hours of the mine blast injury (Figure 8-11). This was the initial treatment received for more than 80 percent of the wounded. The severity of the wounds was distributed as shown in Figure 8-12. It is interesting to speculate that the distribution of the severity of injuries may have changed with the delivery time. Increased delivery times would exacerbate the severity of the injuries but tend to decrease the number of fatal injuries.

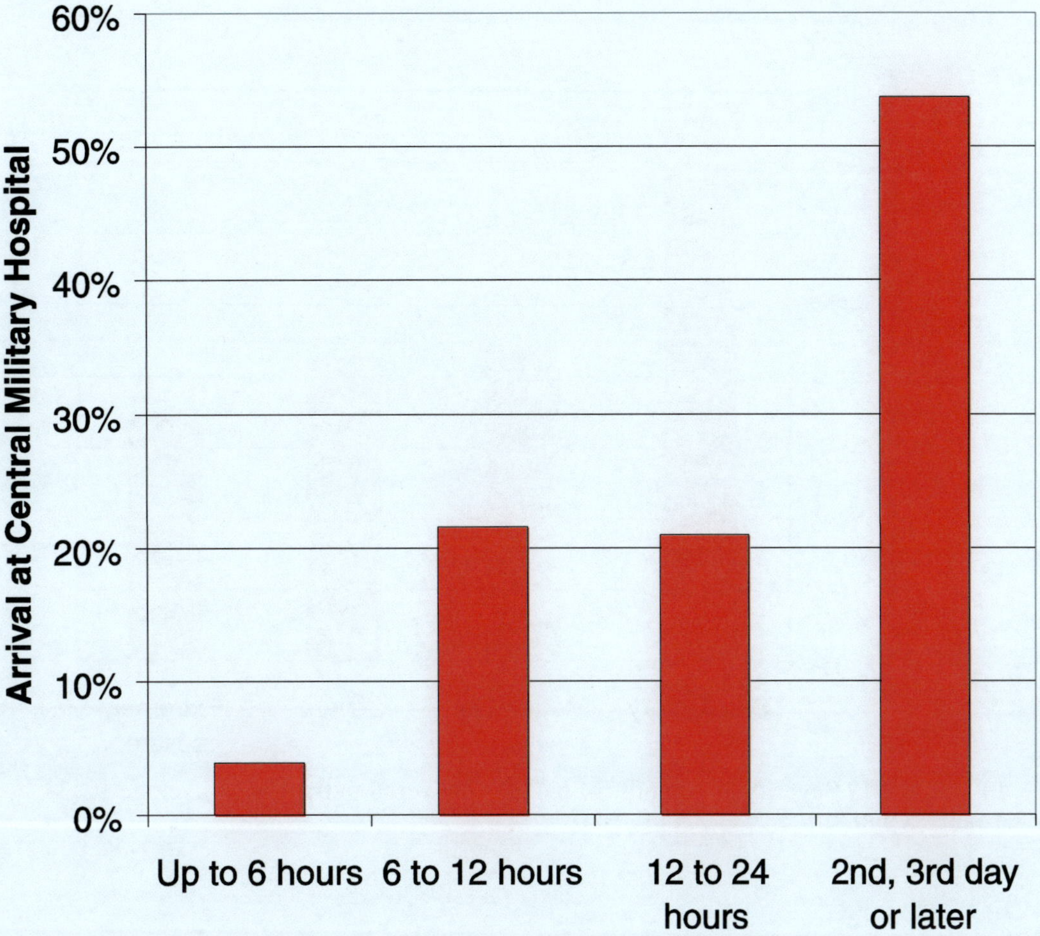

FIGURE 8-11 Time of delivery of wounded to the CMH (average 1983-1984).
SOURCE: Nechaev et al., 1995.

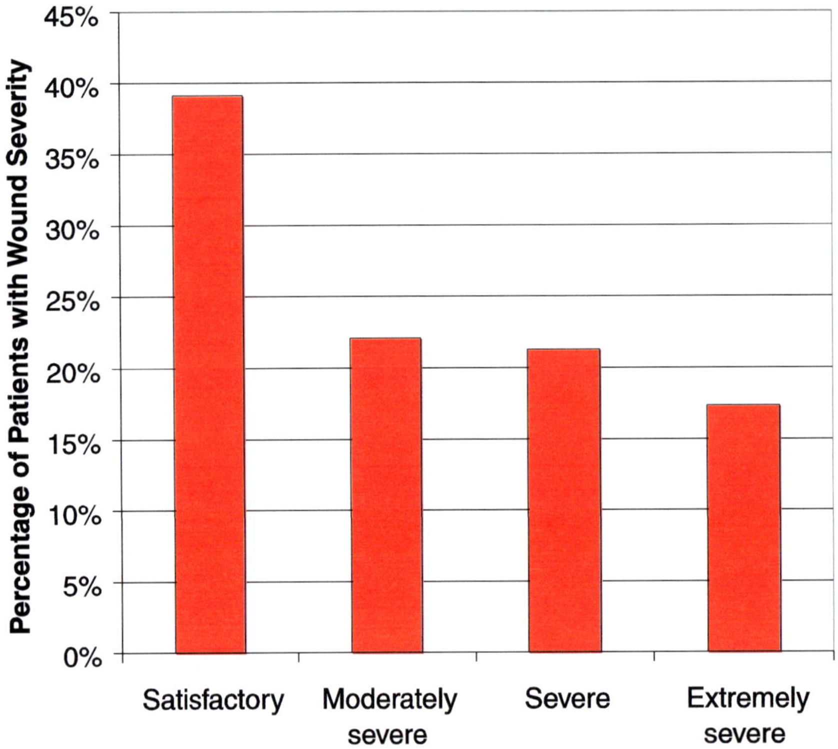

FIGURE 8-12 Severity of wounds for patients delivered to the CMH (average 1983-1984). SOURCE: Nechaev et al., 1995.

This work is almost the only instance of comprehensive epidemiology for BABT trauma. As with all epidemiological studies, there are limitations. The ranges were estimated, and battlefield trauma care may have been significantly different from care by Western militaries. This difference could influence the distribution of injuries. To correlate this study with Western body armor, it might be useful to acquire Russian body armor for calibration with Western vest designs.

Current Epidemiology for Battlefield/Law Enforcement BABT

Both the U.S. Army and the Office of the Director, Operational Test and Evaluation personnel emphasize to the committee that there have been no known U.S. soldier deaths due to small arms and shrapnel that were attributable to a failure of the issued ceramic body armor for threats for which the armor was designed (Rickey, 2010). Based on this, the U.S. military has fielded hard body armor with adequate survivability characteristics to soldiers in combat. Further, since the National Institute of Justice (NIJ) undertook the responsibility to standardize personal body armor for law enforcement personnel in 1973, over 2,900 lives have been saved.[61]

The tragic failure of soft zylon body armors in a small number of well-publicized cases in law enforcement further emphasizes the rarity of either soft or hard armor failures and the success of the testing programs based on clay. What is unknown, however, is the link to human injury for the existing hard armor assessment methodology. It is necessary to determine whether the standard is overly conservative and how to assess trade-offs of weight and mobility against protection from ballistic threats.

Large Animal Experiments for Behind-Armor Blunt Trauma

Animal experiments may be used for the development of injury criteria for blunt trauma, including BABT. However, animals used in such experimentation, typically livestock, have significant differences in anatomy from humans. So, scaling data from animals to humans must be performed with techniques that may be uncertain or nonexistent.

There have been a number of experiments that investigated thoracic penetration and BABT with soft body armor (Prather et al., 1977; Carroll and Soderstrom, 1978; Lidén, 1988), but the committee concentrated on studies that are applicable to BABT with hard body armor. This includes several direct impact studies from high velocity projectiles..

Large animal studies specifically designed to assess injury resulting from nonpenetrating ballistic impact on body armor have been done by U.S., Canadian, Swedish, Danish, Dutch, and French teams, mostly in anesthetized pigs and using various models of body armor and threats ranging from .38 cal to .30 cal (7.62

---

[61] Available online http://www.dupont.com/kevlar/lifeprotection/survivors.html?NF=1.

mm) and .50 cal (12.7 mm). Many of these studies are reported by the NATO Task Group on Thoracic Response to Undefeated Body Armor (Mayorga et al., 2010) and a series of international symposia called Personal Armor Systems Symposia.

Lovelace Foundation

Bowen et al. (1966) report experiments on dogs with the lateral thorax impacted by nonpenetrating missiles as shown in Figure 8-13. Aluminum missiles with variable masses were used, and the impact end was a flat cylinder with a diameter of 7 cm. The impactor masses varied from 63 g to 381 g, impact velocities varied from 18.9 m/sec to 91.4 m/sec, and dog mass varied from 12.2 kg to 23.1 kg. The dogs were positioned so that impacts were produced at the right lateral chest wall near the midthorax. Ribs fractured include the fourth rib through the eighth rib, implying impact locations near the fifth or sixth rib. Animals that survived the immediate postimpact time were sacrificed 30-40 minutes after the impact time.

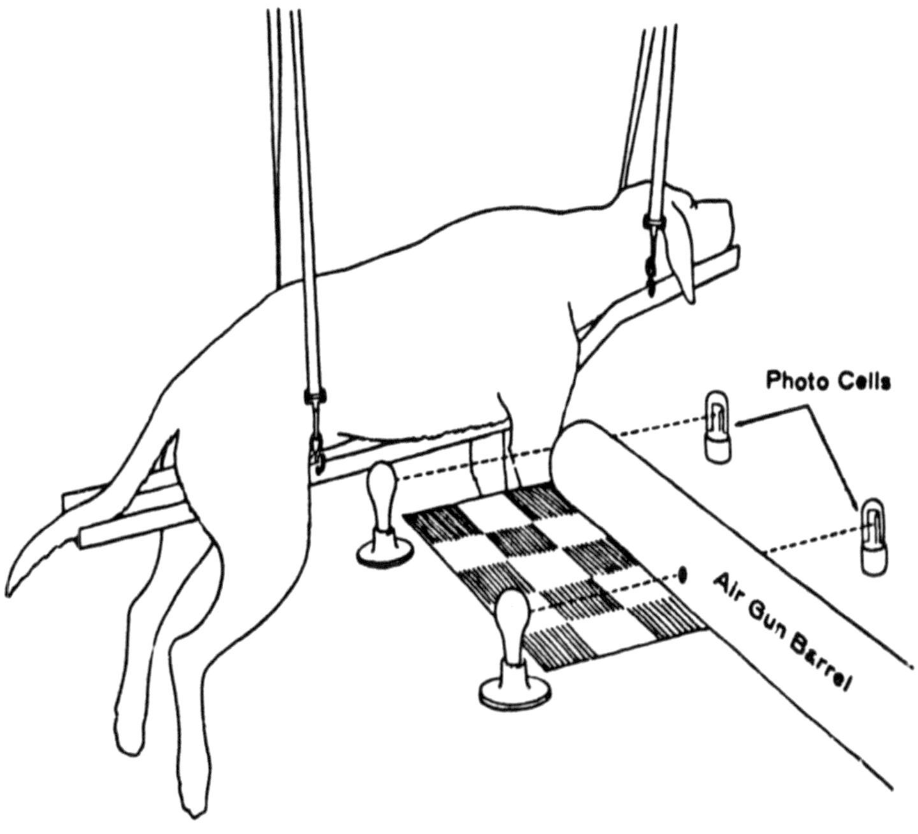

FIGURE 8-13 Lateral dog thorax impacted by nonpenetrating missiles. SOURCE: Bowen et al., 1966.

Scaling techniques were developed for transferring dog values to equivalent human values. These scaling relations, however, are uncertain, because the impact area of the missile was not scaled in the experiments with the body mass of the dog (Bowen et al., 1966). The mass of the impacted lung was compared with the mass of the lung on the contralateral side. This ratio is generally correlated with severity of injury; the threshold value of the ratio of the right mass to the left for fatalities in the watched period is approximately 2.3. In addition, the scaled energy of impact (scaled to a 75 kg man) is well correlated with fatalities. In Figure 8-14, the increased impact lung mass is plotted against the scaled impact energy. The increased lung mass of the animals measured postmortem has some limitations as an injury measure, because bleeding into the lung occurs most effectively while the animal is living. However, over the limited time that the study followed the test animals, the correlation is relatively good.

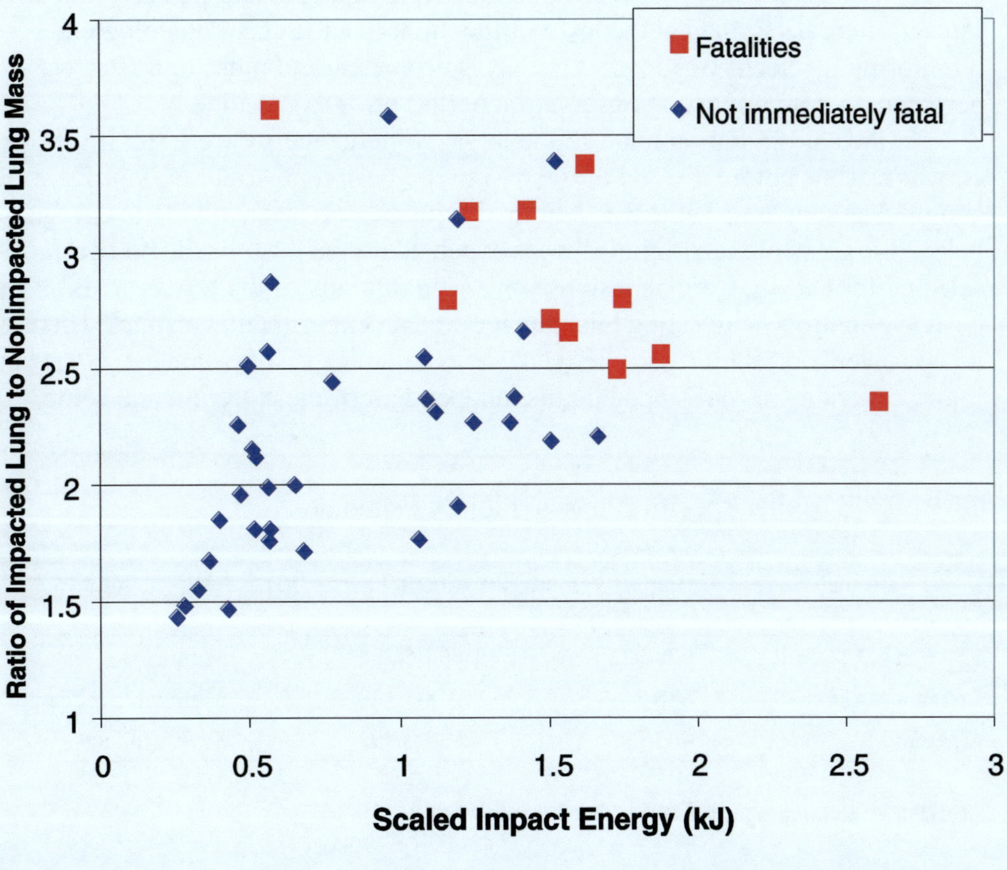

FIGURE 8-14 Impact energy (scaled to a 75 kg man) vs. increased lung mass. SOURCE: Bowen et al., 1966.

The impactor mass of 63 g may represent behind-armor impact in some regimes. For body armor with 24 kg/m^2 areal density, the effective impact diameter of approximately 37 mm reported by Mirzeabassov et al. (2000) implies an additional 84 g of mass owing to the induced motion of the body armor. In addition, the time scale is similar to BABT forcing. Tam et al. (2000) report approximately 2 msec to full displacement of approximately 4 cm with a peak acceleration of approximately 28,000 g.

Danish Army Combat School

Animal experiments were performed to assess the potential for thoracic injury from BABT behind undefeated body armor. As reported by Knudsen and Gøtze (1997), projectiles included a 5.56 mm NATO ball at 921 m/sec, a 7.62 mm NATO ball at 848 m/sec, and 12.7 mm AP rounds with reduced load at 463 m/sec to 595 m/sec. Tests were performed using 20 swine weighing 60 kg each, with one used as a control. The swine were supported in a standing posture, and lateral impacts were performed at the level of the xiphoid cartilage. Physiological monitoring by electrocardiogram (ECG), spirometer, and pulse oximeter was performed for an hour-long observation period prior to sacrifice.

Two of the four animals at the largest kinetic energy were sacrificed for ethical reasons before the end of the 60-minute period posttrauma. Injuries were assigned as shown in Table 8-5. Minimal injuries were seen at kinetic energies below 1.7 kJ, while fatal injuries were expected above 8 kJ for the body armor selected for testing. Cardiac lesions were seen in many of the test animals as well as in the controls, suggesting this damage was an experimental artifact. There was no consistent correlation seen between skin damage and lung damage. No specific injury criterion was developed that is independent of the experimental setup.

TABLE 8-5 Bullet Specifications and Injury Outcome

Round	Impact Velocity (m/sec)	Kinetic Energy (kJ)	Injuries
5.56 mm NATO ball	920-922	1.693-1.700	Minimal
7.62 mm NATO ball	838-861	3.248-3.429	Minimal to moderate
12.7 mm M2	463-595	4.839-7.992	Severe to fatal

SOURCE: Knudsen and Gøtze, 1996.

Oksbøl Trials

As a more extensive follow-up to the Danish Army Combat School tests, trials were performed using extensive instrumentation in Oksbøl, Denmark, in

1999 under the aegis of NATO. The Oksbøl trials used porcine specimens similar to those used earlier in the Danish Army Combat School tests. The protective equipment tested was provided by Denmark, France, the United Kingdom, and the United States. The Danish and French developed body armor systems that defeat the 7.62 × 51 mm threats while the U.K. and U.S. systems defeat the more energetic 12.7 mm (.50 cal) sniper rifle threat. Three designs of body armor were used with each design tested in groups of eight pigs each. There were three control animals. The impact site was the right lateral thorax in the middle of the eighth rib. Instrumentation included pressure and acceleration measurements in the thoracic wall and physiological measurements for half an hour before euthanasia, as stipulated by the animal welfare oversight committee. In retrospect, this was too short a time in which to develop traumatic sequelae.

Observations included recordings from pressure transducers, accelerometers, ECG, blood oxygen saturation, respiratory rate observations, and postexperiment autopsy examinations. The gross and microscopic studies included lungs, liver, and, in some studies, the heart and kidneys (Mayorga et al., 2010). Unfortunately, no intestine, brain, spleen, aorta, or spinal cord studies were reported. Each of four separate experiments (Danish, British, and two French studies) followed more or less the same protocol with respect to animal anesthesia and physiological observations, but the selection of protective armor varied widely, with an attempt in each group to compare one type of armor to another.

The extensive test series was designed to discern the cause of wounding, by separating the effects of the initial large, short-duration pressure peak from the effects of a secondary displacement pressure peak, presumably caused by the deformation of the body wall behind the body armor. The three body armor configurations are shown in Figure 8-15. The first type is a typical body armor with a steep first pressure peak and a more extended later displacement peak (G1). The second type should have a second peak only (G2), and the third should have a first peak only (G3).

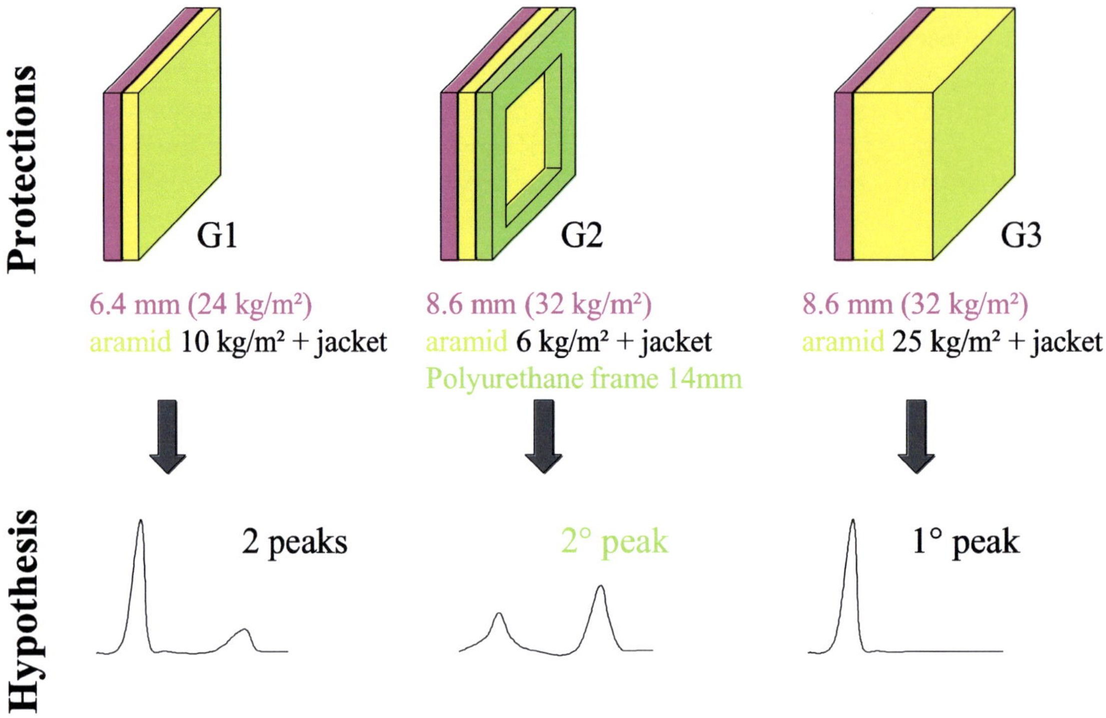

FIGURE 8-15 Body armor for Oksbøl trials. SOURCE: Sarron et al., 2000.

The armor did not perform as expected. As shown in Figure 8-16, the first-peak-only armor (G1) decreased the second peak but also reduced the first peak pressure. Second-peak-only (G2) armor significantly reduced both the first peak and the second peak but never so much that the first peak was less than the second peak. The first-peak-only armor is a difficult engineering problem, implying a need for infinite rigidity in the armor system and thereby significantly limiting resulting BFD. The second-peak-only design also presents a very difficult problem insomuch as it requires a match in impedance between the rear of the armor and the tissue.

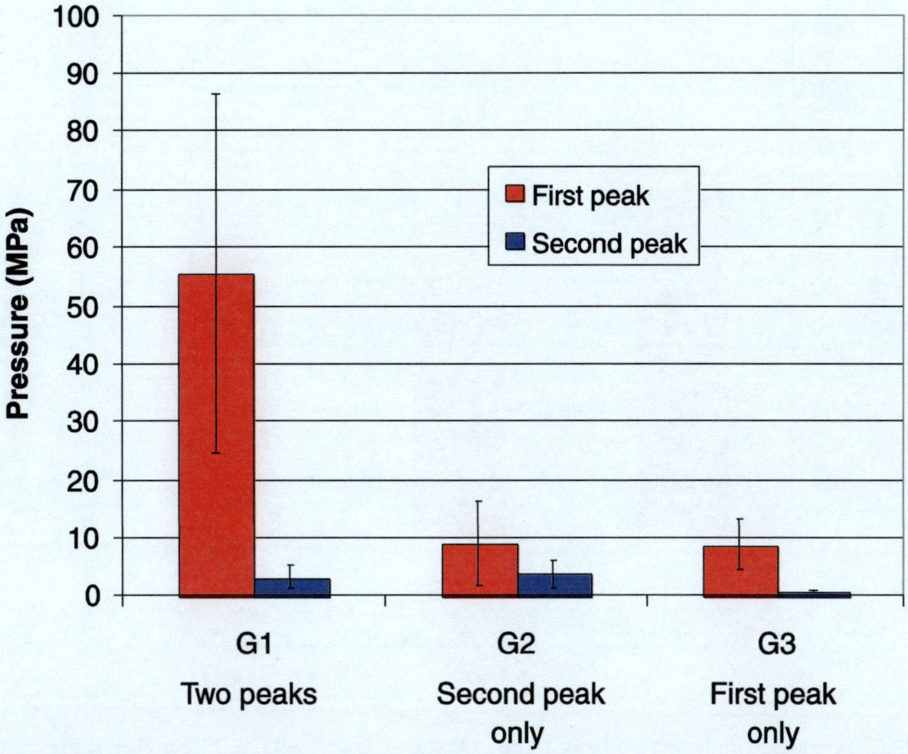

FIGURE 8-16 Average first and second peak pressure, Oksbøl trials. SOURCE: Sarron et al., 2000.

All animals tested were injured; however, injury scaling is difficult. The typical injury criterion used in automobile blunt trauma, AIS, is not specific enough to delineate wounding behavior in the Oksbøl series. All subjects had AIS 2-3 level injury. However, postmortem lung mass, as shown in Figure 8-17 appears to be a more specific injury measure in this experiment as all the animals were sacrificed within 60 min. Different times of expiration and the dynamic effects of hemorrhage may confound lung mass measurements.

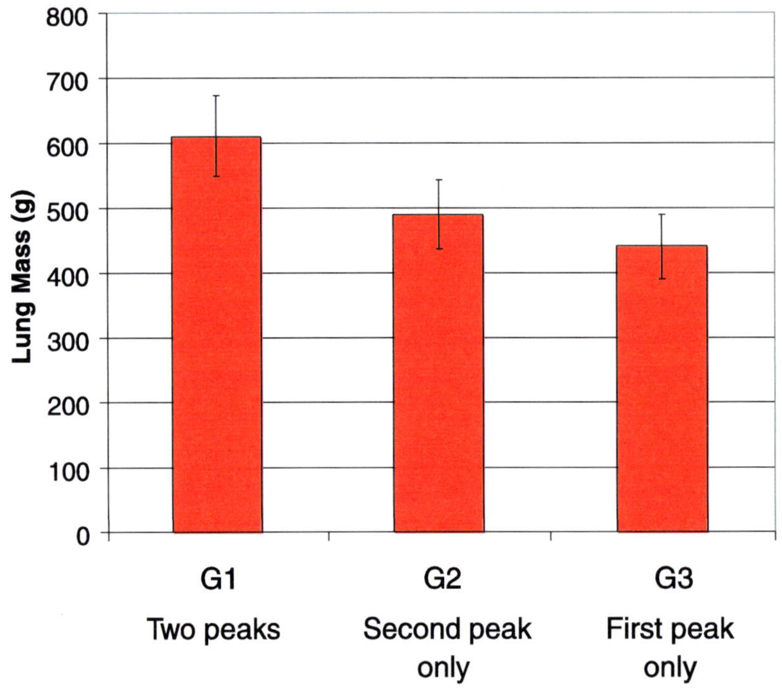

FIGURE 8-17 Average postmortem lung mass, Oksbøl trials. SOURCE: Sarron et al., 2000.

The effect of the first peak only may be best studied by comparison with shock impingement. To get a qualitative idea of the effect of this pressure peak, we can compare the Oksbøl peak uncorrected for the location of the pressure transducer in the tissue with the Bowen curves as shown in Figure 8-18. While the use of this internal pressure is not appropriate for the Bowen curves, it is likely conservative for the assessment of an injury threshold. The Oksbøl experiments had pressure peaks in excess of 30 MPa with durations of 0.3 to 0.4 msec. The 100 percent blast lethality threshold (Bowen et al., 1966) is below this value. This suggests that the local lung damage may be due to transmission of a high-amplitude pressure wave. Indeed, lung mass injury from the Oksbøl tests scales directly with the first peak only.

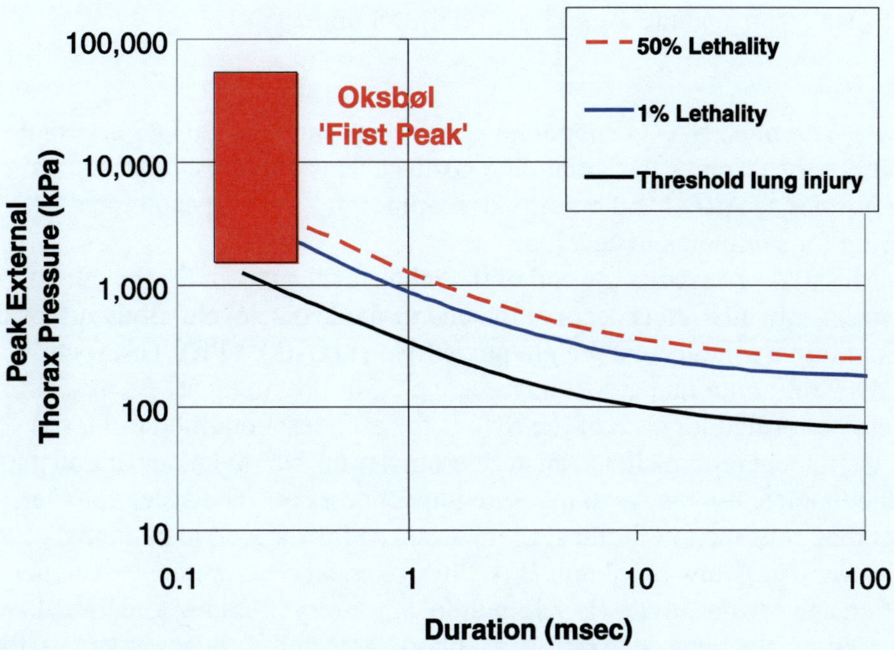

FIGURE 8-18 Oksbøl first peak on Bowen curve. SOURCE: Sarron et al., 2000.

The results showed significant injury and a high mortality for most of the study groups, along with surprisingly significant thorax and lung injuries from hard armor having areal densities up to 24 kg/m^2 with foam backing. Pressure and accelerometer recordings were incomplete as the majority of studies experienced saturation of the instruments with pressures exceeding 34 MPa. Nevertheless, some very important observations and correlations were made for high-velocity ballistic impacts and the subsequent deformations of the back face of body armor.

French Délégation Générale pour L'Armement (DGA)

An extensive test series using porcine subjects was performed by DGA in the 1990s and early 2000s. (Sendowski et al., 1994; Sarron et al., 2001) The model selected was a female swine of mass 60 kg (± 5 kg).

Four impact areas were selected:

- Pulmonary near the seventh right dorsal vertebra. The animals were observed for 2 hr after trauma.
- Lateral cardiac area at the fifth left thoracic vertebra. The animals were observed for 2 hr after the trauma. This condition was chosen to investigate cardiac contusion.
- Mediastinum opposite the apex of the heart. The animals were observed for 15 min prior to euthanasia. This condition was chosen to investigate induced ventricular fibrillation.

- Liver. The animals were observed for 15 min.

As with most NATO-supported animal experiments, the DGA animals were not observed for a substantial time posttest. This prevents the investigation of such injuries as ARDS and others that require long-term physiological monitoring for symptoms to develop.

All shots occurred at the end of the inhale cycle except for the lateral cardiac tests, in which shots occur at the end of the exhale cycle. Body armor used included ultra-high molecular weight polyethylene (UHMWPE). Test rounds included a 7.62-mm round at 829 m/sec and a 5.56-mm round at 989 m/sec. Five tests were performed for each of the rounds for each test condition (40 tests).

Instrumentation included an accelerometer on the rib and a balloon gauge inside the thoracic esophagus to measure impact pressure. The accelerometer, however, did not function for most of the tests, and the systemic pulmonary pressures were very low (~1.8 mm Hg). Physiological measurements included respiration and cardioactivity (ECG), pulmonary artery pressure, abdominal aortic artery pressure, and vena cava pressure. Blood gases and cellular enzymes CPK and LDH were also measured.

Posttest measurements included extensive grading of locations of skin damage. The cutaneous wound was statistically significantly greater for the 7.62 series than for the 5.56 series. The average diameter of the cutaneous wound was found to be about 14 cm.

Pulmonary wounds were assessed from the right thoracic shots. A bruise developed under the pleura surrounded by a region with inflammation. In some tests, emphysema developed at the center of the impact. The percentage of injured lung for the 7.62 mm (13 percent) was statistically different from that for the 5.56 mm series (7.4 percent). Cardiac wounds were assessed using the lateral cardiac area shots and the mediastinal shots. The tests were highly variable, probably due to the shape of the sternum.

Initial apnea duration in 30 tests averaged 15 sec for the 7.62 mm tests and 8 sec for the 5.56 mm tests. This included all shots in the sternal area. Secondary apnea occurred in several of the tests. Deaths seen in these tests are shown in Figure 8-19.

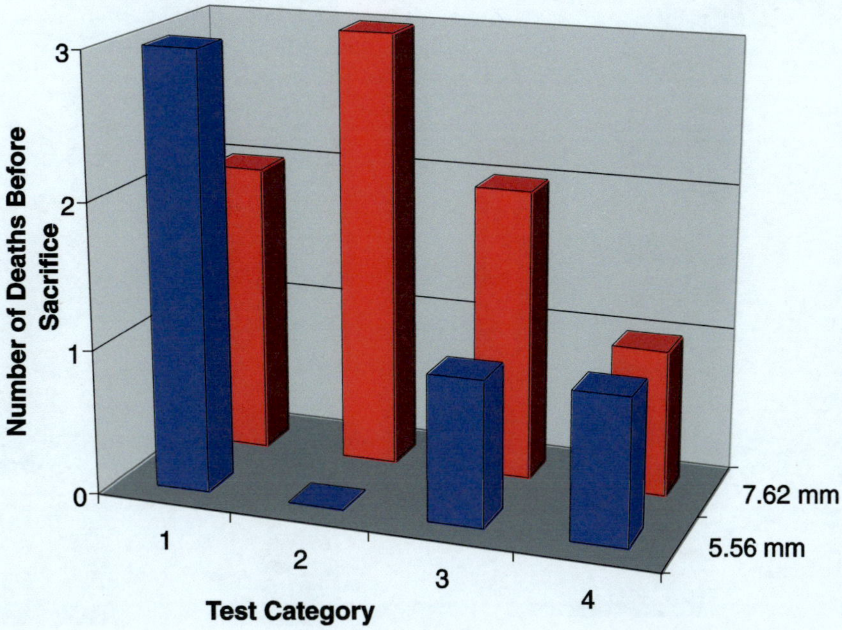

FIGURE 8-19 Animal fatalities during monitoring period. SOURCE: Adapted from Sendowski et al., 1994.

Experiments at the DGA using 7.62 mm test rounds into laminated UHMWPE body armor concentrated on investigating physical measurements of the thoracic wall using a flash X-ray technique with lead markers, as shown in Figure 8-20 (Sarron et al., 2001). This technique provides a good representation of the motion of the chest wall up to a maximum velocity of 30 m/sec. These studies showed a large negative pressure (~3 MPa) in preliminary data, suggesting that a cavitation injury mechanism might be involved. A shock wave arrives before significant displacements, and pressures are not correlated with the local displacements.

From the extensive studies in anesthetized pigs by the French investigators reported in a NATO summary of international studies (Mayorga et al., 2010), a synopsis of the relationships between lung contusion areas, recorded pressures, and deformation measurements can be made. These results are summarized in Figure 8-21 and Figure 8-22. Since Figure 8-20 presents a superimposition of a series of cine-radiographs, the timing of the chest deformation relative to the BFD cannot be related to the maximum deformation or pressure measurements shown in Figures 8-21 and Figure 8-22.

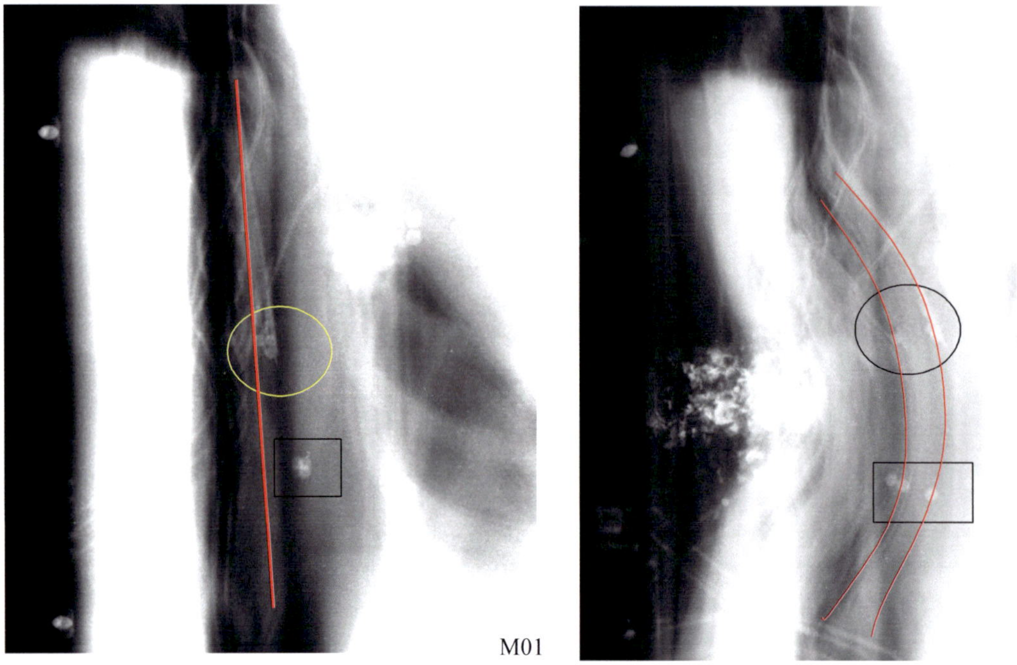

FIGURE 8-20 BABT flash X-ray. The red lines estimate the skin surface and the square and circle enclose reference markers. The left view is before impact. The right view superimposes four time images 0, 1.5, 2, and 2.5 msec after impact. The impact disperses the single marker inside the square and the circle showing the deformation of the thoracic wall with time. SOURCE: Sarron et al., 2001.

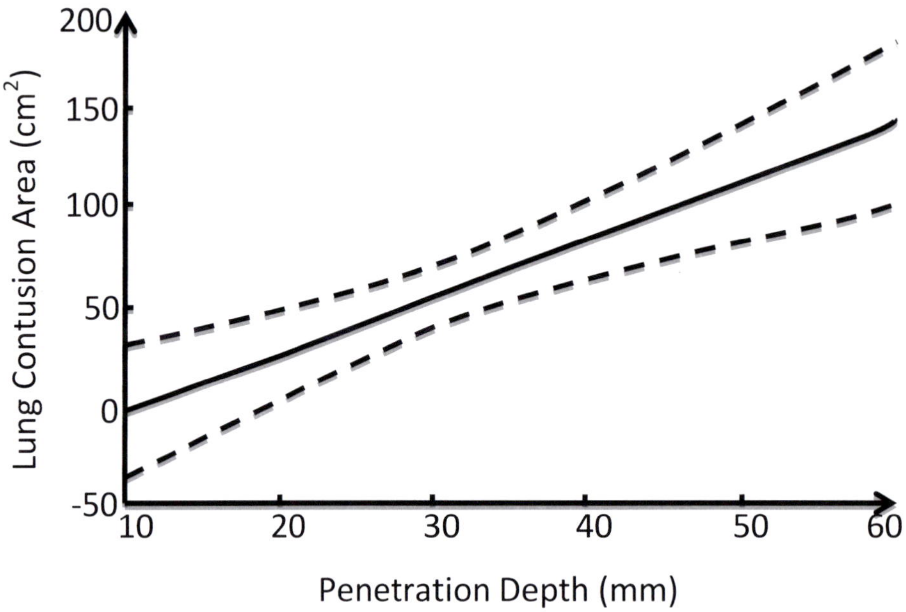

FIGURE 8-21 Relationship between area of lung surface contusion and maximum back-face deformation of body armor. SOURCE: Sarron et al., 2001.

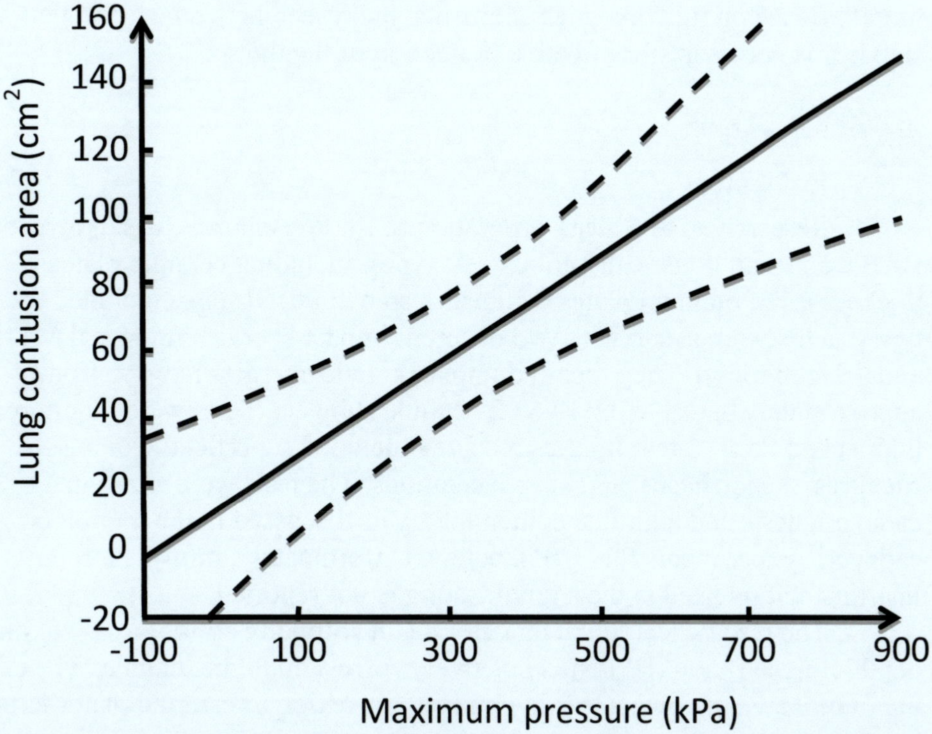

FIGURE 8-22 Relationship between area of lung surface contusion and pressure 6 cm from point of impact. SOURCE: Sarron et al., 2001.

In order to find a correlation between intrathoracic pressure, BABT, and high- vs low-velocity bullets, 20 pigs, protected by a NIJ Level 3 or 4 bulletproof vest, were shot with 7.62 mm NATO bullets (2.4 kJ and 3.2 kJ), and 10 unprotected pigs were shot by air gun with 40 mm rubber projectiles (0.07 to 0.2 kJ) (Prat et al., 2010). Rib fractures occurred in 21 of the 30 animals with no correlation to the projectile kinetics; however, intrathoracic peak pressures showed a good correlation with the volume of lung contusions.

U.S. Army Aeromedical Research Laboratory

Tests were performed by the U.S. Army Aeromedical Research Laboratory at Fort Rucker, Alabama, to evaluate the effect of polyvinyl chloride foam backing at standoffs of 14 mm, 21 mm, and 27 mm (Haley et al., 1996). Seventeen pigs (91 kg) were tested behind ceramic body armor with a 12.7 mm test round. Three control pigs were used, and the animals were monitored for 3 hr after the tests. The subjects were instrumented with accelerometers on the front and back faces of the armor plate, and load cells were placed behind the pig to evaluate the global force. Heart rate and respiration rate were monitored. The standoff foam was found to be rate sensitive, transmitting large pressures to the thorax. Based on the force measurements, the researchers concluded that a 25 mm standoff is necessary for effective protection of the thorax.

Anter Corporation

Researchers at Anter Corporation of St. Petersburg, Russia, investigated BABT on small dogs with various vest types, including ceramic plates with two thicknesses of titanium plates (Mirzeabassov, 2000). Rounds included three bullet types: a 7.62-mm round with two different impact energies, a .45-cal M11911A1, and a 9-mm round. They reported impact kinetic energies ranging from approximately 0.3 kJ to 3.2 kJ on 21 canine subjects. Tests were monitored using high-speed flash X-ray. Injuries received included superficial wounds, rib fractures, hemorrhages, and deep lacerations. The tests were correlated to both cadaveric tests and a limited epidemiology as discussed in the section on cadaveric experiments for BABT below. Unfortunately, information on instrumentation used in the animal testing is not reported.

The researchers found that BFD with soft body armor has a maximum depth of penetration (H) that is positively correlated to the diameter (L) of the maximum area of contact (S). In contrast, the action of ceramic plates tends to lower the depth of penetration relative to the contact diameter (L). For these experiments, the depth of deformation and area of contact were measured using a high-speed flash X-ray, and the volume of the deformation was inferred assuming an ellipsoid of revolution.

Swedish Studies

Swedish studies conducted 32 years after NIJ protocol development demonstrated that the U.S. criterion allowing a 40 mm impression behind the vest was not protective for higher velocity projectiles (Gryth et al., 2007). The tests involved 22 pigs protected by armor and 7.62 × 51 mm (800 m/sec) rifle bullets. The anesthesized animals were monitored for brain, circulatory, respiratory, and blood chemistry changes during the acute period after the ballistic exposure. Extensive anatomical examinations were performed. The principal conclusion was that 50 percent of the animals died in the group that had vests protective to 40 mm. Indeed, 25 percent of animals died in armor that protected to 34 mm impressions. It is important to note that neither the type of clay nor the temperature conditions, well known to be important variables for the clay surrogate, were stated for the Swedish studies. Another important point is that the velocity of the projectiles used was more than twice that of the earlier U.S. handgun studies using .38 cal, 128-grain bullets on goats.

Other Swedish studies involving animals and body armor used larger caliber projectiles in soft-armor-protected pigs and compared the pathological results to depth of penetration in a tissue surrogate consisting of soap to emulate the U.S. clay surrogate. In addition to using 9 mm projectiles, they also used 44 magnum and 12 gauge solid shotgun projectiles. The soft armor protection consisting of various layers of Kevlar with and without foam backing was minimally protective for most of the nine experiments (Lidén et al., 1988). One of the most important studies with hard armor was an electroencephalographic study on pigs impacted with 7.62 mm (800 m/sec) bullets. Although the armor was not pierced, five of the eight pigs showed temporary electroencephalogram changes, and all pigs showed lung injuries (Drobin et al., 2007).

The most recent Swedish studies evaluated the efficacy of adding lightweight material to hard body armor to attenuate the transmission of pressure waves (Sondén et al., 2009). This material is now generally called trauma-attenuating backing (TAB). Twenty-four pigs protected by a ceramid/aramid body armor without (n = 12) or with TAB (n = 12) were shot with a standard 7.62 mm assault rifle. The TAB significantly decreased the size of the lung contusion, decreased hemoptysis, and reduced peak pressures by 91 percent.

Cardiac Trauma Thresholds

The most extensive live-animal, live-fire tests using body armor were conducted in the 1980s in the United Kingdom. Forty-eight anesthetized pigs were instrumented with pressure transducers and accelerometers. High-speed photography as well as cineradiography observations were used to define the kinetics of deformations in the armor and in the chest. In addition, temporal and spatial pressures were measured. Animals were studied for the consequences of 64 J to 363 J delivered mostly to the anterior sternum using an air gun with projectiles of 0.14 and 0.38 kg and velocities between 20 and 74 m/sec (Cooper et al., 1982). The principal findings were that the degree of heart damage was related to the ratio of chest wall displacement to anterior-posterior chest diameter and that this was proportional to the energy of the impact divided by the product of the diameter of the impactor and the body mass. The diameters of the impactors (3.7 and 10 cm) are much larger than those of the live-fire projectiles but are in the range of the diameters of the backface deformations.

In another study, small steel ball fiduciaries were implanted in and around the heart to measure displacements using cineradiography (Cooper et al., 1984). These studies give important basic information regarding the thresholds for cardiac damage from energy delivered to the chest wall and the diameter of the BFD. This information can be included as criteria for body armor design, not only to defeat a high-velocity, high-energy projectile (e.g., 3 kJ to 13 kJ), but also to evaluate risks to personnel from BABT.

Other Animal Ballistic and Blast Experiments

There have been a multitude of both large- and small-animal studies on the effects of the overpressure associated with explosive discharges from nearby bomb explosions, improvised explosive devices, and artillery weapon discharges. In addition, there have been studies of the dispersion of trauma from penetrating wounds in both small and large animals using histopathology to evaluate trauma to organs remote from the projectile wound trajectory. Observations date from the writings of Aristotle, who noted intestinal trauma in deer after blunt injury to the body (Vance, 1923). Notable among experimental studies is one that showed pressure changes and damage to the peripheral and central nervous systems in pigs after a high velocity (1500 m/sec) projectile wound of the thigh (Suneson et al., 1987, 1990). There were permeability increases in small vessels of the brain and sciatic nerve after impacts with energy of 700 kJ associated with pressures of 125 kilopascal (kPa) in the brain and 270 kPa in the abdomen. The speed recorded for the shock wave was 1,400 m/sec, about what would be expected for sound velocity in tissues. Another important live-animal study of blunt trauma used impactors with velocities from 16 to 94 m/sec in anesthetized pigs (Cripps, 1996). Cripps produced small bowel injury at a threshold deformation speed of 40 m/sec. Injury to the colon occurred at all speeds. Microscopic studies of the brain of 15 kg dogs after bullet impacts to the thighs showed clear signs of damage from transmitted pressures (Wang et al., 2004). Other studies in pigs without protective vests focused on the consequences of blast trauma (Axelsson et al., 2000).

Finding: U.S. body armor prevents high-velocity bullets from penetrating the body but may not protect personnel from the shock wave resulting from the initial projectile impact and the trauma induced by the backface deformation. Tests in Europe have shown that adding trauma attenuated backing material to body armor vests may provide some degree of protection by attenuating the transmission of pressure waves.

Finding: Details surrounding the force that is transmitted from the body armor to the person wearing the armor, including the amount, the timing, and the immediate and long-term consequences of this force, are unknown.

An important missing link between the design of body armor and thresholds of injury and death is a lack of knowledge of the kinetic energy thresholds. Animal and human cadaver research experiments in this area are vital to establish these thresholds. These thresholds will guide the development of more effective and possibly lighter protective body armor. Such data are needed to validate blunt trauma prediction models and to provide guidelines for developing a physical surrogate for testing manufactured armor adherence to specifications. A general plan for the conduct of the needed studies is given in Appendix J.

POTENTIAL ADVERSE EFFECTS OF BODY ARMOR IN BLAST EXPOSURES

While body armor protects against ballistic penetrating missiles, it might lead to adverse effects from blast exposures. British Army studies showed a higher incidence of primary blast injury in fatally injured soldiers, 90 percent of whom were wearing body armor, than in civilian bystanders (Mellor and Cooper, 1989). Human experiments involved exposure of vest-wearing volunteers to a blast wave from a chemical explosion to simulate a muzzle blast. Ten test subjects were exposed to the blast wearing different clothing configurations, including Kevlar and ceramic vests. Explosive charges were detonated 3 m from the subjects, who were instrumented with a strain-gauge pressure transducer in the esophagus for intrathoracic pressure measurements. Shock wave pressures outside the body were about 17 kPa, and those on the esophagus were 7-8 kPa (Young et al., 1985). These experiments did not evaluate the relationship of the blast wave frequency and the resonance frequency of the thorax covered by a particular vest. However, the maximum energy transfer occurs when the predominant frequency of the incident shock wave most closely matches the resonance of the thorax (Cooper 1996). The resonance of the human thorax is between 40 and 50 Hz (Von Gierke, 1968). But the stress wave, whether from an air blast or from pressure transduced from a ballistic hit, will have different frequency spectra and a different energy coupling relationship depending on the projectile velocity in the case of a ballistic hit. Higher blast loadings can have an energy spectrum whose components are close to the natural resonance of the human thorax, thereby causing greater injury.

Another mechanism that can account for adverse effects in subjects wearing body armor is the reflection of incoming power at interfaces associated with the vest-air-thorax space. An analysis of the physical phenomenon can be made based on the physics of reflection and transmission of power through or from surfaces of differing density and bulk compressive moduli (i.e., acoustic impedance). The governing equation is the same that applies for propagating electromagnetic fields and acoustic pressure waves. The well-known Fresnel equation for sound gives the ratio of pressure reflected to incident pressure based on the difference in acoustic impedance. The impedance is proportional to the square root of the compression modulus times the density of the material. The power transmitted to the thorax is greatest when there is a match between the medium of the vest and that of the thorax. But if the vest impedance for a given frequency is less than that of the thorax (ribs and associated muscle and skin), then the reflected power will be less than that for a high impedance vest next to a lower impedance thorax.

However, the vests used in the previous studies are no longer used. Recent results (Wood et al., 2010) show that NIJ Level 2 and Level 4 police-issue ballistic vests substantially reduce the peak overpressure of primary blast waves. Attenuation ratios of peak reflected pressure were observed to increase with increased input pressure. The NIJ Level 2 vest showed overpressure attenuation ratios ranging from 3.4 at low input pressure levels to 14.2 at maximum input pressure levels of this study. Similarly, the NIJ Level 4 vests showed an ability to attenuate the peak reflected pressure seen behind the armor vest. Attenuation ratios for the Level 4 vests varied from 9.5 at low input levels to 56.8 at the maximum peak pressure input used in this study. The vests used in the human studies wherein a definite increase in intrathoracic pressure was observed did not have a NIJ level classification. They were described as (1) military field jacket (control), (2) woven ballistic armor (Kevlar vest), (3) ceramic flak vest (6.4 kg), and (4) ceramic flak vest over the Kevlar. Further study is needed using vests that are consistent between studies and that are manufactured according to the NIJ specifications.

In experimental studies with rats and pigs it has been shown that the lung injury increases if foam material is interposed between a blast wave and the thorax. However, if another material with high impedance is placed between the blast and the foam over the thorax, the injury is much less severe (Cooper et al., 1991). These phenomena are explained by the acoustic impedance of the layering of materials of a particular body armor design and the acoustic impedance of the tissues of the thorax and lungs (Cooper et al., 1991). Animal studies confirm that much less pathological damage is sustained in the lungs and intestines from a blast pressure for high impedance body armor than for lower impedance body armor, presumably because less energy coupling occurs (Cooper, 1996; Cripps and Cooper, 1996).

Animal studies confirmed that with higher levels of blast loading, the wearing of cloth ballistic vests resulted in increased lung injury as measured by lung weight increase and death relative to the control group without vests. The experiments on sheep used blast pressures of 115 kPa to 420 kPa (Phillips et al., 1988) and suggest that vests be designed to modulate the blast or ballistic energy spectra so that less coupling to the resonance frequency of the thorax occurs.

Data for current issue military vests are not available. The observations above are relevant to blast effects and also to trauma from ballistic effects as acoustic impedance matching physics is relevant to the transmission of energy to the body. These experimental results from different groups with a variety of armor material and the related physics suggest opportunities to design vests where selected material properties of layers will confine the energy of a projectile to the vest rather than allow transmission to the body.

Finding: The design for future body armor vests should consider blast effects as well as trade-offs between bulk, weight, and protection.

Discrepancies between published measurements of pressure changes in intrathoracic pressure for human subjects exposed to blasts from explosives with and without vests needs to be resolved. In the current threat environment, protection against blasts must be considered at least as important as ballistic impact protection, and the relationship between the two threats needs to be better understood.

Recommendation 8-1: The Army medical and scientific testing communities should adequately fund and expedite the research necessary to experimentally and epidemiologically quantify the physiologic and medical impact of blunt force trauma on the body from both ballistic and blast threats to soldiers.

Cadaveric Experiments for Behind-Armor Blunt Trauma

A very limited number of ballistic behind armor impact studies on cadavers have been performed. For injury association using animal experiments, such cadaveric experimentation is likely crucial as there are significant differences between livestock anatomy and human anatomy, especially in the mediastinal region.

Anter Corporation

In addition to the animal test described earlier, the Russian Anter Corporation conducted 13 cadaver experiments (Mirzeabassov et al., 2000), The tests used various vest types, which included two thicknesses of titanium plates and ceramic body armor. Rounds included 7.62 mm with two impact energies, a .45-cal M11911A1 and a 9 mm round. The researchers reported impact kinetic energies ranging from approximately 0.3 kJ to 3.2 kJ. Details regarding the preparation of the cadavers and conditions under which the tests were conducted are not reported. The authors state some differences in tolerance between animal and cadaver results for the internal organs but do not quantify them.

Instrumented Cadaveric Specimens

Researchers instrumented 17 cadaveric specimens, including 6 females and 11 males with a mean age of 73 years, as reported by DeMaio et al. (2001). The study investigated various body armors with different velocity regimes. Instrumentation included accelerometers at the sternum, T7, carina, and ligamentum arteriosum, as well as pressure transducers in the right and the left heart ventricle, and the left chest. Measurements of the impact pressure between the armor and chest wall were reported as not reliable. Pressurization of the lungs and cardiovascular system was used to make the cadavers more realistic human surrogates. Posttest evaluations assessed exterior wounds, sternal and rib fractures, cardiac bruising, and other injuries, including pulmonary injuries and spinal fractures.

Three body armor systems were tested: a soft vest with 9 mm test rounds, a light plate with a 7.62 mm round at two representative velocities, and a heavy plate with a 7.62 round at one representative velocity. Three injury levels were defined: survivable (with minimal trauma), immediately survivable (nonlethal if treated within 1 hr), and lethal (fatal even with treatment within 1 hr). Injuries received ranged from light surface friction to deep, extensive bilateral open chest wounds. The most severe injuries arose from complete plate penetration; an estimated three quarters of the cases where the round went fully through the plate were estimated to be lethal.

Use of the data from the instrumentation for this study poses several difficulties. The accelerometers were attached to the sternum using suture material. However, accelerations were seen in excess of 1,000 *g* in some tests. The effective mass of a typical shock accelerometer (nominal mass ~ 1 g) at this acceleration level exceeds 1 kg. To ensure repeatable results, a rigid connection is required. Second, the measurement of uniaxial accelerations on viscoelastic components within the thorax such as the carina and ligamentum arteriosum is questionable. Results will vary significantly based on local details of mounting and local viscoelastic behavior of the compliant structures of the carina and ligamentum arteriosum. While it may be possible to use these measurements for qualitative estimation of the gross arrival of local tissue deformation owing to the significantly increased density of the accelerometer compared with the surrounding tissue, intra specimen comparisons are not reliable. So, this work is most useful for qualitative injury performance of the articles tested. No injury risk function has been developed in terms of measured dynamic variables.

BABT Injuries Behind Hard Armor

A highly deforming hard body armor study to estimate the mechanical correlates with BABT injury in nine cadavers and two anthropomorphic test dummies used a range of velocities including low-severity impacts, medium-severity impacts, and high-severity impacts based upon risk of sternal fracture (Bass et al., 2006). Thoracic injuries ranged from minor skin abrasions (AIS 1) to severe sternal fractures (AIS 3+) and were well correlated with impact velocity and bone mineral density. Eight male cadavers were used to develop a criterion for injury risk. A 50 percent risk of AIS 3+ injury corresponded to a peak impact force of 24,900 ± 1,400 N. This study also investigated spinal impacts behind body armor in a single test. Correlation of the injuries to the sternum and spine in the same specimen under the same threat round velocity suggests that sternal impact may not be the worst case for behind-hard-armor impact. Preliminary data from a single specimen with a matched sternal and spinal impact behind the body armor suggest that additional spinal impact research would be of significant value. This study, however, did not assess impacts to the ribs or more general loading conditions with body armors of different characteristics.

Wayne State University

Human cadavers provide realistic models for studying injury biomechanics in blunt ballistic impacts. Past experiments looked at the use of projectiles with masses and velocities much less than those of handgun and rifle threats; nevertheless these experiments do provide data on the effects of behind-armor deflections of given forces and kinetic energy. Bir (2000) reported results of low velocity (20-250 m/s) and high mass (20-200 g) projectiles used to study force–time, deflection–time and force–deflection responses on the chest of 13 human cadavers The chest wall deflection was about 5.5 cm for a kinetic energy of 112 J and a deflection of about 2.5 cm for energy of 28 J for the two conditions of 0.14 kg mass at 40 m/sec and a 0.14 kg mass at 20 m/sec. The measured forces for these two conditions were 10 kN and 3.4 kN, respectively. Whereas the velocity realms differ substantially from those of bullet threats, these data can be used to evaluate the consequences of forces and kinetic energy behind body armor impacted by high-velocity projectiles if the area or volume of the vest indent is similar to that from the projectiles used in these experiments.

Finding: Although there are several studies using animal and cadaveric experiments to study behind-armor blunt trauma (BABT) injuries for hard body armor, the current work does not allow the development of a thoracic BABT injury criterion. Additional animal and/or cadaveric experimentation are required to develop a BABT injury criterion.

Finding: There is a need for a robust and widely used ballistic trauma injury classification scale. Although there is a number of existing injury scales, including a widely used scale for automobile injuries, the abbreviated injury scale promulgated by the Association for the Advancement of Automotive Medicine, none is well suited to ballistic trauma. Data on which to base a satisfactory injury scale will require the collection of military epidemiological data on a large scale.

Finding: The fidelity of anatomical, physical, and mathematical finite-element models simulating the human thorax, heart, lungs, liver, and kidneys is limited at the present time. Thus, damage to such organs as the intestines, spinal cord, brain, or vascular system from transmitted pressures associated with blunt trauma cannot be predicted.

Recommendation 8-2: The Army should perform high-speed ballistic tests using human cadavers and large animal cadavers to provide responses to deforming hard armor impacted by velocities likely to be encountered in combat. These tests should be extensively instrumented to determine dynamic deformation characteristics in the human and animal torsos to provide data that can be correlated with clay response at the same rates (or with alternative media or other test methodology) and with epidemiology and medical outcomes in the soldier. The studies should ensure that velocity and backface deformation regimes replicate those for current and future desired body armor testing protocols.

The recommended testing should be performed as soon as practical to address the following goals:

- *Near term*. Determine local three dimensional displacement time histories of animal and human cadavers to correlate with clay or other emerging test methodologies.
- *Intermediate term*. Determine pathophysiological effects of behind armor injury and correlate with acute injury and potential injury cascades.
- *Long term*. Incorporate injury outcome and mechanical response into emerging test methodologies and ongoing assessments of pathophysiological behind-armor effects in order to develop protective concepts.

Rationale for Large-Animal, Live-Fire Experiments

Large animal studies are needed to evaluate damage to organs remote from the site of the blunt trauma for both acute effects and late effects. The experimental evidence for such remote effects comes from battlefield observations, previous large and small animal studies, and medical reports on civilian gunshot accidents.

The earliest observations of the effects of penetrating injuries on the nervous system remote from the site of penetration were case studies from the Civil War of temporary and sometimes long term motor and sensory paralysis (Mitchell et al., 1864). During World War I and World War II, autopsy studies revealed evidence of brain pathologies caused by blast, but more detailed study of soldiers' symptoms later concluded that most of the disorders were of a psychological rather than a physical nature (Mott, 1919). Livingstone et al. (1945) were among the first to propose that the transfer of kinetic energy from a pressure wave might damage the nervous system. Blast exposure is not the same as a ballistic impact from high-speed projectiles, but internal biophysical phenomena as they relate to the central nervous system trauma may be similar.

Damage caused by the transmission of kinetic energy from the point of impact on the torso to remote body organs in humans has been observed on a number of occasions (Chamberlin, 1966; Carroll and Soderstrum, 1978; Sperry, 1993; Akimov et al., 1993; Cannon, 2001; Krajsa, 2009) and are corroborated by Civil War case studies (Mitchell et al., 1864). A report on human trauma from BABT in law enforcement personnel emphasized that protection from penetration does not preclude significant thoracic trauma (Wilhelm and Bir, 2008; Courtney and Courtney, 2010). When remote organs such as brain and intestines were included in examinations of animals clad in body armor then subjected to live-fire tests, there was evidence of substantial injury. A notable result of the small-animal studies was definite evidence of blood–brain barrier dysfunction subsequent to a high-speed bullet impact distant from the brain. Those studies used Evans Blue dye injected into the blood pool before the test.

Other observations in people exposed to blast waves indicated diminished cognitive caability and long-term encephalographic changes as well as complex neuropsychiatric symptoms (Cernak and Noble-Haeusslein, 2010). It should be noted that a causal connection between pressure waves from explosives or nonpenetrating blunt trauma and cognitive or psychiatric disorders is not the topic of this report, although this subject of blast-induced traumatic brain injury remains an area for intense scrutiny, as exemplified by the conclusions of a recent National Institutes of Health workshop (Hicks et al., 2010). An advanced imaging method study of soldier brains after blast trauma showed neuronal damage from blast exposure, but the cases were complicated by associated nonpenetrating head trauma (MacDonald et al. 2011)

The need to evaluate transmitted pressure waves in a variety of battlefield threats is an important reason for recommending large-animal live-fire experiments.

In addition, far more extensive data are needed than were collected in the past. Long-term testing involving large animals will require extensive use of pressure transducers, cineradiography, metabolic imaging, and neurochemical cerebral spinal fluid and blood assays that are appropriately instrumented, as described in Appendix J.

Recommendation 8-3: The Army should perform live large-animal, live-fire tests to simulate the behavior of current and proposed new body armor against expected threats.

INSTRUMENTED DETERMINATION OF BACKFACE DEFORMATION—RESEARCH DIRECTIONS

A dummy or surrogate for human response is generally used to provide a reliable and inexpensive test methodology for research. This surrogate allows repeatable characterization of the performance of ballistic protective gear. Existing ballistic impact simulators may be divided into three classes, as shown in Figure 8-23. The first class, "bulk tissue simulants," is made up of a single layer of material that allows measurement of deformation responses—for example, posttest residual clay penetration depth or dynamic gelatin penetration depth. The second class, "instrumented response elements," generally includes a simplified thoracic wall that simulates the motion of the surface and may incorporate multiple layers. The third class is "instrumented detailed anatomical surrogates." These generally include some form of thoracic viscera and are designed to investigate a wide range of blunt trauma kinetic energies and projectile diameters.

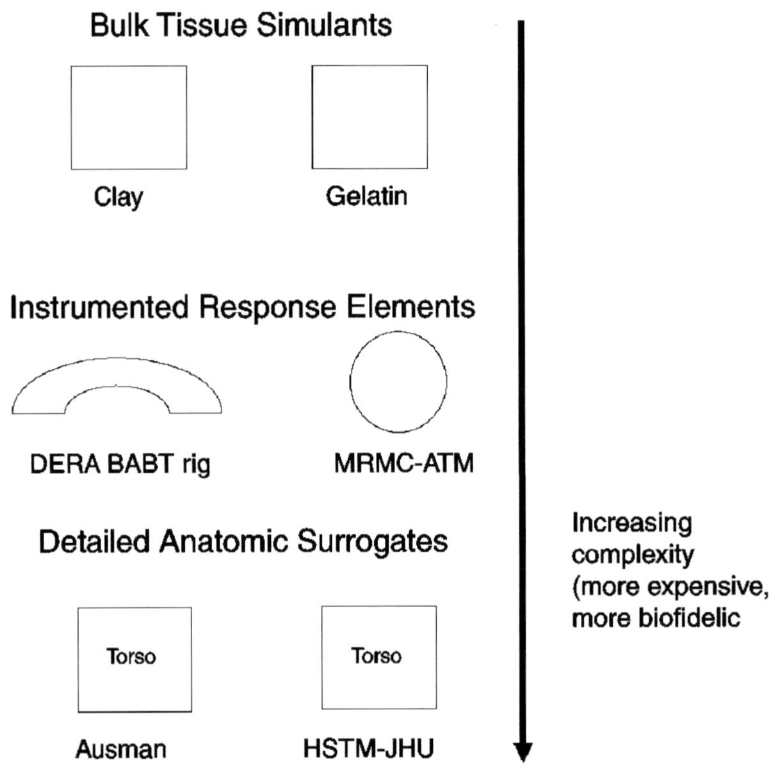

FIGURE 8-23 Examples of BABT assessment devices and methodologies.

To develop BABT test methodologies, all three types may be useful. There are trade-offs as devices run from simpler to more complex, usually less expensive to more expensive, but the more complex devices generally have the potential to assess more detailed injury criteria where appropriate. For instance, it is difficult to evaluate complex BABT interactions, especially for very lightweight body armor systems, without detailed anatomical surrogates. However, it is advisable to use a relatively inexpensive bulk tissue simulant or instrumented response element for production testing, because in multiple tests there is a potential for penetrating events that can destroy a test device worth thousands or tens of thousands of dollars. Each class of device may have an advantage for a given test condition. The three classes of devices are discussed below.

Bulk Tissue Simulants

Bulk tissue simulants are characterized by a single ideally isotropic and homogeneous layer. Although a number of bulk tissue simulants have been studied (e.g. Mirzeabassov et al., 2000), the Army has focused on simulants originally proposed in Prather et al. (1977). See Chapter 4 for a discussion of gelatin, plasticine, and clay simulants.

Instrumented Response Elements

There are at least six instrumented response elements used in current ballistic research environments. These are described below.

DERA BABT Simulator

The U.K. Defense Evaluation and Research Agency (DERA) has developed a test device intended to evaluate the injury effect of behind-armor blunt trauma, called the DERA BABT rig (Tam et al., 2000). The device is similar to a half-cylinder silicone rubber chest wall (GE Silicones RTV 428) developed by Cooper et al. (1996) enclosed in a framework providing for rotation and vertical positioning. The physical model was derived from a finite-element model that included high-rate blast and blunt response of the thorax. The DERA rig instrumented response element also allows for varying the thickness of response element materials.

The assumed BABT injury mechanism for this physical model is that injuries are a function of chest wall motion, including displacement amplitude, velocity, acceleration, and deformation profile. The plausible mechanism for porcine organ damage is the impact pressure wave and subsequent displacement. The test device uses a novel laser deformation sensor array system to measure the time history of displacement, as shown in Figure 8-24. The deformation sensor is kept outside the bullet trajectory and may provide velocity and acceleration response of the back face of the test device.

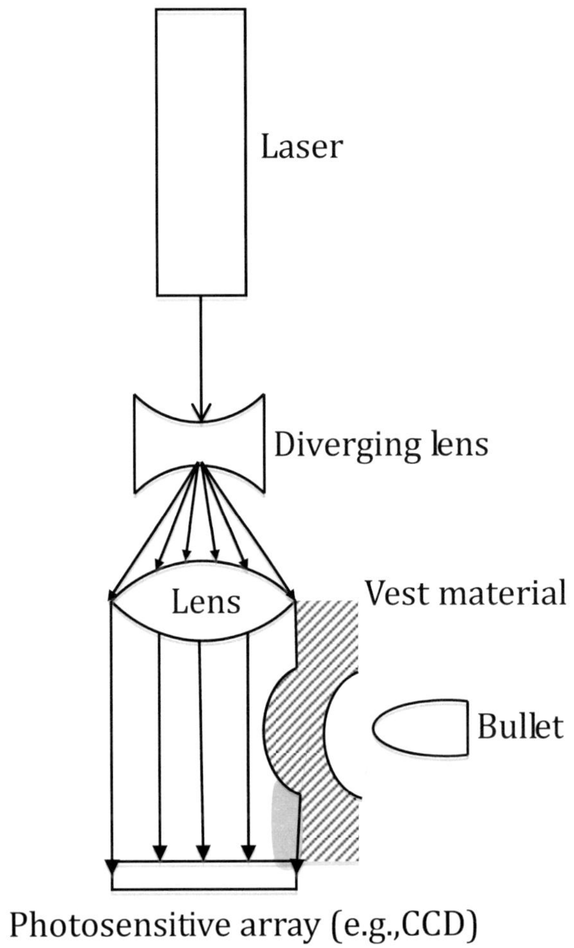

FIGURE 8-24 DERA BABT simulator displacement sensor system. SOURCE: Adapted from Tam et al., 2000.

The body armor has been tuned to lateral eviscerated pig baton data and has been tested using a 12.7-mm AP round against the U.K. Enhanced Body Armor and a 7.62-mm NATO ball against Improved Northern Ireland Body Armour. Validation studies included both lateral and anterior shots. Peak accelerations were comparable between the BABT rig and lateral pig shots. These accelerations were approximately 20,000 *g* for the 7.62-mm test round and the Improved Northern Ireland Body Armour. However, the response of the anterior porcine chest wall varied somewhat from the rig behavior. As the rig was developed using lateral impacts, this is not surprising.

Comparative experiments reported by Cannon et al. (2000) involved six eviscerated and six intact pigs using 7.62-mm rounds at 3 kJ behind commercial ceramic body armor. The velocity of the wall deformations of the eviscerated model were similar to that of the physical model (9.6 m/sec for eviscerated vs. 15.2 m/sec for physical). Peak displacement of 12.6 mm was seen in the eviscerated pig, and peak displacement of 7.8 mm was seen in the intact pigs. Local peak acceleration was approximately 13,000 *g*. Mean time to peak displacement was found to be approximately 2.7 msec. Viscous criterion of 0.29 m/sec was calculated for both. The DERA BABT device has been used in cooperative testing with Natick Soldier Center. The tests include body armor with a ceramic plate and a UHMWPE laminate. The test round was a 7.62-mm M80 ball round at 838 m/sec nominal velocity.

There are two significant drawbacks to this system. First, the system has been tuned to a specific thoracic wall velocity. Outside this regime, it is unlikely to respond appropriately to the rear face impact. Second, the simulator is shaped like a cylinder, and the laser system relies on this cylindrical shape to operate properly. As the human body is not cylindrical, it is difficult to use to evaluate actual body armor. To enhance the system, a different displacement measuring system could be developed using a multiple-laser time of flight system.

DREV Torso Injury Assessment Rig

The Defense Research Establishment Valcartier (DREV) has developed a thoracic injury assessment rig that is similar to the DERA BABT simulator (Bourget et al., 2002). For the DREV simulator both the material and the geometry have been altered to have a similar mechanical response to nonlethal baton impacts (Bir, 2000). Otherwise, the DREV test rig is similar in character and performance to the DERA BABT rig.

Anter Corporation

The private Anter Company in St. Petersburg, Russia, has patented a multilayer thorax simulator, as shown in Figure 8-25 (Mirzeabassov, 2000). This simulator is also covered under U.S. Patent 5850033. The simulator is proposed for both penetrating and blunt injuries. It includes an outside skin layer, a layer of muscle-simulating material, and a layer of bone-simulating material. Stiff paper is located on either side of the muscle stimulant, which consists of 10 mm layers of unvulcanized rubber. The brittle, strain-sensitive paper indicates the level of strain inside the simulant to approximate the temporary cavity during penetrating injury and to determine the extent of local deformation from blunt trauma.

There is no electronic instrumentation for this simulator; penetration or injury is indicated by paper between layers. The construction is reusable, and the paper indicator layers may be replaced. To collect additional information, these layers could be augmented with pressure-sensitive film. The basis for this injury model is experimentation using small dog subjects. Validation rounds include 7.62 mm with two impact energies, a 5.56-mm M16A1 and a 5.45-mm AK74. Validation included local thoracic deformation imaged using flash X-ray.

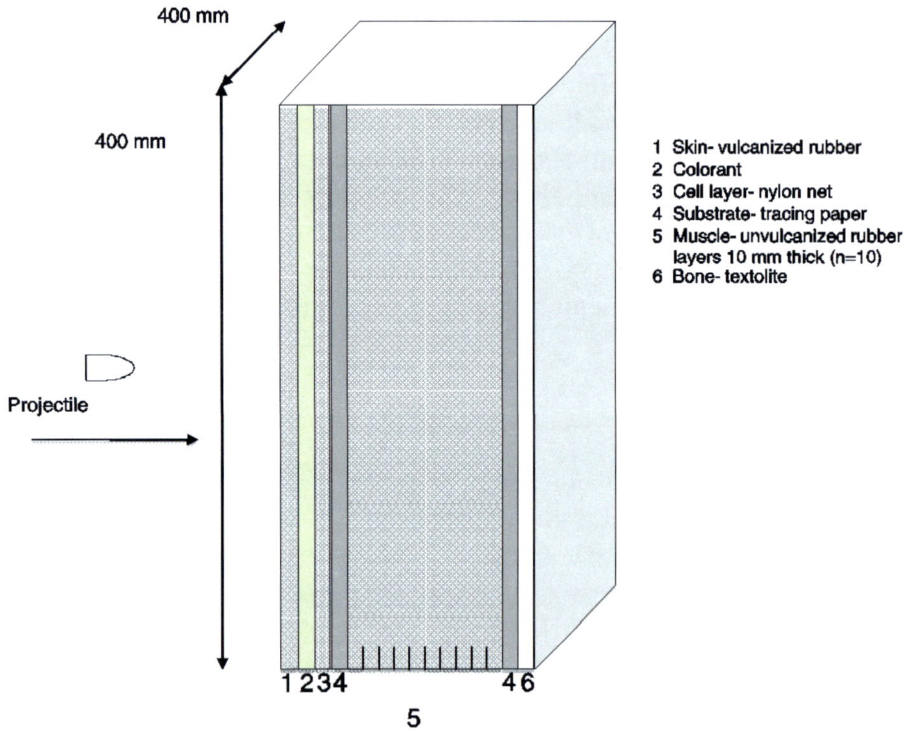

FIGURE 8-25 DERA tissue viscoelastic stimulant concept as described by Mirzeabassov et al., 2000.

Hybrid III Dummy

The standard automobile frontal crash test dummy, the Hybrid III, has been used by several investigators for high rate impact (Figure 8-26). The Hybrid III dummy has been used for mine blasts and high rate impacts from blasts in structures (Bass et al., 2001a and Bass et al., 2001b). The rib structure of the Hybrid III consists of six ribs constructed of steel overlying a viscoelastic damping material. The ribs are connected in the front to a sternal bib, and a single displacement sensor is standard instrumentation. This single displacement sensor has significant limitations (Butcher et al., 2001). Enhanced instrumentation is available with an array of string potentiometers. Use of the Hybrid III dummy has several advantages: It is widely used and is manufactured in several different sizes representative of various standard anthropometries.

There are substantial drawbacks to the use of the Hybrid III as a ballistic BABT test device. The dummy has been validated only for low-velocity impacts.

Further, the local response of the Hybrid III is likely not biofidelic, even for these low-rate impacts (Kent et al., 2006).

FIGURE 8-26 Hybrid III 50th percentile male dummy. SOURCE: Courtesy Humanetics Innovative Solutions.

Anthropomorphic Test Module

The U.S. Army Medical Research and Materiel Command developed an anthropomorphic test module (ATM) as an instrumented response element for BABT injury assessment, as shown in Figure 8-27.[62] The shoulders of the torso are not instrumented but provide an anthropometrically correct platform for mounting body armor as worn by soldiers. The response of the ATM element is measured using a multiple accelerometer array implanted within a polymer with approximately cylindrical form. The initial peak impact response is mitigated using a rubber pad over the surface of the response element as shown in the center of Figure 8-27.

[62]Michael Leggieri, U.S. Army Medical Research and Materiel Command, briefing to the committee at Aberdeen, Md., on August 11, 2010.

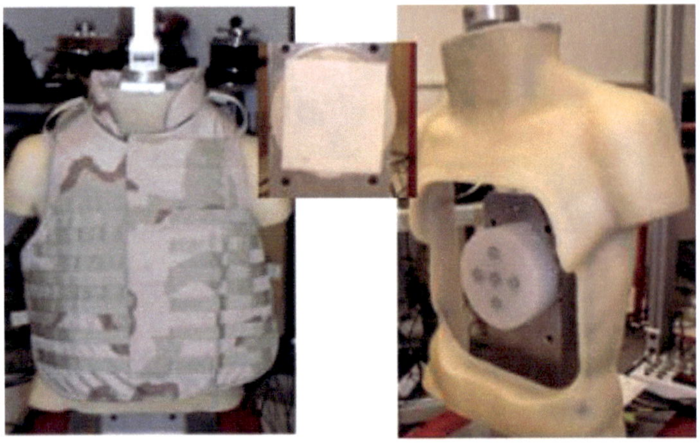

FIGURE 8-27 Left: ATM with mounted body armor. Center: ATM instrumented response element with padding. Right: oblique view of response element within torso. SOURCE: Michael Leggieri, Director, DoD Blast Injury Research Program Coordinating Office, U.S. Army Medical Research and Materiel Command, "Blunt Trauma Research to Support a New Body Armor Blunt Trauma Performance Standard and Testing Method," presentation to the committee, August 11, 2010.

Essential elements of the ATM methodology include these:

- The development of a detailed anatomical finite-element model of the human torso (Figure 8-28).
- Identification of global mechanical response and injury response in animal and cadaver tests for impact with hard projectiles.
- Development of a simple response element with mechanical response that is correlated with the animal and cadaver tests.
- Validation of the human finite-element model with the animal and cadaver tests.
- Correlation of the response element deformation with finite-element model calculations and hence animal and cadaver injury response.

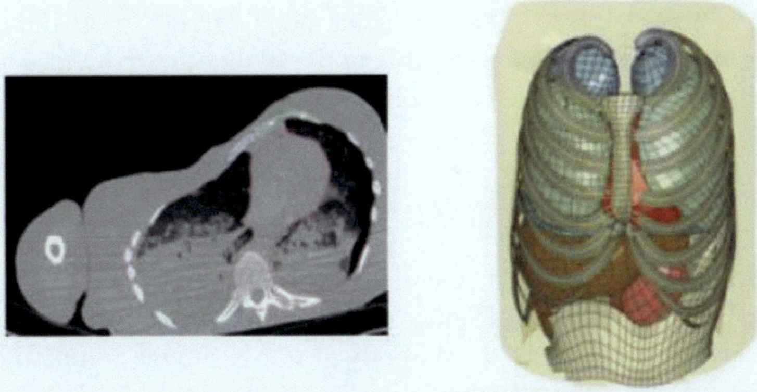

FIGURE 8-28 Left: human CT scan. Right: finite-element model, ribs and internal viscera. SOURCE: Michael Leggieri, Director, DoD Blast Injury Research Program Coordinating Office, U.S. Army Medical Research and Materiel Command, "Blunt Trauma Research to Support a New Body Armor Blunt Trauma Performance Standard and Testing Method," presentation to the committee, August 11, 2010.

The model is based on 30 moderate-rate porcine and 12 low-rate blunt impactor cadaver tests. Generally, test impact velocities did not reach those typical of high-rate impact behind hard body armor. A limited model validation was performed using the animal and cadaver tests for the finite-element model. The response included global response of the model and surrogates for a limited number of instrumentation locations.

Though the data provide good correlation with automobile impact corridors (Kroell et al., 1974), the principal limitation is the lack of robust validation data for rates typical of impact behind hard body armor at typical rifle round velocities. Further limitations of the model involve the beam formulation of the ribs, which must be derived from validation data.

Research questions raised include response to penetration; repeatability of measurements, including the potential for material properties changes in the rubber with repeated shots; and model validation. These issues are the subject of ongoing research.[63]

The ATM model and associated finite-element models are more complex than the typical response elements considered previously and likely represent a transition device between instrumented response elements and detailed anatomical surrogates.

Instrumented Detailed Anatomical Surrogates

The final class of instrumented surrogates, detailed anatomical surrogates, includes anatomical features that are similar to humans, generally including internal organs. When validated, these may in principal be appropriate for the

[63]Ibid.

investigation of complex behind-armor phenomena. Owing to the complexities of design and instrumentation, however, it is unlikely that these devices will directly form the basis for large-scale testing of body armor.

AUSMAN

AUSMAN is a reusable mechanical surrogate developed by the Australian Department of Defense, Defense Science and Technology Organization. The torso surrogate is shown in Figure 8-29, and body armor is shown mounted on the torso in Figure 8-30. AUSMAN consists of a 21-kg polymeric skeletal system enveloping a simulated cardiopulmonary system and liver and incorporates an anthropomorphic rib structure and a realistic spine. The entire thorax is encased in polymer, with gel coupling between internal viscera and the rib structure. The lungs are simulated by open cell foam as are a heart and liver. In addition, the design includes a mount for a Hybrid III head and neck to allow neck injury assessment.

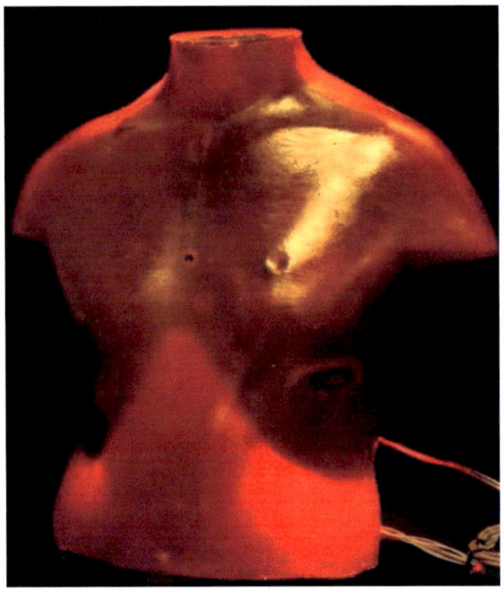

FIGURE 8-30 AUSMAN upper torso.

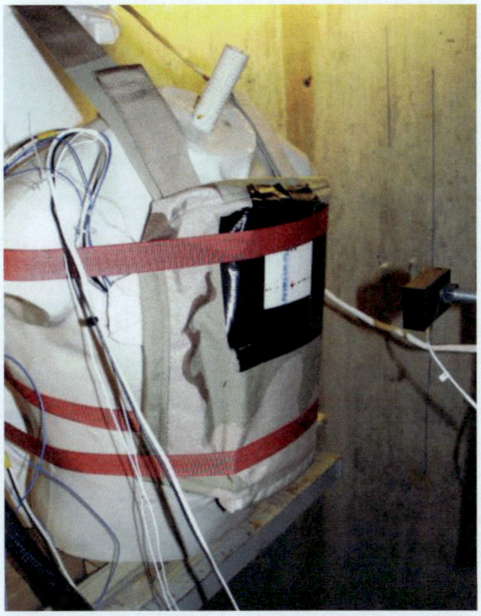

FIGURE 8-30 AUSMAN thorax with body armor in place, prior to testing. SOURCE: Reprinted with permission of Cameron Bass.

Mechanical response is measured by accelerometers in the sternum and midthoracic spine with pressure transducers in the lungs. AUSMAN has limited validity for blast, although an early version was correlated against sternal ballistic response in cadavers (Bass et al., 2006). The blast version of the dummy surrogate is currently under development.

Swedish Anthropometric Dummy

The Swedish National Defense Research Institute (Jonsson et al., 1986) has developed a model thorax to be used for blast, ballistic blunt impact, and missile impact. The thorax is enclosed by a rubber tube with a thickness of 6 mm and an elliptical shape that is 35 cm laterally and 25 mm in the anterior/posterior direction. Internal viscera are simulated by a water filled cavity that includes foam rubber surrogate lungs shaped as cylinders. These lungs have a diameter of 10 cm and a length of approximately 20 cm. They are sealed using a thin shell of rubber to prevent water infiltration. The lungs are positioned in the chest using wire mesh and strings that are rigidly mounted at each end of the thorax model. Each simulated lung includes a pressure transducer located at the center of the lung structure.

Experiments on this model were performed using a shock tube, a pendulum device, and ballistic impact behind body armor. Test rounds included 7.62 mm at 870 m/sec in both soft point and full metal jacket, a 9-mm submachine gun at 430 m/sec, and a 37-mm antiriot baton round at 76 m/sec. Body armor for the rifle rounds was 10 mm thick polyethylene trauma pack, 17-ply Kevlar 29, and a 10-mm ceramic plate. For the 9-mm round, the body armor used a 2.3-mm steel plate in place of the 10-mm ceramic plate. For the baton round, no body armor was used. Impacts were performed at the mediastinum with energies that range from 0.2 kJ to 5.6 kJ.

Intrathoracic pressure was measured for impact with the various rounds tested and compared with lateral rabbit thorax impact experiments using the pendulum impactor. An injury scale for intra-lung pressure was developed using pendulum impacts with velocities of approximately 5 m/sec that are stroke limited to 10 percent thoracic compression. Researchers found that the blast pressure peaks were far larger in the blast experiments than in the impact experiments for similar levels of lung damage. So, they concluded that dummies validated for blast pressure will not be calibrated for blunt impacts using the same pressure measurements. Assessments using this dummy predict a low risk of lung injury from a 7.62-mm test round at 3.2 kJ behind the body-armor combination of trauma pack, ceramic plate, and Kevlar-29.

The researchers observed small (~10 percent of peak) oscillations in the pressure signal at approximately 3 kHz. They concluded that the oscillations were related to structural forcing. There are significant questions regarding impact injury with mediastinal forcing using injury indices developed in a different velocity regime for lateral impacts. Further development of this surrogate is unknown.

Instrumented Model Human Torso

The Johns Hopkins Applied Physics Laboratory has been developing two research models for assessment of ballistic trauma: a frangible device and a reusable device.

Frangible JHU Model. The physical structure of the proof of concept frangible model includes rib, sternum, and spinal structure. Internal organs are represented as homogeneous solid viscera of urethane material, and the model has a skin and subcutaneous fat layer with an interposed sensor pad between the skin and fat layer. An emphasis was placed on incorporation of biofidelic materials. The bone material was chosen to be frangible with a failure modulus similar to human values. The design is intended to allow rapid replacement of frangible components.[64]

Instrumentation includes a sensor pad composed of an array of piezoelectric elements that are sensitive to bending. Additional piezoelectric arrays and resistive flexure grids are used inside the rib structure and near the heart.

Data from the anterior piezoelectric sensor array were sampled at 10 MHz and the remaining arrays were sampled at 25 kHz. This lower data rate does not appear to be an intrinsic limitation in data rate of the piezoelectric sensor arrays.

---

[64]Personal communication between Matt Bevin, Johns Hopkins University, and committee member Dale Bass circa 2002.

Preliminary testing was performed with soft body armor (Type II). Rounds tested include 9-mm at 330 m/sec and .357-Magnum JSP at 420 m/sec. Analysis of the anterior piezoelectric element only was reported. The researchers draw a correlation between measured peak voltage of the array and entering kinetic energy. There was no reported analysis of additional arrays.

An advantage of this model is that it has the potential for extending sensor instrumentation. The calibration of the sensors might be difficult, however, because of the difficulty integrating bending elements across linear and planar structures without substantial error (Bass et al., 1998). In addition, the piezoelectric materials used for measurement of compressive force are generally extremely sensitive in bending. Further, the model is frangible. It is difficult to produce a cost-effective model with large frangible components.

Reusable Johns Hopkins University Human Surrogate Torso Model. The reusable Human Surrogate Torso Model developed by the Johns Hopkins University has two versions, including 5th and 50th percentile human anthropometry, and has a detailed anthropomorphic skeletal structure with overlying skin and internal organs, including heart, lungs, stomach, and intestinal mass. The model is constructed so that the ribs have the fracture and bending properties of bone, and the organs are constructed of silicone polymers. Material properties of these organs have not yet been compared with human material properties at high rates of deformation. Sensors used in the torso include sternal and spinal accelerometers and pressure sensors at various locations. Additional instrumentation may include a vertebral load cell, surface pressures and load transducers, strain gauges for bone simulants, and possible displacement sensors.

The Human Surrogate Torso Model has been used in NIJ soft body armor tests to characterize mechanical response (Roberts et al., 2007; Merkle et al., 2008). Tests include 9-mm threats at 436 m/sec incoming velocity. The responses have further been correlated with clay response. The device has no validation against an injury metric and is still in development.

Developmental Testing Requirements

In sum, there are several existing test devices that are potentially suitable for use in the development of a test methodology for ballistic BABT. None is currently suitable for use as a test device for BABT with hard body armor without further development or experimentation.

For such developmental testing, two aspects must be considered. The first is the biofidelic response of the surrogate, and the second is the validation of the mechanical correlate from this model with an injury model. Each surrogate class has advantages. For example, many of the instrumented response elements and all of the anatomical surrogates have anthropomorphically appropriate thoracic form, reducing the risk of misleading appliqué response in clay or gelatin for body armor meant to be worn. A substantial drawback to instrumented surrogates is the cost of surrogates and sensor replacement. However, careful design can likely minimize both handling and sensor maintenance costs.

Finding: Instrumented response elements are in a primitive state for the evaluation of ballistic behind-armor blunt trauma for hard body armor against rifle round threats. Although several devices have associated instrument response and injury criteria that have been validated against a small range of loading conditions, there is no test device suitable for use without further development and validation.

Finding: Instrumented anatomical surrogates are not detailed enough to assess ballistic behind-armor blunt trauma for hard body armor with rifle round threats.

Recommendation 8-4: The Army should develop finite-element simulation models of human and live-animal thoracic response to behind-armor blunt impact. The validation of this simulation should be hierarchical from the small scale to the large scale. This includes the dynamic local response of constituent materials such as skin, bone, muscle, lung, liver, and other tissues; the regional response of the tissues under loading; and the global response of the whole torso. It should also include deformations from soft and hard body armor impacted with appropriate threats.

Recommendation 8-5: The Army medical community should enhance the current trauma registries to provide a program of injury epidemiology for ballistic impact, including behind-armor blunt trauma. This should include collection of both injury and noninjury events and should be similar to the federal crash databases used by the Department of Transportation—for example, the Fatality Analysis Reporting System and the National Automotive Sampling System for traffic injuries/fatalities, including injuries induced by both penetrations and backface deformations.

Recommendation 8-6: Using experimentally determined links to injury, response, and epidemiology, the Army should ensure that the clay or other alternative test methodology for hard body armor has humanlike dynamic response and is suitable for the development of behind-armor blunt trauma injury criteria.

Recommendation 8-7: To achieve improvements in behind-armor blunt trauma (BABT) research methodology in the medium term, the Army should develop instrumented thoracic simulators as response elements (sensors). Necessary preludes to this effort include the following:

- Establishing BABT phenomenology and injury criteria using human cadavers, animal models, and field injury epidemiology coupled with well-validated finite-element simulations.
- Establishing human BABT mechanical response for the range of design conditions for personal protective body armor. This should include impact on soft and hard body armor of anticipated threats.

Because of the high kinetic energy imparted to vests from current threats, laboratory testing with surrogates must remain well below ballistic V_0 using well-characterized armor systems to avoid extensive damage to structures and instrumentation. On the other hand, complex phenomenology can be investigated using appropriately validated devices that have the potential to reduce the overall cost of assessing BABT. One approach would be to initially use instrumented response elements in parallel with current methodologies focusing on dynamic displacement and/or force-response sensing capabilities.

Recommendation 8-8: In the long term, beyond simple clay torso surrogates and one-layer torso simulants, the Army should use the road map in Figure 8-31 to investigate the use of detailed anatomical surrogates (such as cadavers, instrumented models, etc.) as research devices to evaluate behind-armor blunt trauma.

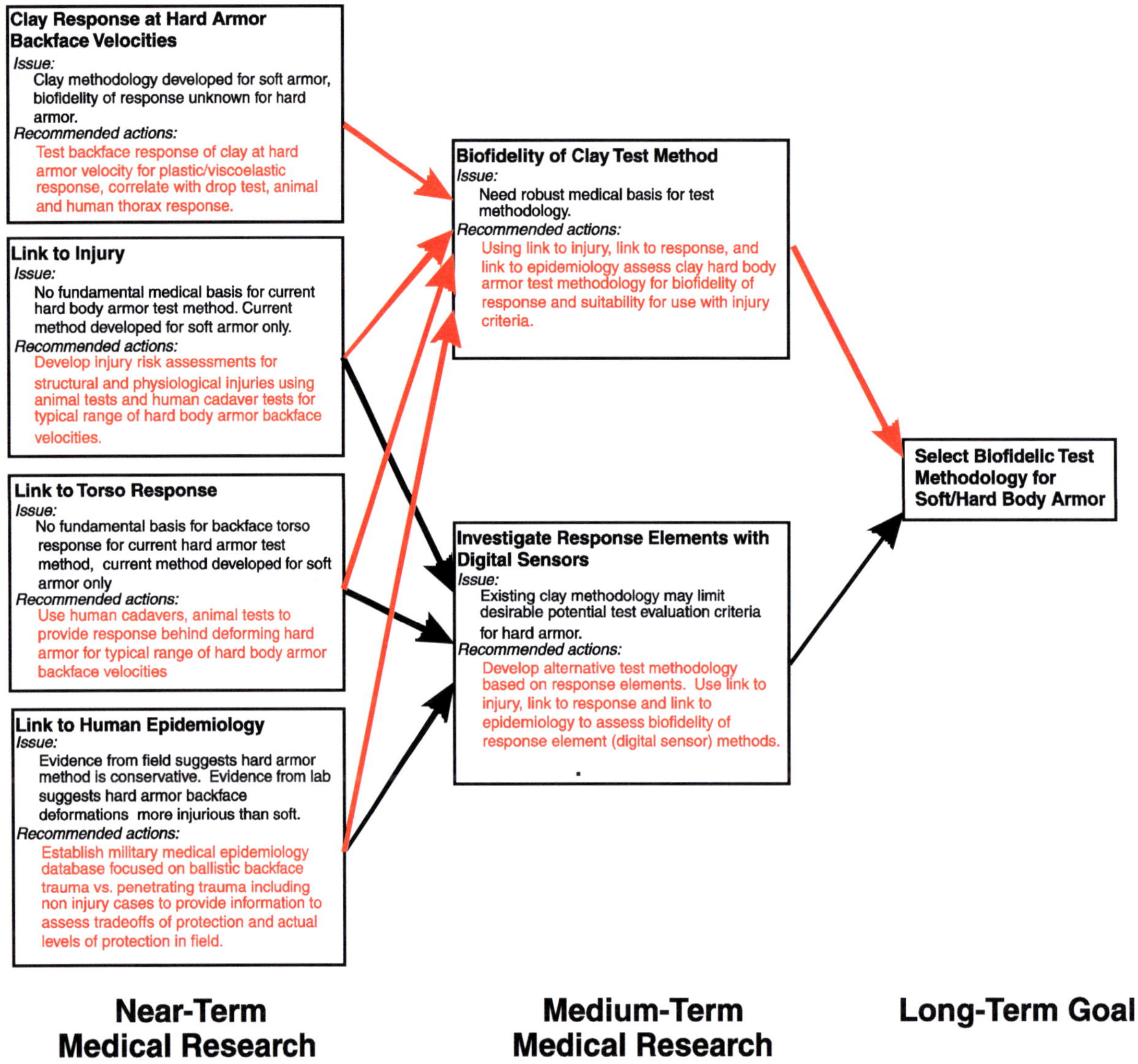

FIGURE 8-31 Road map showing suggested near-term and medium-term research needs, and a long-term goal to provide the fundamental medical basis for injury risk assessment behind helmets and hard body armor.

MEDICAL RESEARCH NEEDS

The principal biomedical issues relating to the development of a test methodology for soft and hard body armor testing with a strong basis in biomedical response and injury include these:

- The response of the clay currently used in the backface impact methodology has limited biomedical basis in human body response. As the human torso responds differently for impacts at different rates, the current clay response methodology has no biomedical basis for hard body armor impacted with high-velocity rifle rounds.
- The link to human injury in the current clay methodology was developed for the behind-armor impact of soft armor and has a limited biomedical basis even for soft body armor. The current methodology has no link to human injury for hard body armor impacted with high-velocity rifle rounds.
- The backface torso response and the effects of BABT on organs remote from the point of trauma are likely dependent on impact rates for both soft and hard body armor.
- There are only very limited links to human epidemiology for injuries from BABT when rifle rounds impact hard body armor worn in combat. This fact combined with the fact provided by DoD that there are no known fatalities from design threats suggests that it is unknown whether body armor is overdesigned against current threats. This has substantial implications for battlefield mobility, thermal loads, and other important issues in combat.

In the face of real-world constraints on dollars and manpower, Figure 8-31 provides a prioritized, time-phased road map for the near-term and medium-term medical research that is needed to reach the long-term goal of developing a test methodology for soft and hard body armor based soundly in biomedical response and injury.

Near-Term Actions

As shown in Figure 8-31, several near-term actions are needed to address the biomedical issues enumerated above and provide a strong biomedical basis for future body armor testing:

- The backface response of clay must be tested for its plastic and viscoelastic characteristics and correlated with relevant drop tests as well as animal and human thoracic response. See Recommendation 4-2.
- Injury risk assessments for structural and physiological injuries must be developed using animal tests and human cadaver tests for a typical range of hard body armor backface velocities. Experience from the

limited number of previous animal, cadaveric, and surrogate studies should be assessed to help guide these studies. See Recommendations 8-2 and 8-3.

- Tests involving human cadavers or animals must be conducted to determine response behind deforming hard armor for a typical range of hard body armor backface velocities. Experience from the limited number of previous animal, cadaveric, and surrogate studies should be assessed to help guide these studies. See Recommendation 8-2.
- A military medical epidemiology database must be established that focuses on ballistic backface trauma vs. penetrating trauma, including non-injury cases to provide information to assess tradeoffs of protection and actual levels of protection in field, See Recommendations 8-1 and 8-5.

Medium-Term Actions

The near-term actions shown in Figure 8-31 should provide input to medium-term medical research to produce a biofidelic test methodology for soft and hard body armor.
Key medium-term actions are an assessment of the biofidelity of the clay test method and development of instrumented response elements. Specifically,

- Assess the biofidelity of the clay or other alternative test method using near-term research results. See Recommendation 8-6.
- Develop an alternative to the clay methodology using digital sensors with a thorax response element. See Recommendation 8-7.
- Develop detailed anatomical surrogates. See Recommendation 8-8.

The long-term goal is to improve body armor by choices of materials and system configurations so the weight and protection are optimized. To reach this goal, a practical biofidelic test methodology for both soft and hard body armor is essential.

REFERENCES

Akimov, G.A, M.M. Odinak, S.A. Zhivolupov, B.S. Glushkov, and T.I. Milovanova. 1993. The mechanisms of the injuries to the nerve trunks in gunshot wounds of the extremities. (experimental research) (article in Russian). Voennomeditsinskii Zhurnal (Military Medical Journal) 80(9):34-36.

Axelsson, H., H. Hjelmqvist, A. Merdin, J.H. Persson, and A. Suneson. 2000. Physiological changes in pigs exposed to a blast wave from a detonating high-explosive charge. Military Medicine 165:119-126.

Axelsson, H. and J.T. Yelverton. 1996. Chest Wall Velocity as a Predictor of Nonauditory Blast Injury in a Complex Wave Environment. The Journal of Trauma: Injury, Infection, and Critical Care 40(3S): 31S-37S.

Baker, S.P., B. O'Neill, W. Haddon Jr, and W.B. Long. 1974. The injury severity score: A method for describing patients with multiple injuries and evaluating emergency care. Journal of Trauma 14(3):187-196.

Barwood, M., P. Newton, and M. Tipton. 2009. Ventilated vest and tolerance for intermittent exercise in hot, dry conditions with military clothing. Aviation, Space, and Environmental Medicine 80(4):353-359.

Bass, C.R., S.M. Duma, J.R. Crandall, W.D. Pilkey, N. Khaewpong, R.H. Eppinger, and S. Kuppa. 1997. Interaction of air bags with cadaveric upper extremities. Journal of Passenger Cars, SAE Transactions 106:3644-3650.

Bass, C.R., J.R. Crandall, C. Wang, W.D. Pilkey, and C. Chou. 1998. Open-loop chestbands for dynamic deformation measurements. Journal of Passenger Cars, SAE Transactions 107(6):1380-1387.

Bass, C.R. 2000. Development of a procedure for evaluating demining protective equipment. Journal of Mine Action, Version 4.2. Available online at http://maic.jmu.edu/journal/4.2/Focus/Bass/bass.htm. Last accessed February 24, 2011.

Bass, C.R., B. Boggess, M. Davis, E. Sanderson, G. Di Marco, and C. Chichester. 2000. A methodology for evaluating demining personal protective ensembles for AP landmines. UXO/Countermine Conference, New Orleans, La.

Bass, C.R., M. Bolduc, and S. Waclawik. 2002. Development of a nonpenetrting, 9 mm, ballistic trauma test method. Personal Armor Systems Symposium, Colchester, U.K.

Bass, C., R. Salzar, S. Lucas, M. Davis, L. Donnellan, B. Folk, E. Sanderson, and S. Waclawik. 2006. Injury risk in behind armor blunt thoracic trauma. International Journal of Occupational Safety and Ergonomics 12(4):429-442.

Bass, C.R., K.A. Rafaels, and R.S. Salzar. 2008. Pulmonary injury risk assessment for short-duration blasts. The Journal of Trauma 65(3):604-615.

Beekley, M.D., J. Alt, C.M. Buckley, M. Duffey, and T.A Crowder. 2007. Effects of heavy load carriage during constant-speed, simulated, road marching. Military Medicine 172(6):592-595.

Bensel, C. K. and J.M. Lockhart. 1975. The Effects of Body Armor and Load-Carrying Equipment on Psychomotor Performance. ADA014978. Natick, Mass.: U.S. Army Natick Research and Development Labs, Clothing Equipment and Materials Engineering Lab.

Bir, C. 2000. Comparison of Injury Evaluation in Non-lethals, Non-Lethal Technology and Academic Research Symposium, Portsmouth, N.H.

Bowen, I.G., E.R. Fletcher, D.R. Richmond, F.G. Hirsch, and C.S. White. 1966. Biophysical mechanisms applicable in assessing the responses of the thorax energized by blast overpressure or by nonpenetrating missiles, DASA-1857. Washington, D.C.: Defense Atomic Support Agency, Department of Defense.

Bowen, I.G., E.R. Fletcher, and D.R. Richmond. 1968. Estimate of man's tolerance to the direct effects of air blast. Technical Progress Report, DASA-2113. Washington, D.C.: Defense Atomic Support Agency, Department of Defense.

Butcher, J., G. Shaw, C. Bass, R. Kent, and J. Crandall. 2001. Displacement Measurements in the Hybrid III Chest. SAE Technical Paper 2001-01-0118. SAE 2001 World Congress, Detroit, Mich.

Cannon, L., W. Tam, P. Dearden, and P. Watkins. 2000. Thoracic wall response to nonpenetrating balistic injury: The validation of a physical model of behind armor blunt trauma. 12th Conference of the European Society of Biomechanics, Dublin.

Cannon, L. 2001. Behind armour blunt trauma—An emerging problem. Journal of the Royal Army Medical Corps 147:87-96.

Carroll, A. and C. Soderstrom. 1978. A new nonpenetrating ballistic injury. Annals of Surgery 188(6):753-757.

Carstensen, E.L., D.S. Campbell, D. Hoffman, S.Z. Child, and E.J. Ayme-Bellegarda. 1990. Killing of Drosophila larvae by the fields of an electrohydraulic lithotripter. Ultrasound in Medicine & Biology 16(7): 687–698.

Carstensen, E.L., S. Gracewski, and D. Dalecki. 2000. The search for cavitation *in vivo*. Ultrasound in Medicine and Biology 26(9): 1377-1385.

Cavanaugh, J. M. 1993. The biomechanics of thoracic trauma, in Accidental Injury, Biomechanics and Prevention, eds. A.M. Nahum and J.W. Melvin, Springer-Verlag.

Clare, V., J. Lewis, A. Mickiewicz, and L. Sturdivan. 1975. Blunt Trauma Data Correlation. EB-TR-75016. Aberdeen Proving Ground, Md.: Edgewood Arsenal.

Clemedson, C.J. and H.I. Hultman. 1954. Air embolism and the cause of death in blast injury. Military Surgeon 114:424-437.

Cooper, G., B. Pearce, M. Stainer, and R. Maynard. 1982. The biomechanical response of the thorax to nonpenetrating impact with particular reference to cardiac injuries. The Journal of Trauma: Injury, Infection, and Critical Care 22(12):994-1008.

Cooper, G., R. Maynard, B. Pearce, M. Stainer, and D. Taylor. 1984. Cardiovascular distortion in nonpenetrating chest impacts. The Journal of Trauma: Injury, Infection, and Critical Care 24(3):188-200.

Cooper, G.J. 1996. Protection of the lung from blast overpressure by thoracic stress wave decouplers. The Journal of Trauma 40(3S):105S-110S.

Cooper, G., B. Pearce, A. Sedman, I. Bush, and C. Oakley. 1996. Experimental evaluation of a rig to simulate the response of the thorax to blast loading. The Journal of Trauma 40(3S):38S-41S.

Cooper G.J., D.J. Townend, S.R. Cater, and B.P. Pearce. 1991. The role of stress waves in thoracic visceral injury from blast loading: Modification of stress transmission by foams and high-density materials. Journal of Biomechanics 24(5):273-285.

Cooper, G. and A. Jonsson. 1997. Protection against blast injury. Scientific Foundations of Trauma, ed. G. Cooper. New York, N.Y.: Butterworth-Heinemann.

Courtney, M.W. and A.C. Courtney. 2010. Working toward exposure thresholds for blast-induced traumatic brain injury: Thoracic and acceleration mechanisms. Neuroimage 54(S1): 55S-61S.

Crandall, J.R., C.R. Bass, S.M. Duma, and S. Kuppa. 1998. An evaluation of 5th percentile female hybrid III biofidelity and injury criteria measurement methods during out of position occupant tests with a driver airbag. Journal of Passenger Cars, SAE Transactions 107(6):1154-1161.

Cripps, N.P.J. and G.J. Cooper. 1996. The influence of personal blast protection on the distribution and severity of primary blast gut Injury. The Journal of Trauma 40(3S):206S-211S.

DeMaio, M. 2001. Closed chest trauma: Evaluation of the protected chest, Presentation to the NATO Task Group on Behind Armor Blunt Trauma, Copenhagen.

Desaga, H. 1950. Blast injuries. In: German Aviation Medicine, World War II, vol. 2. pp. 1274-1293. Washington, D.C.: US Government Printing Office.

Drobin, D., D. Gryth, J. Persson, D. Rocksen, U. Arborelius, L. Olsson, J. Bursell, and B. Kjellstrom. 2007. Electroencephalogram, circulation, and lung function after high-velocity behind armor blunt trauma. The Journal of Trauma 63(2):405-413.

Eppinger, R., J. Marcus, and R. Morgan. 1984. Development of dummy and injury index for NHTSA's thorax side impact protection research program. Proceedings of the 28th Stapp Car Crash Conference, SAE Paper Number 840885. Warrendale, Pa.: Society of Automobile Engineers.

Fung, Y., R. Yen, Z. Tao, and S. Lui. 1988. A hypothesis on the mechanism of trauma of lung tissue subjected to impact load. Journal of Biomechanical Engineering 110:50-56.

Gryth, D., D. Rocksen, J. Persson, U. Arborelius, D. Drobin, J. Bursell, L. Olsson, and T. Kjellstrom. 2007. Severe lung contusion and death after high-velocity behind-armor blunt trauma: Relation to protection level. Military Medicine 172(10):1110-1116.

Harman, E., P. Frykman, C. Pandorf, W. Tharion, and R. Mello. 1999. Physiological, Biomechanical, and Maximal Performance Comparisons of Female Soldiers Carrying Loads Using Prototype U.S. Marine Corps Modular Lightweight Load-Carrying Equipment (MOLLE) with Interceptor Body Armor and U.S. Army All-Purpose Lightweight Individual Carrying Equipment (ALICE) with PASGT Body Armor. ADA365448. Natick, Mass.: U.S. Army Research Institute of Environmental Medicine, Military Performance Division.

Haley, J.L., N.M. Alem, J.B. McEntire, and J.A. Lewis. 1996. Report Number 96-14 Force Transmission Through Chest Armour During the Defeat of a .50

Calibre Round. Fort Rucker, Ala.: U.S. Army Aeromedical Research Laboratory.

Hawkins, J., L. Koonin, S. Palmer, and C. Gibbs. 1997. Anesthesia-related deaths during obstetric delivery in the United States, 1979-1990. Anesthesiology 86(2):277-84.

Jonsson, A., E. Arvebo, and B. Shantz.. 1986. Intrathoracic pressure variations in an anthropomorphic dummy exposed to air blast, blunt impact, and missiles. The Journal of Trauma 28(1S):125S -131S.

Kent, R., C.R. Bass, W. Woods, R. Salzar, S. Lee, and J. Melvin. 2006. The role of muscle tensing on the force-deflection response of the thorax and a reassessment of frontal impact thoracic biofidelity corridors. Journal of Automobile Engineering 220(7):853-868.

Knapik, J.J., K.L Reynolds, and E. Harman. 2004. Soldier load carriage: Historical, physiological, biomechanical, and medical aspects. Military Medicine 169(1):45-56.

Knudsen, P.J. and H. Gøtze. 1996. Behind armour blunt trauma. Biomedical Experiments, Unpublished manuscript.

Knudsen, P.J and H. Gøtze. 1997. Behind armour blunt trauma. Trials report 02/97. Oksbøl, Denmark: Danish Army Combat School.

Krajsa, J. 2009. Causes of pericapillary brain hemorrhages caused by gunshot wounds. Ph.D. Dissertation (in Czech). Brno, Czech Republic: Institute of Forensic Medicine, Masaryk University.

Kroell, C.K., D.C. Schneider, and A.M. Nahum. 1971. Impact tolerance and response of the human thorax. SAE 710851. 15th Stapp Car Crash Conference.

Kroell, C. K., D.C. Schneider, and A.M. Nahum. 1974. Impact tolerance and response to the human thorax. . SAE 741187. Proc. 18th Stapp Car Crash Conference, pp. 383-457.

Lewis, E.A. 2005. Dissertation: A Novel Behind Armour Blunt Trauma injury. Cranfield, Bedfordshire, U.K.: Cranfield University.

Lidén, E., R. Berlin, B. Janzon, B. Schantz, and T. Seeman. 1988. Some observations relating to behind-body armour blunt trauma effects caused by ballistic impact. The Journal of Trauma 27(1S): 145S-148S.

Link, M.S., P.J. Wang, B.J. Maron, and N.A. Estes. 1999. What is commotio cordis? Cardiology in Review 7(5):265-269.

Lui, B., Z. Wang, H. Leng, Z. Yang, and X. Li. 1996. Studies on the mechanisms of stress wave propagation in the chest subjected to impact and lung injuries. The Journal of Trauma 40(3S):53S-55S.

Martin, P., G. Hall, R. Crandall, and W. Pilkey. 1998. Measuring the Acceleration of a Rigid Body. Shock and Vibration 5(4):211-224.

Martinez, B. 1999. A Health Hazard Assessment for Blast Overpressure Exposures: Blast Overpressure Research Program. Technical Report J2997.74-99-106. San Diego, Calif.: Jaycor.

Maynard, R., D. Coppel, and K. Lowry. 1997. Blast injury of the lung. Scientific Foundations of Trauma, ed. G. Cooper. New York, N.Y.: Butterworth-Heinemann.

Mayorga, M. 1997. The pathology of primary blast overpressure injury. Toxicology 121(1):17-28.

Mayorga, M., I. Anderson, J. van Bree, P. Gotts, J-C. Sarron, and P. Knudsen. 2010. Thoracic Response to Undefeated Body Armor. 1 Neuilly-sur-Seine, France: North Atlantic Treaty Organization, Research and Technology Organisation.

Mellor, S.G., G.J. Cooper. 1989. Analysis of 828 servicemen killed or injured by explosion in Northern Ireland 1970-84: The hostile action casualty system. British Journal of Surgery 76(10):1006-1010.

Merkle, A., E. Ward, J. O'Connor, and J. Roberts. 2008. Assessing behind armor blunt trauma (BABT) under NIJ Standard-0101.04 conditions using human torso models. The Journal of Trauma 64(6):1555-1561.

Miller, P.R., M.A. Croce, P.D. Kilgo, J. Scott, and T.C. Fabian. 2002. Acute respiratory distress syndrome in blunt trauma: Identification of independent risk factors. The American Surgeon 68(10):845-851.

Mirzeabassov, T., D. Belov, M. Tyurin, and L. Klyaus. 2000. Further investigation of modelling for bullet-proof vests. Personal Armor Safety Symposium, Colchester, U.K.

Mitchell, S.W., G.R. Morehouse, and W.W. Keen, Jr. 1864. Reflex paralysis. Circular No. 6, Surgeon General's Office, U.S. Government. Found in Yale Library Reprints, 1941.

Mott, F.W. 1919. War Neuroses and Shell Shock. Henry Frowde and Hodder & Shoughton. London, England: Oxford University Press.

Nechaev, E., A. Gritsanov, N. Fmoin, and I. Minnullin. 1995. Mine blast trauma. St. Petersburg, Russia: Russian Ministry of Public Health and Medical Industry.

Ness, L.S. 2011. Jane's Ammunition Handbook 2011-2012, 20th Edition, edited by A.G. Williams. Alexandria, Va.: Jane's Information Group.

Phillips, Y.Y., T.G. Mundie, J.T. Yelverton, and D.R. Richmond. 1988. Cloth ballistic vest alters response to blast. The Journal of Trauma 28(1S):149S-152S.

Prat, N., F. Rongieras, E. Voiglio, P. Magnan, C. Destombe, E. Debord, F. Barbillon, T. Fusai, and J.C. Sarron. 2010. Intrathoracic pressure impulse predicts pulmonary contusion volume in ballistic blunt thoracic trauma. The Journal of Trauma 69(4):749-755.

Prather, R., C. Swann, and C. Hawkins. 1977. Backface Signatures of Soft Body Armors and the Associated Trauma Effects. ARCSL-TR-77055. Aberdeen Proving Ground, Md.: Edgewood Arsenal.

Raeman, C.H., S.Z. Child, D. Dalecki, C. Cox, and E.L. Carstensen. 1996. Exposure-time dependence of the threshold for ultrasonically induced murine lung hemorrhage. Ultrasound in Medicine & Biology 22:139-41.

Roberts, J.C., A.C. Merkle, P.J. Biermann, E.E. Ward, B.G. Carkhuff, R.P. Cain, and J.V. O'Connor. 2007. Computational and experimental models of the human torso for non-penetrating ballistic impact. Journal of Biomechanics 40(1):125-136.

Rossle, R. 1950. Pathology of blast effects. German Aviation Medicine, World War II, Vol. 2. U.S. Air Force Surgeon General, pp. 1260-1273.

Ryan, J.M., N.M. Rich, R.F. Dale, B.T. Morgans, and G.J. Cooper. 1997. Ballistic trauma: Clinical relevance in peace and war. New York, N.Y.: Oxford University Press, Inc.

Sarron, J.C., C. Destombe, H. Gøtze, and M. Mayorga. 2000. Physiological results of NATO BABT experiments. Personal Armour Systems Symposium Colchester, U.K. 5-8 September.

Sarron, J.C., J.P. Caillou, C. Destombe, T. Lonjon, J. Da Cunha, P. Vassout, and J.C. Allain. 2001. Displacements of thoracic wall after 7.62 non penetrating impact. Presentation to the NATO BABT Task Group, Copenhagen.

Sendowski, I., P. Martin, J. Muzellec, and J. Breteau. 1994. Rear side effects of protective vests in non-perforating model. Tolerance Norms Research, Institute of Tropical Medicine of the Armed Forces Health Service. General Directorate for Armament, France.

Sondén, A., D. Rocksen, L. Riddez, J. Davidsson, J. Persson, D. Gryth, J. Bursell, and U. Arborelius. 2009. Trauma attenuating backing improves protection against behind armor blunt trauma. The Journal of Trauma 67(6):1191-1199.

Sperry, Y.K. 1993. Scleral and conjunctival hemorrhages arising from a gunshot wound of the chest. Journal of Forensic Sciences 38(1):203-209.

Stapp, J. 1949. Human exposures to linear deceleration, Air Force Technical Report No. 5915. Wright-Patterson Air Force Base, Ohio: U.S. Air Force Material Command.

Suneson, A., H. Hansson, and T. Seeman. 1987. Peripheral high-energy missile hits cause pressure changes and damage to the nervous system: Experimental studies on pigs. The Journal of Trauma 27(7):782-789.

Suneson, A., H.-A. Hansson, and T. Seeman. 1990. Pressure wave injuries to the nervous system caused by high-energy missile extremity impact, Part II. Distant effects on the central nervous system: A light and electron microscopic study on pigs. The Journal of Trauma 30(3):295-306.

Tam, W., M. Dorn, and P. Gotts. 2000. A new method for quantifying behind armour blunt trauma. Proceedings of the Personal Armor Systems Symposium, Colchester, U.K.

van Bree, J. and P. Gotts. 2000. The 'Twin Peaks' of BABT. Personal Armor Systems Symposium, Colchester, U.K.

Vance, B.M. 1923. Traumatic lesions of the intestine caused by nonpenetrating blunt force. Archives of Surgery 7(1):197-212.

Viano, D.C. and I.V. Lau. 1985. Thoracic impact: A viscous tolerance criterion. Tenth International Conference on Experimental Safety Vehicles, Oxford, England, pp. 104-114.

Von Gierke, H.E. 1968. Response of the body to mechanical forces – an overview.
Annals of the New York Academy of Sciences 152(1):172-186.

Wang, Q., Z. Wang, P. Zhu, and J. Jiang. 2004. Alterations of myelin basic protein and ultrastructure in the limbin system at the early stage of trauma-related stress disorder in dogs. The Journal of Trauma 56(3):604-610.

Weller-Ravell, D., R. Addatto, and J.B. Boreman. 1975. Blast injury of the chest: A review of the problem and its treatment. Israeli Journal of Medical Sciences 11(2-3):286-274.

Wilhelm, M. and C. Bir. 2008. Injuries to law enforcement officers: The backface signature injury. Forensic Science International 174(1):6-11.

Wood, G.W., B.Z. Panzer, J.K. Shridharani, K.A. Matthews, and C.R. Bass. 2010. Attenuation of blast overpressure behind ballistic protective vests. Personal Armor Science Symposium (PASS-2010).

Young, A.J., J.J. Jaeger, Y.Y. Phillips, J.T. Yelverton, and D.R. Richmond. 1985. The influence of clothing on human intrathoracic pressure during airblast. Aviation, Space, and Environmental Medicine 56(1):49-53.

9

Future Improvements in Testing Methodology

This chapter discusses the notion that the body armor testing community should retain and improve on the current body armor testing methodology that has evolved from the early work of the Prather and other studies as the way into the future. It describes the four main body armor community stakeholders: the users; the technologists; the medical researchers; and the production testers. It also describes how the current methodological basis for testing can be improved by aligning the body armor testing community stakeholders and by developing a common understanding of the dynamics and measurements of behind-armor phenomena to link medical research to product testing criteria.

BUILDING ON THE PRATHER STUDY

Chapter 3 described the original "Prather study" (Prather et al., 1977) and subsequent work that formed the underpinning of modern body armor testing. Mr. Prather told the committee that the original work of his team was intended to provide a quick turnaround process that could be the starting point for conducting soft body armor testing by the Army. It was never intended to become the body armor testing gold standard for military and police forces nationally, especially not for hard body armor.[65] Nonetheless, it has in fact evolved into a standard approach internationally even though the character of the threats as well as the composition and construction of body armor have changed.

The original work in the late 1970s provided an efficient method for testing body armor without live animals using a surrogate that allowed determining the adequacy of a given soft body armor to prevent a certain magnitude of backface deformation (BFD) that would cause serious harm to warfighters or law enforcement personnel.

Perhaps the greatest accomplishment of this test approach as used currently is the assertion by the Army Program Executive Office Soldier (PEO Soldier) that no soldier is known to have died on the battlefield as a result of the penetration of body armor by

[65]Russell Prather, Survice Engineering Company, "The Lightweight Body Armor Program - A History," presentation to the committee, August, 10, 2010.

rounds the body armor was designed to defeat. Owing to the limited basis for the Prather work, however, essential trade-offs of protection with weight are unknown. At the present time and for the next few years this methodology needs to be retained and the incremental improvements and refinements over the past three decades need to continue until an acceptable alternative method can be introduced for both development and testing of new armor for realistic threats.

Finding: The committee finds that the current body armor testing methodology that has evolved from the early work of Prather et al. (1977) should be retained and improved on while investigating alternative methods.

Synopsis of Near-Term Improvements

Chapter 3 concluded with a summary of strengths and weaknesses of the Prather approach, which is repeated here as Table 9-1. The committee focused on retaining the strengths while providing insights into overcoming the weaknesses.

TABLE 9-1 Strengths and Weaknesses of the Prather Methodology

Strengths	Weaknesses
Ease of use	Clay constituents have changed considerably since original study
Immediate results	Clay variability (handling, thixotropy, temperature effects, etc.)
Relatively low cost	Current methodology requires elevated clay temperatures
Large historical database of results	All variability in testing results is assumed to be design flaws in the armor
Apparent success in field for soft body armor	Method has limited medical validation for soft body armor
Apparent success in field for hard body armor	Method has no medical validation for hard body armor
	Pass/fail criterion

In Chapter 4 the report discussed and provided findings and recommendations for mitigating the weaknesses associated with the variability of clay and its formulations, especially important for production testing. The road map of the Chapter 4 findings and recommendations is reiterated as Figure 9-1. It includes immediate actions to improve the current clay methodology and longer term activities to develop techniques for optimizing body armor development and testing manufactured vests that are alternatives to the clay surrogate testing methods.

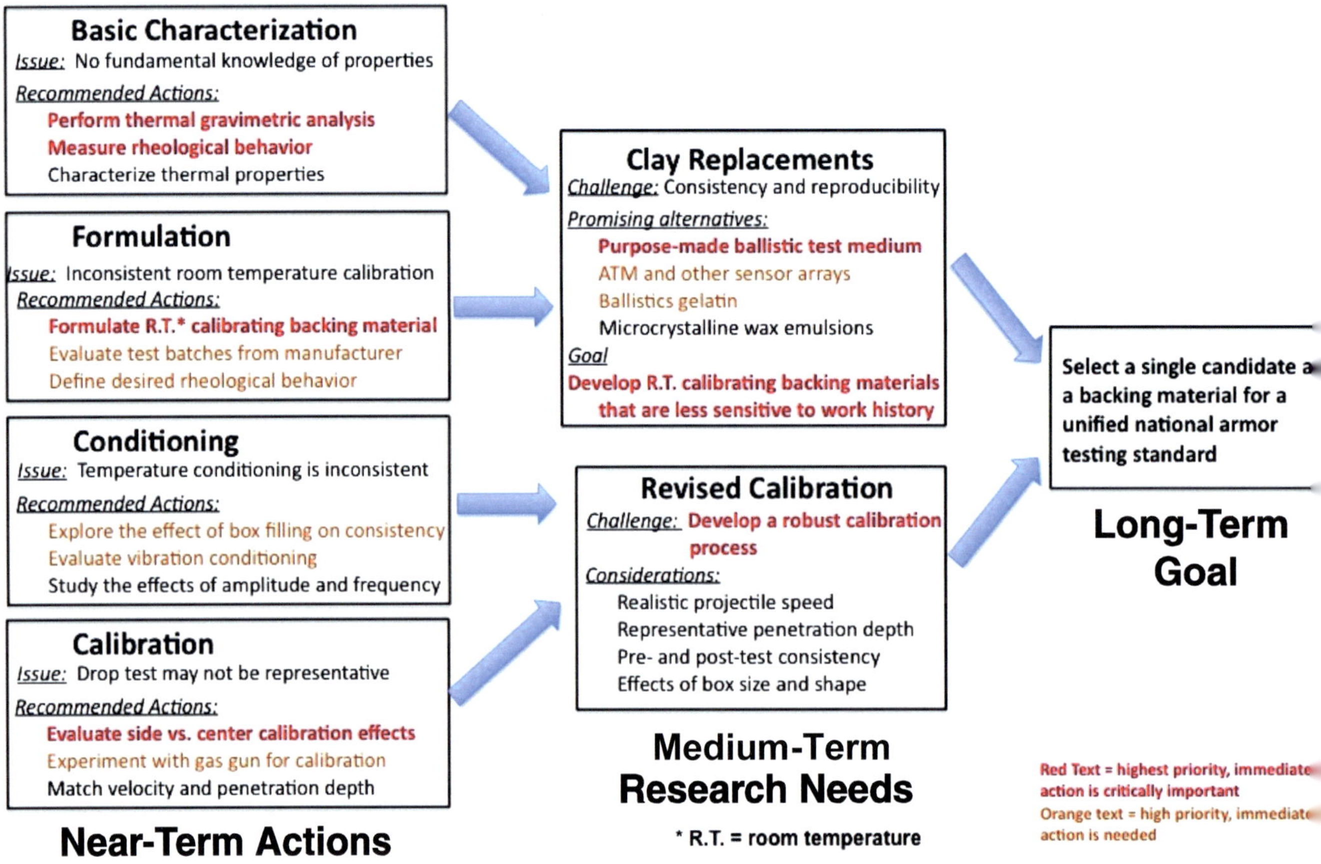

FIGURE 9-1 Road map showing suggested near-term actions, medium-term research needs, and a long-term goal to develop a more consistent backing material and a more reliable process for evaluating hard armor. The color coding shows "highest priority" items in red text with "high priority" actions in orange.

To overcome the weaknesses of the Prather methodology, Chapter 5 provides findings and recommendations on the instrumentation to measure indents in the clay recording medium. This understanding should lead to refinements to the current body armor testing process and provide a platform for evaluating new body armor designs to defeat future threats while minimizing the ergonomic penalties of vest bulk and weight.

Chapter 8 described how related medical studies of blunt force trauma should be performed to better understand the interactions between the armor and the back face, including the amplitude and speed of back-face displacement, so that body armor can be perfected that minimizes physical injury. The road map of the Chapter 8 findings and recommendations is recapitulated in Figure 9-2.

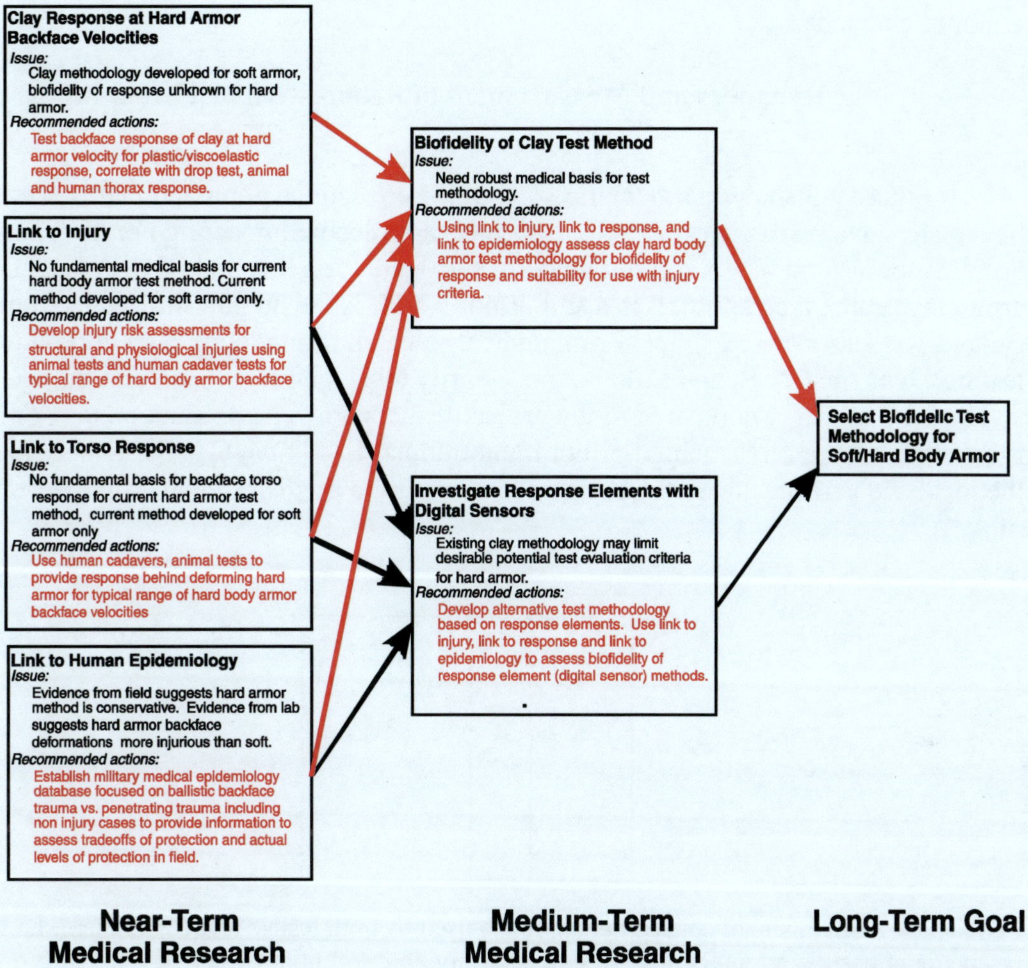

FIGURE 9-2 Flow chart showing suggested near-term and medium-term research needs, and a long term goal to provide the fundamental medical basis for injury risk assessment behind helmets and hard body armor.

LINKING MEDICAL RESEARCH DATA TO PRODUCT TESTING CRITERIA

Medical researchers and production testers have adopted the original Prather approach as a conceptual basis for medical experimentation and for body armor testing. However, each community has evolved different processes to create and measure behind-armor phenomena.

Dynamics and Measurement of Behind-Armor Forces

Figure 9-3 shows a schematic view of the conceptual approach that is used by both testers and researchers. The figure depicts a projectile impacting normally onto the body armor in front of a recording medium surrogate for a human body. A is a hard armor (typically, a ceramic), B is a soft armor, and C is the human surrogate (animal, cadaver, or sensor-based simulator in medical research or modeling clay in production testing). Note that A, B, and C do not necessarily touch and that the lateral dimensions of A, B, and C are large compared to the projectile diameter. The production tester's BFD or the medical researcher's behind-armor blunt trauma (BABT) in C is the final outcome of the localized damage due to the ballistic load imparted by the projectile on the front face of A.

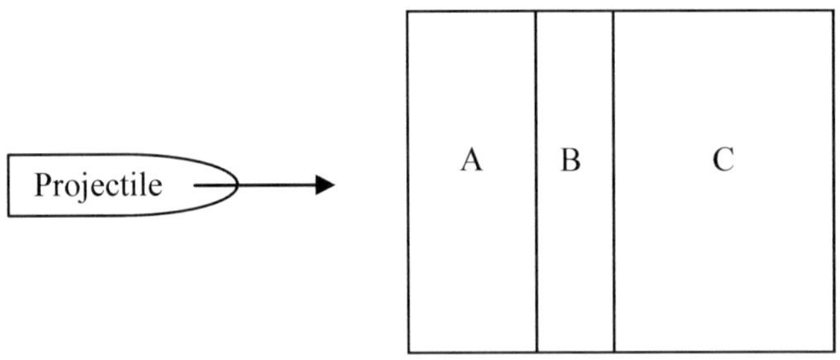

FIGURE 9-3 Schematic of conceptual approach used by both testers and researchers showing a projectile impacting normally onto hard body armor (A), soft body armor (B), and a recording medium surrogate for a human body (C).

The motion or deformation of the B/C interface can be considered in two parts:

- An early-time motion, and subsequent dynamic deformation, due to the stress wave that originates from the point of impact at front of the hard armor and propagates to the B/C interface, followed by various wave reflections from different interfaces and boundaries.
- A late-time, or final, BFD due to the transmission of the projectile itself through the hard armor (A) and the soft armor (B), but without perforation through the soft armor.

The following simple exercise can be used to determine the characteristic times associated with each of these two parts.

$$\text{Wave transit time} = T_W = \frac{\text{Thickness of A}}{\text{Acoustic speed in A}} + \frac{\text{Thickness of B}}{\text{Acoustic speed in B}}$$

$$\text{Projectile transit time} \approx T_P \approx \frac{\text{Thickness of A} + \text{Thickness of B}}{\text{Half the projectile velocity}}$$

In the latter expression, half the projectile velocity is used because the projectile comes to rest within the armor. As such, half the projectile velocity is a good estimate of the average projectile velocity.

During the time between T_W (microseconds) and T_P (milliseconds), the interface B/C and the human surrogate C are subjected to dynamic forces and deformations that depend on several factors: projectile velocity, area of impact, and thicknesses of A and B.

Establishing a correlation between BABT and BFD is crucial and requires that two key issues be addressed. First, the impact conditions (projectile mass and velocity) in the medical research and the production testing efforts need to be identical and to reflect the actual threat faced on the battlefield. Second, the motion and stress at the B/C interface, which correspond to BABT and BFD, need to be measured and correlated as a function of time (microseconds to milliseconds). Without addressing both these issues, it will be difficult to make a meaningful comparison between test results in medical research and production testing.

The practical issue facing the body armor community is that since the medical-research and production-testing stakeholders use different projectiles and different recording media (C) as human surrogates, their base data for BABT and BFD are not easily comparable. Some ideas on aligning the two different processes will be discussed below.

Aligning Recording Media Data

The committee appreciates that medical researchers and production testers have different goals. Medical researchers are trying to develop very specific insights in how behind-armor forces cause trauma to a specific organ, groups of organs, or other localized portions of the human body. Higher cost recording media such as electronic sensors or organ surrogates are the norm in this type of research. Production testers are concerned with holistic approaches that can test many plates or helmets in a short period of time. Since the Prather study, production testers have used inexpensive modeling clay as a recording medium. Ideally, both the medical and production testing community would use the same object C. That may be possible in the long run, as described in Chapter 4, with the development of better alternatives to modeling clay, including inexpensive, disposable sensors. However, the realities of the different levels of detail needed by these two stakeholder communities as well as the practical aspects of cost and time make it likely that at least for the short to medium term the different recording media currently used will persist.

As a result of these practical differences, the recording medium C that stands in for a human being (Figure 9-3) will be different for medical research (a large animal, cadaver, or sensor-based simulator) and for production testing (a clay box). Therefore, the dynamic motion and deformation at the B/C interface discovered by the two efforts will be different. Better correlation between the two can probably be best accomplished by monitoring in real time (with the requisite time resolution) the motion and deformation at the B/C interface in the two cases, using yet-to-be-realized identical sensors.

One possible near-term way to obtain valuable information on this BFD using the current clay methodology was recommended in the Phase II study report (NRC, 2010, p. 24) as follows: "To better understand and measure the forces that create the backface deformation, the Army should experiment with inserting microscopic temperature and displacement sensors into the clay near the site of the backface deformation." One possible idea is a set of microscopic sensors embedded in a frangible wire grid system on thin paper, as shown in Figure 9-4. A sensor grid system such as shown in the figure could be placed immediately behind armor at the B/C interface and immediately ahead of medical recording sensors for medical research. A similar grid could be placed immediately behind armor and immediately ahead of modeling clay during production testing.

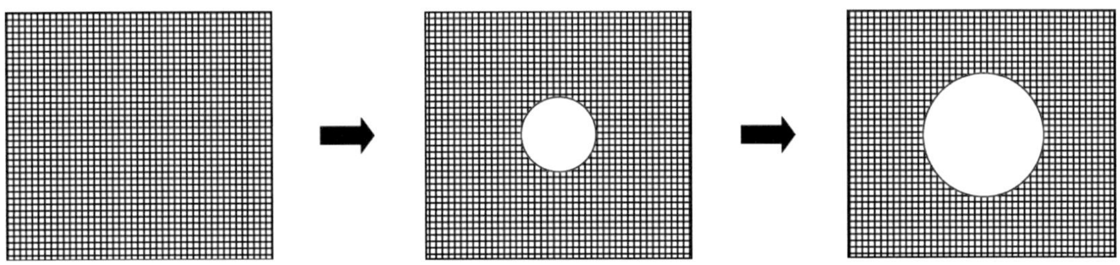

FIGURE 9-4 Schematic of the dynamic measurement method.

As behind-armor forces contact the wire grid, regardless of its makeup, the wires sequentially break and the dynamic contact area can be measured using resistive, conductive, or capacitive methods at high rate. In principle, the data can be recorded at 1 MHz or greater. An approach such as this could allow common near-term measurement of behind-armor forces and could compensate largely for the variability inherent to the recording media. It would also allow medical researchers and production testers to document that they are in fact creating comparable behind-armor displacements—even if the latter are using rifled firearms and the former are using smooth-bore or rifled gas guns to launch projectiles.

Considering the inherent variability of the clay, the application of a thin layer of sensors is unlikely to have a significant impact on BFD formation. Assuming that experiments with sensors on clay using medical recording media confirms this to be the case, a practical application of measurements derived by this method could be a better correlation between BABT injury data developed in the medical community and BFD criteria used by production testers.

The author of the Prather study stated that the original BFD criterion of 43 mm was very conservative for typical lower-rate deformations behind soft body armor.[66] That is, it would almost certainly require a dynamic force significantly greater than that which produces a 43-mm BFD to cause serious injury to a human. Prather and colleagues (1977) did not address higher-rate deformations behind hard body armor.

It is possible that if medical researchers used dynamic measurements as described above, they might be able to better determine the balance between increasing levels of BABT and the risk of human injury. The behind-armor impact that produces an acceptable risk of human injury could be replicated in clay by the production testers. Such research could lead to the adoption of a less conservative BFD criterion for production testing. A less conservative BFD could, in turn, allow the technology community to develop armor for soldiers that provides adequate survivability but is lighter, enhancing soldier mobility.

Finding: Recording medium data from medical research and production testing need to be correlated using identical sensors having the requisite time resolution. The results need to be shared among the stakeholders.

This could be a significant practical improvement that results from better alignment of stakeholder communities, including users, technologists, medical researchers, and production testers.

Recommendation 9-1: The Director of Operational Testing and Evaluation should take the lead in aligning the production testing, medical research, and body armor/helmet technology development communities so that the data outputs from their various processes can be easily correlated. This will lead to a better understanding of the relationships among body armor testing performance, human/animal survivability, and

[66]Russell Prather, Survice Engineering Company, "The Lightweight Body Armor Program - A History," presentation to the committee, August, 10, 2010.

other trade-offs. Specifically, two policies should be adopted and applied: (1) specify acceptable ranges for projectile weights and velocities used to generate behind-armor dynamic forces during testing and research and (2) investigate the use of standardized sensors behind armor to measure the amount of dynamic force that is produced during testing and research.

The recommendation means relating data from comparable projectiles and recording media used in production testing and medical research as well as relating the use of recording media such as animals, cadavers, or sensor-based simulators to the use of modeling clay for production testing. The data should be shared among stakeholders to promote a more detailed understanding of the relationship between body armor performance, human survivability, and other trade-offs. Importantly, implementation of this recommendation will provide data- based evidence for adopting credible BFD pass/fail testing criteria for body armor and helmets. As stated in the Phase II report (NRC, 2010), the 1977 Prather study with its BFD criteria has to its credit resulted in the fielding of successful—but almost certainly heavier than necessary—body armor. The current pass/fail testing criteria, based on the early Prather work, should be retained until Recommendation 9-1 has been completed and reviewed by all stakeholders.

SYNCHRONIZING THE STAKEHOLDERS

The actions contained in Figures 9-1 and 9-2 require the coordinated activities of the entire body armor community. Stakeholders include not only the at-risk warfighters and law enforcers but also the organizations and individuals involved in fabrication design, fabrication technology, materials testing, production quality assurance, performance criteria development, and performance verification, as well as those involved in linking medical damage thresholds to body armor performance. The stakeholders and some of their interactions are shown in Figure 9-5.

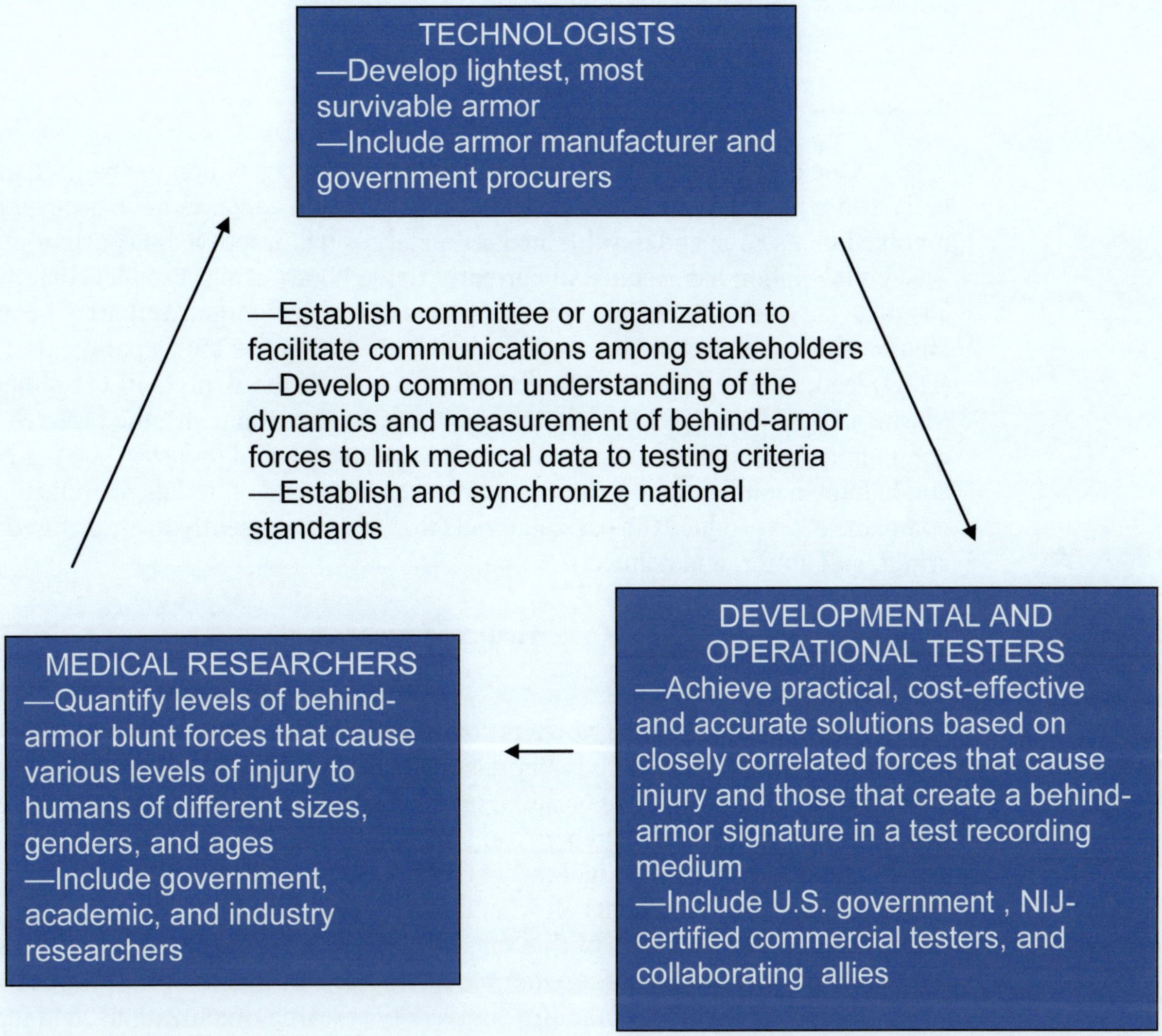

FIGURE 9-5 Schematic of stakeholder relationships. Testing and fielding of the most effective armor requires close communications among the stakeholders.

Military and Law Enforcement Personnel

The principal stakeholders are the warfighters and civil servants at risk for gunshot wounding who need protection with limited ergonomic or other penalties. The stakeholders also include military and civilian medical personnel as well those who test the adequacy of fielded vests against evolving threats.

Technologists

One of the reasons body armor has been so successful on the battlefield is that body armor manufacturers and government acquisition agencies have been actively involved in research and development of materials that improve body armor performance. This collaboration has resulted in currently fielded body armor products that provide adequate survivability against specific threats at relatively light weights. The principal members of this stakeholder group, for which the Army has the Department of Defense (DoD) lead, are the acquisition community (to include the Army's PEO Soldier and the Marine Corps' Program Manager Infantry Combat Environment), the research community (including the Army Research Laboratory), and industry, such as body armor and helmet manufacturers, defense munition experts, and materials manufacturers (the companies that produce the base materials that are subsequently manufactured into hard armor, soft armor, and helmets).

Medical Researchers

Medical and other researchers continue to push the boundaries of quantifying and correlating BABT forces with injury data collected from experiments with animals, cadavers, and simulators. Importantly, medical research conducted by government, academic, and industry researchers can focus on the different levels of injury that the same behind-armor forces might inflict on humans of different sizes, genders, and ages. These data can lead to insights on what injury BABT force is likely to cause on various humans. Important members of this stakeholder group include medical research organizations such as the U.S. Army Medical Research and Materiel Command and the Armed Forces Institute of Pathology; university research organizations; and industry contract researchers.

Testers—Developmental and Operational

The production testers take the body armor items that have been provided from industry or researchers and test them following prescribed processes and standards to ensure they are effective and suitable for military use. The products are tested to determine if they meet basic specifications and to learn how they function while worn by the soldier in a simulated or actual battlefield environment. Production testing is based on practical, cost-effective, and accurate processes. These processes need to ensure that the armor being tested will prevent specified threat rounds from penetrating it and keep the measured BFD below the level that could cause serious injury for the soldier while achieving the lowest practical weight burden. The organizations that have been doing this type of testing include DoD organizations such as the Office of the Director, Operational Test and Evaluation (DOT&E) and the Aberdeen Test Center; the National

Institute of Justice (NIJ) and the National Institute of Standards and Technology, which have developed the national standards for testing procedures; NIJ-certified commercial testing facilities; and the military organizations of U.S allies that have adopted U.S. standards for testing their own armor.

Aligning Stakeholders

As stated above, each of the four stakeholder groups has worked hard to improve its part of the body armor community. However, the committee feels that faster advances could be made if there was better coordination and communications among the stakeholders.

The Phase II report (NRC, 2010) recommended that the ad hoc clay working group be empowered and adequately resourced to gather information, influence research, and develop working-level consensus across body armor testing organizations. The report recommended that, after the clay working group had reached a reasonable consensus, DOT&E and NIJ should convene a nationally recognized group to review all appropriate considerations and develop recommendations that could lead to a single national body armor testing standard to achieve more uniform testing results.

The committee observed that the clay working group has made reasonable progress in developing working-level consensus but that not all stakeholders are participating. A nationally recognized coordination committee, as recommended in Phase II, would facilitate the alignment of activities among the stakeholders.

Members of such a military-industry-academia committee would be drawn from the stakeholder groups and would facilitate the passing of information to all stakeholders. It could also assist in rationalizing priorities for research and allocating responsibilities among the more senior organizations. It could provide an efficient means for informal feedback on policy and procedures that are related to two or more of the stakeholders. It could also oversee the development and modification of national standards that could unite various stakeholder groups.

The proposed military-industry-academia coordinating committee could, as appropriate, host occasional conferences to bring interested parties together to present papers and otherwise share ideas on topics of interest to multiple stakeholders. Subcommittees could be set up by the coordinating committee to communicate among the production testing community, NIJ-certified commercial testers, manufacturers, acquisition experts, and others to quickly provide feedback on proposed changes to testing procedures. This could ensure that all testers, government and commercial, as well as product manufacturers and end users are able to provide feedback that will minimize unintended consequences from proposed changes and ensure uniformity in the procedures once adopted.

The coordinating committee could also be used to coordinate biomedical and biomechanical research or to organize conferences from time to time, assuring more of a systems engineering framework for body armor technology and standards development. Members of the coordinating committee would be cognizant of relevant international efforts—in particular, the North Atlantic Treaty Organization-organized working groups. The overall need is for the coordinating committee to provide oversight and facilitate the exchange of information between stakeholder groups.

Finding: The nationally recognized coordination committee recommended in the Phase II report is needed to align and accelerate efforts of technologists, production testers, and biomedical researchers in behind-armor blunt trauma/ backface deformation - related research for both body armor and helmets.

Establishing National Standards

The Phase II report (NRC, 2010) provided insights into some of the problems arising as a consequence of not having uniform standards. For example, the NIJ, with assistance from the National Institute of Standards and Technology, has developed requirements for the conditioning and validation of the clay box used in body armor testing. Those requirements are described in the NIJ body armor standard. However, in many cases the requirements are general and not prescriptive in that they do not define the methods, tools, materials or details involved in building clay blocks, ensuring that the blocks are uniform, conditioning the clay blocks, and measuring deformations resulting from the validation drop tests or the tests of the armor.[67]

The NIJ standard is used by DoD testing organizations to guide the development of their procedures for clay handling and deformation measurement. These procedures are used in the testing program to determine if various types of body armor are adequate for military applications. Other non-DoD and private testing laboratories also use NIJ standards to guide their procedures to test the body armor used by police forces and other organizations throughout the country. These standards have also been adopted by the military forces of some other countries to guide their body armor testing.

Over time, the NIJ standard has undergone multiple revisions, and depending on the circumstances, different versions of the standard have been adopted by various testing organizations at different times in their histories. As a result, it is possible at this time that identical body armor plates tested by different organizations could achieve dissimilar and not easily comparable results. In the extreme case, a plate could be deemed acceptable at one testing facility and unacceptable at another. It would therefore be a considerable improvement to have one standard or only a few, so that identical plates would be likely to achieve the same test results regardless of the test facility.

The NIJ standards offer a broad set of procedures that will evolve over time owing to changes in technology and other considerations. The Aberdeen Test Center has made decisions on specific procedures and refinements to be used in the testing of body

---

[67]NIJ requirements were originally developed to support independent civilian law enforcement organizations. Those organizations wanted maximum flexibility to allow decentralized procurement for various types of body armor on the market. The NIJ philosophy was to permit the various testing laboratories flexibility in technical approaches as long as the standard was met. Civilian law enforcement has long since adopted the NIJ approach. The military subsequently adopted the NIJ standards for body armor testing but felt that the large quantities of body armor that were being purchased in a centralized manner could benefit from more prescriptive requirements. Ideally, there is a middle-ground solution where the NIJ standard(s) could (1) meet the needs of both civilian law enforcement and the military and (2) facilitate reproducibility of results at different laboratories by addressing more detail than was originally envisioned.

armor plates in an effort to standardize both commercial and government production testing.

Eventual adoption of fewer standards—or even a single national standard —would require detailed analysis of key issues such as the threats that are being protected against and the rationale for differences in testing processes. Ideally, developing consensus across all the organizations involved in body armor testing will be more effective than simply mandating national standards.

As an important step in this process, the committee agrees that the ad hoc clay working group approach that was started by and is currently chaired by DOT&E can serve as organizational nucleus and a way ahead for DoD. The working group began as an assembly of individuals with expertise in clay properties, clay calibration, clay working techniques, and future efforts in body armor testing. Two efforts in particular could help lead to a single national body armor testing standard:

- Collaborating on and investigating clay properties, formulation, calibration, and working techniques.
- Collaborating on alternatives to the existing test procedures and standards.

Reducing the number of national standards for body armor testing requires examining issues other than just the recording medium. An encompassing standard for testing would include the following:

- Rationalization of instruments and procedures to achieve consistency in measuring the indents in backing materials.
- Application of statistics and other mathematical tools to improve standardization, such as determining test sample size.
- Alignment of body armor and helmet testing procedures.

Finding: The original ad hoc clay working group could be expanded to form Department of Defense's portion of the national body armor testing standardization committee that was recommended in the Phase II report.

The mission for an expanded working group would include not only the group's original tasks but also the areas touched on above. Membership would consist of experts from each stakeholder group. Examples of the tasks to be performed by the standardization committee include these:

- Gather and document information that defines and explains the reasons for the different testing procedures used by various organizations.
- Determine areas where alignment of processes among organizations makes sense.
- Determine areas where different missions, customer requirements, resources, and other organizational considerations provide a reasonable rationale for different testing procedures to be retained, at least in the short term.
- Oversee additional analysis that is required to make recommendations on procedure and process changes. (It would be useful here to design

experiments, gather data, and perform analyses that would lead to informed recommendations to the chains of command of the participating organizations.)
- Achieve consensus from all stakeholders that will, ideally, lead to the drafting of a single national testing standard or at least fewer such standards.

Once the testing standardization committee achieves a consensus, it should take actions to gain ratification of the appropriate national standards for the testing of body armor and helmets for both military and police forces. After ratification, the committee could conduct a periodic review to determine if existing standards need to be updated.

Recommendation 9-2: The Director of Operational Test & Evaluation and the National Institute of Justice (NIJ), in collaboration with the military services, unified commands, government testing organizations, NIJ-certified testing laboratories, medical researchers and governmental and commercial material developers should convene a national body armor testing standard committee to review all appropriate considerations and develop recommendations that could lead to updated national body armor configurations and testing standards for body armor and helmet testing.

This final recommendation is conceptually the same as Recommendation 15 in the Phase II report (NRC, 2010). However, it has been expanded to include helmet testing. Since helmets and body armor plates have different requirements, there will likely be different testing standards for them for the foreseeable future.

REFERENCES

NRC (National Research Council). 2010. Phase II Report on Review of the Testing of Body Armor Materials for Use by the U.S. Army. Washington, D.C.: National Academies Press.

Prather, R., C. Swann and C. Hawkins. 1977. Backface Signatures of Soft Body Armors and the Associated Trauma Effects. ARCSL-TR-77055. Aberdeen Proving Ground, MD: U.S. Army Armament Research and Development Command Technology Center.

Appendixes

Appendix A

Biographical Sketches of Committee Members

MG (ret.) Larry G. Lehowicz is the manager of the Experimentation, Test and Training Sector Group at Quantum Research International. Prior to that, he was the corporate vice president for business development, engineering, logistics, and strategic solutions at Science Applications International Corporation. Previously, he was vice president of Quantum Research International. He retired from the U.S. Army as a major general and commander of the U.S. Army Operational Test and Evaluation Command, an organization dedicated to ensuring that warfighting systems, information management systems, and other military equipment are prepared for combat use. Gen. Lehowicz served as deputy chief of staff for combat development at the Army Training and Doctrine Command, and he was assistant division commander of the Tenth Mountain Division. He has a B.S. in geology from Kent State University and an M.B.A. from Syracuse University. He is also a graduate of the U.S. Army War College. General Lehowicz was the chair of the National Research Council's Committee on Assessment of Test Infrastructure Requirements to Support Testing of Defense Directed Energy Systems. He served previously as the vice-chair of the Committee on Army Unmanned Ground Vehicle Technology and was a member of the Committee on Alternative Technologies for Anti-Personnel Landmines.

Cameron R. Bass is director of the Injury Biomechanics Laboratory in the Biomedical Engineering Department at Duke University. He is a recognized expert in blast and ballistic injury risk modeling with over 15 years' experience in biomechanics. This includes substantial experience developing biomechanical injury models of blast, ballistic, and blunt trauma. Following postdoctoral experience (on an NSF fellowship) developing injury biomechanics models for blunt impact at the University of Virginia, Dr. Bass established a military and high-rate biomechanics program at the University of Virginia Center for Applied Biomechanics, which he ran from 1995 to 2008. One initial focus of the program was cranial, thoracic and spinal injuries from behind-armor blunt trauma (BABT), which led to the development of a BABT head injury assessment methodology being used at the U.S. Army Research Laboratory to evaluate ballistic protective helmets and other biomechanically based injury risk functions. In recent years Dr. Bass's program has focused on the assessment of brain and thoracic trauma from primary blast and high-rate blunt trauma. Dr. Bass has developed animal and human cadaver models for assessing blast injuries, including the first large animal model, which demonstrated diffuse injury to axons from short-duration blasts that do not cause fatality from pulmonary trauma. Dr. Bass has over 50 peer-reviewed publications in the areas of blast and blunt injury biomechanics and tissue biomechanics. He was awarded a Ph.D. in 1994 from the University of Virginia.

Thomas F. Budinger, NAE and IOM, is professor in the graduate division of the University of California, Berkeley; senior medical scientist at the Lawrence Berkeley National Laboratory (LBNL); and professor emeritus at the University of California, Berkeley, and San Francisco Medical Center. Dr. Budinger has authored numerous papers on specific research topics including biomedical electronics, aging and cardiovascular physiology, bioastronautics, image processing and reconstruction, nuclear magnetic resonance, positron emission tomography, reconstruction tomography, and inverse problem mathematics. He is coauthor of the text Ethics *of Emerging Technologies: Scientific Facts and Moral Challenges.* He received the Gold Medal from the American Roentgen Ray Society in 2009 and the Hal Anger Memorial Lectureship from the Society of Nuclear Medicine in 2010. Dr. Budinger graduated magna cum laude in chemistry from Regis College and received an M.S. in physical oceanography from the University of Washington. He subsequently received an M.D. from the University of Colorado and a Ph.D. in medical physics from the University of California, Berkeley.

Morton M. Denn is Albert Einstein Professor of Science and Engineering and director of the Levich Institute at the City College of New York. He is past professor and chair of chemical engineering at the University of California, Berkeley, and head of materials chemistry at the LBNL. He served as editor of the *Journal of Rheology* and received the Bingham Medal from the Society of Rheology in 1986 and the Founders Award from the American Institute of Chemical Engineers in 2008. He was elected to the NAE in 1986. Dr. Denn received a B.S.E. from Princeton University and a Ph.D. from the University of Minnesota, both in chemical engineering. His expertise is relevant to this study in polymer rheology, including process dynamics of materials.

William G. Fahrenholtz is a professor of ceramic engineering at Missouri University of Science and Technology at Rolla. Before that, he was assistant professor of ceramic engineering and a research investigator in the Graduate Center for Materials Research, University of Missouri-Rolla. Dr. Fahrenholtz also worked as a research assistant professor of chemical engineering at the University of New Mexico, where he researched ceramic-metal reactions and composite formation by reactive hot pressing and reactive metal penetration, examined processing methodologies, characterized microstructures, studied reaction sequences, and evaluated mechanical properties. His current research interests include processing and characterization of ceramics, ultra-high-temperature ceramics, reaction-based processing of ceramics and ceramic-metal composites, cerium oxide coatings for corrosion protection of aluminum, and thermodynamics. Dr. Fahrenholtz received his B.S. and M.S. in ceramic engineering from the University of Illinois at Urbana-Champaign and a Ph.D. in chemical engineering from the University of New Mexico.

Ronald D. Fricker, Jr., is an associate professor at the Naval Postgraduate School. He's current research is focused on the performance of various statistical methods for use in biosurveillance, particularly epidemiologic surveillance, and statistical process control methodologies more generally. His recent research includes developing new spatiotemporal algorithms for biosurveillance, useful for both early event detection and situational awareness, and methods for optimizing the performance of biosurveillance systems. His other recent research includes assessing the effects of individual augmentation deployment on naval personnel retention, researching federal support to state and local organizations for domestic terrorism preparedness, and investigating the use of pesticides by U.S. forces during the Gulf War. Dr. Fricker holds a Ph.D. and an M.S. in statistics from Yale University, an M.S. in operations research from The George Washington University, and a bachelor's degree from the U.S. Naval Academy. Upon graduation from the Academy, he served as a surface warfare officer in the U.S. Navy. He has published in *Statistics in Medicine*, the *Journal of the Royal Statistical Society*, *Environmental and Ecological Statistics, the Journal of Quality Technology, Naval Research Logistics, Teaching Statistics,* and *CHANCE*. Professor Fricker is on the editorial boards of Statistics*, Politics and Policy* and the *International Journal of Quality Technology and Engineering.* He has served as the chair of the section on Statistics in Defense and National Security (SDNS) of the American Statistical Association (ASA). Prior to the creation of SDNS, he was a member of the Committee on Statisticians in Defense and National Security, serving as both the chair and vice-chair. He has also served as membership chair for the Quality and Productivity Section and as publicity chair for its Section on the Physical and Engineering Sciences.

Yogendra M. Gupta is currently a Regents professor in the Department of Physics and the Director of the Institute for Shock Physics at Washington State University. He has been studying condensed matter response to shock wave and high pressure loading since 1970, with particular emphasis on examination and understanding of microscopic processes, and has supervised the work of over 90 graduate students and research associates since joining the Washington State University in 1981. He received his B.Sc. (physics., math, and chemistry.) 1966 and an M.Sc. (physics) in 1968 from the Birla Institute of Technology and Science, Pilani, India, and a Ph.D. (physics) in 1972 from Washington State University, Pullman, Washington. He is the author of over 250 publications, has made over 300 invited and contributed presentations; and holds two patents. These include the invited articles "Shock Waves" in the *Encyclopedia of Physics* (Van Nostrand Reinhold) and "Shock Waves in Condensed Materials" in the *Encyclopedia of Science and Technology* (McGraw-Hill). Dr. Gupta has been a member of the External Review Committee for the Physics Capability Review of the Atomic Weapons Establishment of the United Kingdom; the Technology Area

Review and Assessment Panel for Applied Research and Technology Development for the Department of Defense; the American Physical Society Panel on Public Affairs; and the University of California (UC) President's Science and Technology Panel for Oversight of the UC-managed DOE national laboratories, at present he is a member of, the External Review Committee for the Pulsed Power Sciences Center, Sandia National Laboratories, and is the chairman of the Subcritical Experiments Evaluation Committee, DOE/NNSA to name a few.

Dennis K. Killinger is the Distinguished University Professor and professor of physics at the University of South Florida and is an expert in laser and optical remote sensing/lidar, applied laser spectroscopy, laser physics, and free space optical laser communication. He received a B.A. from the University of Iowa, an M.A. from De Pauw University, and a Ph.D. in physics from the University of Michigan. He conducted research on radar analysis and microwave atmospheric propagation while employed as a research physicist at the Naval Avionics Facility and was on the research staff in quantum electronics at Lincoln Laboratory, Massachusetts Institute of Technology, conducting research in the development of new solid-state lasers and their application as spectroscopic LIDAR probes of the atmosphere. In 1987, he joined the Physics faculty at the University of South Florida and is director of the Laboratory for Atmospheric LIDAR and Laser Communication Studies and past technical director of the Technology Deployment Center, working on technology transfer for the university and regional industries. Dr. Killinger is a fellow of the Optical Society of America, a senior member of the IEEE, past associate editor of *Applied Optics* and *Optics Letters*, past member of the NAS/NRC Committee on Optical Science and Engineering, and has served as chairman of several international conferences on lasers and applied spectroscopy. He has published over 200 technical papers, reports, and conference papers, and five books or book chapters.

Vladimir B. Markov is president of Advanced Systems & Technologies, Inc., and past vice president and director of applied optics at MetroLaser, Inc., in Irvine, California. He specializes in the research, design, and development of devices and systems in the areas of laser physics and real-time holography, multibeam interaction, and holographic sensors. Throughout his career, Dr. Markov's activity has been strongly associated with development of the fundamental properties of three dimensional holograms, real-time holography, optical image processing, and holographic nondestructive testing. He was actively involved and participated in development of such areas as nonlinear optical holography, including optical wave front conjugation, lasers with controlled parameters, especially with phase conjugation mirrors, multibeam interaction, and holographic sensors. In the area of holographic interferometry, Dr. Markov developed the technique for studying the vibrational characteristics of large-scale pressurized plastic pipes, a method that for the first time allowed detecting, studying and developing the technology to arrest fast-running cracks propagating in this type of pipe. At the same time, using a more conventional approach, he developed an opto-electronic holographic nondestructive system for defect detection in various items, including components used in the electronic industry and museum objects. Some of his recent activities include the development of a matrix (16 × 16 beams) laser Doppler velocimeter system capable in detection of defects in airframe components (for the U.S. Navy and the Air Force); a novel long-range active laser tracking system for space situation awareness; a miniature laser system for crack precursor detection; a multibeam laser vibrometer; and a novel method for wave function sensing. Dr. Markov has more than 100 papers published in refereed journals, two books, and patents. He is a fellow of SPIE, a member of OSA, and is on the editorial boards of *Optics and Laser Technology* (U.K.) and the *Journal of Holography and Speckle.*

James D. McGuffin-Cawley is chair of the Department of Materials Science and Engineering at Case Western Reserve University. He received a B.S from Alfred University in 1978 and a Ph.D. from Case Western Reserve University in 1984. After 2 years at what is now NASA Glenn Research Center, he began his academic career in the College of Engineering at Ohio State University, where he spent 6 years. Dr. McGuffin-Cawley returned to Case in 1991 as the Great Lakes Associate Professor of Ceramic Processing and became a full professor in 1996. He was named Arthur S. Holden Professor of Engineering in January 2007. Dr. McGuffin-Cawley's research has included work on mass transport and corrosion of ceramics, especially in aerospace applications. Most recently, he has concentrated on advanced processing strategies for producing components from ceramics.

Russell N. Prather is an engineering analyst at Survice Engineering Company. He retired from the Army Research Laboratory, where his career focus was on personnel vulnerability, body armor research, and wound ballistics. He has extensive knowledge about wound ballistics, bioresponse to trauma, mechanical engineering, applied mathematics and statistics, electronics engineering, physics, anatomy, photography, and explosives. Mr. Prather's experience includes planning and conducting experiments to study the effectiveness of protective materials; identifying damage caused by transient deformation of protective materials and developing counteractive measures; assessing the effectiveness of personnel protection devices in response to new threats; and evaluating the ability of new weapons systems and ammunition to incapacitate. Mr. Prather served on the Joint Interservice Body Armor Committee, 1969-1972; Personnel Armor System Working Group, 1971-1975; International Association of Chiefs of Police, 1978-1986; and Joint Working Group for Protective Eyewear, 1986-1992; he represented the Army Research Laboratory, Expert Group on Ballistic Test Methods, in support of NATO Working Group 5, 1994-2000. He was awarded a B.S. in engineering-physics from Loyola College in, Baltimore in 1968 and in 1981 was awarded a master's degree in engineering administration from the George Washington University.

Sheldon M. Wiederhorn (NAE) is a senior fellow emeritus at the National Institute of Standards and Technology (NIST) and continues to carry out a research program on the mechanical properties of ceramic materials. His current interests are to use the atomic force microscope to investigate the atomistics of crack growth in glasses and ceramic materials with the objective of learning more about the crack growth process and its relation to the microstructure of glass. At the National Bureau of Standards, now NIST Dr. Wiederhorn carried out a program on the mechanical reliability of brittle materials. He was one of the first to apply fracture mechanics techniques to study the fracture of ceramic materials. A consequence of his research was the development of techniques to assure the structural reliability of brittle ceramic materials. Techniques pioneered by Dr. Wiederhorn and his colleagues are now used to assure the reliability of glass windows in airplanes, space vehicles, and related applications. Dr. Wiederhorn is best known for the experiments that he developed to characterize subcritical crack growth in glasses. The results of these studies illustrate the complexity of subcritical crack growth, which consisted of stress-enhanced chemical reactions between water and stressed bonds at the tips of small cracks in glass. A natural conclusion of his study was that the failure of glass was caused by the slow growth of cracks to a critical size, which determined the time-to-failure. Dr. Wiederhorn received a B.S. in chemical engineering from Columbia University in 1956 and a M.S. (1958) and Ph.D. (1960) in Chemical Engineering from the University of Illinois. He has received many awards for his research and leadership at the NIST. These include both a Silver (1969) and a Gold Medal (1982) from the Department of Commerce and the Samuel Wesley Stratton Award, (1977) from the National Bureau of Standards. He is also a fellow of the American Ceramic Society (1970) and has received a number of important awards for his research from that society, including the Jeppson Award (1994) for outstanding research on ceramic materials. He is now a distinguished lifetime member of the American Ceramic Society (1998). In 1991, Dr. Wiederhorn was elected a member of the National Academy of Engineering.

Alyson Gabbard Wilson is statistics research associate at the Institute for Defense Analyses. She is past associate professor in the Department of Statistics at Iowa State University and Scientist (Level 5) in the Statistical Sciences Group at Los Alamos National Laboratory. Dr. Wilson received a Ph.D. in statistics from Duke University, and a M.S. in statistics from Carnegie-Mellon University, and a B.A. in mathematical sciences from Rice University. She is a fellow of the American Statistical Association and a recognized expert in statistical reliability, Bayesian methods, and the application of statistics to problems in defense and national security. Prior to joining Iowa State in 2008, Dr. Wilson was a project leader and technical lead for Department of Defense programs in the Statistical Sciences Group at Los Alamos National Laboratory (1999-2008). In this role, she developed and led a $3 million portfolio of work in the application of statistics to the reliability of conventional and nuclear weapons. Before she moved to Los

Alamos, Dr. Wilson was a senior statistician and operations research analyst with Cowboy Programming Resources (1995-1999), where she planned, executed, and analyzed U. S. Army air defense artillery operational evaluations. From 1990 to 1991, she was a mathematical statistician at the National Institutes of Health. Dr. Wilson has served on numerous national panels, including the National Academy of Sciences (NAS) Oversight Committee for the Workshop on Industrial Methods for the Effective Test and Development of Defense Systems (2008-2010), the Sandia National Laboratories' Predictive Engineering Science Panel (2008-2013), the NAS Panel on Methodological Improvement to the Department of Homeland Security's Biological Agent Risk Analysis (2006-2008), and the NAS Panel on the Operational Test Design and Evaluation of the Interim Armored Vehicle (2002-2003). She was on the organizing committee for the Department of Energy Office of Science Workshop on Mathematical Issues for Petascale Data Sets (2008), and an invited participant in the Chief of Naval Operations Distinguished Fellows Workshop on Critical Infrastructure Vulnerability (2008), the DOE/OS Workshop on Mathematical Research Challenges in Optimization of Complex Systems (2006), and the DOE Simulation and Modeling for Advanced Nuclear Energy Systems Workshop (2006). In 2006, Dr. Wilson chaired the American Statistical Association President's Task Force on Statistics in Defense and National Security.

Appendix B

Committee Meetings

This appendix lists the presentations to the committee at its meetings, fact-finding sessions, and a site visit during the course of the Phase III study.

FIRST COMMITTEE MEETING, AUGUST 9-11, 2010 ABERDEEN, MARYLAND

ATC Update on Clay Actions to Date
Shane Esola, Aberdeen Test Center (ATC)

Working, Ageing, and Temperature Effects on Roma Plastina #1
William Perciballi, Armor Works

Development of a Standard Ballistics Testing Clay
Isaac Peng, Chavant

Pragmatics of Body Armor Testing—Manufacturer Views
Dave Reed, Ceradyne

Commercial Body Armor Testing Perspectives
Donn Dunn, H.P. White Laboratory

Future Handling and Processing of Clay
Christian Action, Action International

Phenomenology and Material Response to High Velocity Impact
Yogendra Gupta, Phase III Committee Member

Pluses and Minuses of Clay and Future Alternatives
James Zheng, PEO Soldier

Gelatin as Future Testing Alternative
Robert Kinsler, Army Research Laboratory (ARL)

ATC Road Map on Phase II Recommendations
MAJ William Lash, ATC

U.S. Army ATC Welcome and Command Overview
COL Jeffrey Holt, U.S. Army, Commanding Officer, ATC

Helmet Testing Procedures and Demonstration
MAJ William Lash, ATC

Statistics Issues Related to Helmet Testing
Linda Moss, ARL

NIST Perspectives on Helmet Testing and Standards Development
Kirk Rice, National Institute of Standards and Technology (NIST)

Helmet Testing Perspectives–USMC PM-ICE
Lt. Col. Kevin Reilly, U.S. Marine Corps, PM-Infantry Combat Environment

ARL/SOCOM Alternative Head Forms for Helmet Testing
Robert Kinsler, ARL
Dixie Hisley, ARL

Biokinetic Head Form and Traumatic Brain Injury
James Zheng, PEO Soldier

Commercial Helmet Testing Perspectives
Jim Martin, Chesapeake Labs

Prather Study Findings
Russell Prather, Phase III committee member

Experimental Study of Behind-Armor or Blunt Thoracic Trauma in High-Rate Loading Conditions and Follow-on Studies
Dale Bass, Phase III committee member

BABTA and Other Recent Approaches to Blunt Trauma Measurement
Michael Leggieri, U.S. Army Medical Research and Materiel Center

PEO Perspective on Blunt Trauma Research
James Zheng, PEO Soldier

Analysis of Blunt Trauma Casualties
Edward Mazuchowski, Armed Forces Institute of Pathology

Perspectives on Electronic Sensors as a Future Alternative to Clay
Adam Fournier, ATC
Andrew Merkle, Johns Hopkins University

SITE VISIT, AUGUST 30-31, 2010
ABERDEEN PROVING GROUND, MARYLAND

Committee members in the instrumentation working group visited facilities of H.P. White Laboratory, Chesapeake Testing Laboratory, ATC, and ARL.

DATA-GATHERING SESSION, OCTOBER 12, 2010
ARLINGTON, VIRGINIA

Committee members on the statistics working group met with statisticians from the Army Evaluation Center and the representatives from the Office of the Program Manager- Soldier Protective Equipment, U.S. Special Operations Command, DOD Inspector General, and the Office of the Director, Operational Test and Evaluation (DOT&E) to discuss issues related to the use of statistical protocols.

SECOND COMMITTEE MEETING, OCTOBER 13-15, 2010
WASHINGTON, D.C.

Summary of Instrumentation Working Group Site Visits
Larry Lehowicz, committee chair

Summary of Statistics Working Group Pentagon Discussions
Larry Lehowicz, committee chair
Alyson Wilson, committee member
Ronald Fricker, Jr., committee member

Update on Clay Experimentation and Specification Activities
Shane Esola, ATC

Differences in BFD Measurement Standards
Richard Sayre, DOT&E

Improvements in Helmet Measurement
Robert Kinzler, ARL

DATA-GATHERING SESSION, OCTOBER 18-19, 2010
KECK CENTER, WASHINGTON, D.C.

Committee members on the methodology working group conducted a data-gathering session at the Keck Center. Questions on prospective post-Prather testing methodologies were posed via telephone to Dixie Hisley, ARL; Andrew Merkle, Johns Hopkins University; Stephen Vatner, New Jersey Medical School; and Weixin Shen, L-3/TRACOR.

Appendix C

Additional Phase III Tasks

The statement of task for the study (Box 1-1) listed tasks for each of the three phases of the study. Included as the last task for Phase III of the study was a request for the committee to document in its final report "other issues regarding body armor testing that the committee found relevant."

In response, this final report addresses the following additional tasks not specifically set forth in the task statement:

1. Provides a roadmap for reducing the variability of clay processes and for migrating from clay to future solutions.

2. Considers the use of statistics to permit a more scientific determination of sample sizes to be used in body armor testing. Specifically, the committee was requested to review a statistically based protocol that had been developed by Office of the Director, Operational Test and Evaluation with assistance from Army statisticians and testers.

3. Develops ideas for revising/replacing the Prather study methodology.

4. Within the time and funding available, reviews and comments on methodologies and technical approaches to military helmet testing.

5. Considers the possibility of combining various national body armor testing standards.

Appendix D

Report Sections Cross-Referenced to the Statement of Task

Original Statement of Task for Phases I, II, and III (See Box 1-1)	Phase III Report Chapter or Appendix
The National Research Council will convene specialists in committee form to consider the technical issues relating to the testing of body armor. To do this the National Research Council shall conduct a three-phase study:	
In **Phase I** a committee will comment on the validity of using laser-profilometry/laser-interferometry techniques to determine the contours of an indent made by a ballistic test in a nontransparent clay material at the level of precision established in the Army's procedures for testing personal body armor. If laser-profilometry/laser-interferometry is not a valid method, the committee will consider whether a digital caliper can be used instead to collect valid data.	Phase I letter report was submitted on December 30, 2009. Findings are listed in Appendix K. Chapter 5
The committee will also provide interim observations regarding the column drop performance test described by the Army for assessing the part-to-part consistency of a clay body used in testing body armor.	Appendix K Chapter 4
The committee will prepare a letter report documenting the findings from its Phase I considerations. This is a 6-week effort beginning November 1, 2009, and ending mid December 2009.	Phase I letter report was submitted on December 30, 2009. Findings are in Appendix K.

In **Phase II** a committee will consider in greater detail the validity of using the column drop performance test described by the Army for assessing the part-to-part consistency of a clay body within the level of precision that is identified by the Army test procedures.	Phase II letter report was submitted on April 22, 2010. Recommendations are in Appendix L.
The committee will prepare a letter report documenting the findings from its Phase II considerations. This is a 6-week effort beginning November 1, 2009, and ending early February, 2010.	Phase II letter report was submitted on April 22, 2010. Recommendations are in Appendix L.
In **Phase III** a committee will consider test materials, protocols, and standards that should be used for future testing of personal armor by the Army.	
The committee will also consider any other issues associated with body armor testing that the committee considers relevant, including issues raised in the Government Accountability Office report *Warfighter Support, Independent Expert Assessment of Body Armor Test Results and Procedures Needed Before Fielding* (GAO-10-119).	Throughout the report and summarized in Appendix F.
The committee will prepare a final report. This is a 14-month effort beginning November 1, 2009, and ending January 2011.	This report constitutes the Phase III final report.
The **final report** will document the committee's findings pertaining to the following issues that are of particular immediate concern to DOT&E.	
The best methods for obtaining consistency of the clay, and for conditioning and calibrating the clay backing used currently to test armor.	Chapter 4
The best instrumentation (e.g., laser scanning system, digital caliper, etc.) and procedures to use to measure the backface deformation (BFD) in the clay.	Chapter 5
The appropriate use of statistical techniques (e.g., rounding numbers, choosing sample sizes, or test designs) in gathering the data.	Chapter 6 Appendixes H, I, and M

The appropriate criteria to apply to determine whether body armor plates can provide needed protection to soldiers; this includes the proper prescription for determining whether a test results in a partial or complete penetration of body armor, including, as appropriate, the soft armor underlying hard armor.	Chapter 2 Appendix F
The **final report** will also document the committee's findings regarding any other issues regarding body armor testing that the committee found relevant. The study team will have access to all data with respect to body armor testing that the team needs for the conduct of the study.	Appendix C. See additional taskings below.

Additional Phase III Taskings Received From DOT&E (See Appendix C)	**Phase III Report Chapter or Appendix**
Provide a roadmap to reduce variability of clay processes and for how to migrate from clay to future solutions.	Chapters 4 and 9
Consider the use of statistics to permit a more scientific determination of sample sizes to be used in body armor testing. Specifically, the committee was requested to review a statistically based protocol that had been developed by DOT&E with assistance from Army statisticians and testers.	Chapter 6
Within the time and funding available, review and comment on methodologies and technical approaches to military helmet testing.	Chapter 7
Develop ideas for revising/replacing the Prather study methodology.	Chapters 8 and 9
Consider the possibility of combining various national body armor testing standards.	Chapter 9

Appendix E

Ballistic Body Armor Insert Composition and Defeat Mechanisms

Ceramics are used in personal armor systems to defeat small-caliber threats such as pistols, rifles, and machine guns. Ceramics are relatively light compared to more traditional armor made of metallic alloys. Properties that contribute to the excellent performance of ceramic armor include high hardness, low density, high elastic constants, and high compressive strength. But as stand-alone items, ceramics would not be particularly good because of their low tensile strength and sensitivity to small mechanical defects such as pores and cracks. Hence, they are used in combination with other materials such as polymers and metals as laminar composites, which enhances their excellent protection properties.

A hard body armor plate typically includes a layer of dense boron carbide or silicon carbide backed by a layer of metal, polymer, or a composite. The purpose of this combination is to convert the projectile's kinetic energy into plastic "work" and stored elastic energy and to broaden the area of contact of between the plate and the body during an impact event. The laminar ceramic/polymer composite is encased in tightly woven canvas. Sometimes additional layers of canvas or other materials are enclosed within the wrapping, depending on the particular manufacturer.

This composite armor package defeats the incoming missile by several mechanisms. On initial impact, the missile is held up by the ceramic surface, which behaves as an elastic barrier. The time the missile is held up is known as the dwell time—the longer the dwell time, the more effective the protective system. During the dwell time, the bullet flattens, flows plastically, and erodes from its tip. At the same time a compressive elastic wave and a shear wave are generated at the point of impact; they propagate radially, reflecting from the back surface and propagating back into the material. The compressive wave converts into a tensile wave upon reflection and acts as an initiator of many cracks within the ceramic. The magnitude of the wave passing into the backup plate depends on the elastic impedance mismatch of the ceramic and the backup material: The closer the match, the less reflection there is of the elastic waves.

A plastic zone develops beneath the contact site. The high pressure under the tip of the bullet and the constraint of the surrounding ceramic material tend to suppress macroscopic fracture and permit plastic deformation to occur. Plastic processes in the ceramic include microcrack formation, amorphization, phase transformation, and twinning and dislocation generation. These are very general mechanisms of plastic deformation that occur in most crystalline ceramic materials. Plasticity is enhanced at the contact point by further constraining the material by inducing compressive stress over the entire outside layer of the ceramic. This can be done by tighter wrapping of the cloth that encases the laminar composite. In addition to a plastic zone, a cone crack develops at the point of contact and propagates into the solid and eventually completely through the ceramic plate. The highly fractured zone of ceramic material, which generates from the rear surface of the ceramic plate, forms primarily within the boundary of the cone crack. This cone crack plays an important role in transferring momentum from the bullet to the backup plate. The cone is filled with plastically deformed and crushed ceramic material before the impact event is complete.

For the bullet to penetrate into the ceramic plate, ceramic material that is under the bullet flows around the bullet and sprays into the air on the impact side of the plate. As the ceramic material flows away from the front of the bullet, it breaks into small particles of ceramic (10-100 μm). These particles erode the bullet as the crushed ceramic flows past the bullet and sprays into the air on the impact side of the armor. In the most favorable scenario, the bullet is completely eroded away and—if within the design parameters of the insert—eliminated as a fatal threat to the person wearing the vest.

Finally, and for purposes of this committee, the incoming momentum of the bullet has to be transferred to the target. This is first done by momentum transfer to the cone of crushed and deformed ceramic. The force is picked up by the backup plate, which catches the moving ceramic cone. As the base of the cone is very much larger than the apex (1 mm vs. 25 mm radius), the pressure at the base is about 1,000 times less at the base than at the apex. The backup plate then deforms, further absorbing the impact force of the bullet. The final transfer of momentum is to the person wearing the protective vest. This absorption of force ends up in blunt trauma injury, sometimes severe enough to topple the person but not to kill him. It should be noted that the momentum transfer of a bullet is only a hundredth that of severe head contact in American football.

Appendix F

Committee Responses to the Government Accountability Office Report

A primary motivation for this study was the Government Accountability Office (GAO) report GAO-10-119 (GAO, 2009). In the report, the GAO recommended that "the Army should provide for an independent ballistics evaluation of the First Article Testing results," that "the Army should assess the need to change its procedures based on the outcome of the independent experts' review and document these and all other key decisions made to clarify or change the testing protocols," and that "the Army provide for an independent external peer review of Aberdeen Test Center's body armor testing protocol, facilities and instrumentation" (GAO, 2009, p. ii).

The committee has addressed questions raised by the GAO in its report, and its responses to seven specific items that were recommended for evaluation and action (GAO, 2009, pp. 37-38.) are summarized here. The committee took its charge to conduct an independent assessment very seriously.

The GAO report was thoughtful and pointed out a number of important issues. There was give and take between GAO, DoD, and the Army. On some issues, DoD and the Army agreed with GAO. On others, there was disagreement. The committee has listened to arguments on many sides of this debate. Its findings and recommendations are based on discussions not only with government experts but also with body armor manufacturers and commercial testers certified by the National Institute of Justice (NIJ).

The committee was precluded from using classified data. As a result, specific first article testing data based on real-world threats were not available to it. However, it spent many hours reading background information and being briefed by experts throughout the body armor community and believes that its findings and recommendations inform most of the important issues raised in the GAO report.

The committee's overarching analysis of the report is that it was generally on target. Over time the Director, Operational Test and Evaluation (DOT&E) and the Army have adopted a number of the GAO recommendations. The committee has studied the report closely, discussed the issues with experts from throughout the body armor community, and agreed with several recommendations; it has taken a number of the issues to additional detailed levels with its findings and recommendations.

SUMMARY OF RESPONSES TO EVALUATION ITEMS AND ACTION ITEMS RECOMMENDED FOR INDEPENDENT ASSESSMENT

Evaluation Item 1: The rounding of back-face deformation measurements

The Phase I report (NRC, 2009, p. 17) suggested that the DOT&E and the Army adopt a common standard for rounding and indicating the appropriate number of significant digits. The specific observation was that the Army should consider using the American Society of Testing and Materials (ASTM) standard ASTM E29-08, Standard Practice for Using Significant Digits in Test Data to Determine Conformance with Specifications (ASTM, 2008). During briefings of the Phase I report, both DOT&E and the Army said they agreed with the recommendation.

Evaluation Item 2: Not scoring penetrations of material through the plate as a compete penetration unless broken fibers are observed in the Kevlar backing behind the plate

The Phase I committee was briefed by the COL Jeffrey Holt, Commander of the Aberdeen Test Center (ATC). He said that the command wanted a definition for a "complete penetration" that minimized subjectivity on part of the tester and manufacturer. As a result, ATC adopted a convention that any test round that penetrated the armor plate and continued on to completely break any Kevlar fiber on the back of the shoot pack would be considered a complete penetration. The committee felt this provided a consistent description for the body armor community and virtually eliminated subjectivity when scoring a plate as having been completely penetrated.

As defined in the new DOT&E protocol, "a complete penetration of the test plate sample occurs on any fair record test shot impact in which the projectile, any fragment of the projectile, or any fragment of the armor material is ejected from the rear of the plate and passes into the first ply (minimum of one complete yarn broken) of the soft armor (ballistic package) located behind the test plate sample when it is placed into the soft armor test panel. The first ply of the soft armor (ballistic package) shall serve as a witness plate." (DOT&E, 2010, p. 4)

Evaluation Item 3: The use of the laser scanner to measure back-face deformations without a full evaluation of its accuracy as it was actually used during testing, to include the use of software modifications and operation under actual test conditions

The Phase I committee was provided a copy of ATC Internal Operating Procedure No. 001: Measurement of Backface Deformation (BFD) Using Faro® Quantum Laser Scan Arm and Geomagic® Qualify® for Hard and Soft Body

Armor, and found that it adequately describes the appropriate and consistent use of the laser scanning system (Huber, 2009; NRC, 2009).

Additionally, the Phase I report stated that the "laser scanning system (including the testing protocols, facilities and instrumentation) as currently implemented by the Army (or for similar equipment) if used in accordance with the Army's procedures, is a valid approach for determining the contours of an indent in a nontransparent clay material at a level of precision adequate for the Army's current ballistic testing of body armor." (NRC, 2009, p. 13)

During Phase III the committee looked for improvements to procedures that could result in more consistent measurements made with laser scanners. There are several findings in the Phase III report that if implemented by DOT&E and the Army, could improve consistency in laser scanning. For example, there is software that allows for smoothing the raw digital data captured by the Faro. The software has settings that allow for various levels of resolution. The committee was shown that just by switching from one level of resolution to another resulted in a 1 mm difference in measurement for the same backface deformation (BFD). Manufacturers can be burdened, in this example, by a 1 mm penalty simply because an operator has selected a particular setting. The committee finds that a control study should be conducted to determine the most reasonable and consistent Faro smoothing settings to be used while measuring BFDs in body armor testing. Similarly, any software selections or system changes that could cause a relevant change to BFD measurements should be studied. Participation in this study should include, at a minimum, ATC, NIJ-certified private testing labs, and body armor manufacturers.

Another finding in the Phase III report that relates to this issue is that there needs to be a single measurement standard to determine the ability of any system to measure a representative BFD regardless of different labs, measurement instruments, and operators. Specifically, a standard model or a BFD artifact (as initially recommended by the National Institute of Standards and Technology) should be developed and used as part of this analysis. Previous work in this area by ATC and by the National Institute of Standards and Technology had established the importance of using an artifact to mimic the measurement process.

The committee believes that because the BFD measurement process is a two-step process (measurement of the preshot surface and measurement of the postshot BFD), both of these steps should be made with a single national standard artifact. It is suggested that an artifact be made that mimics the preshot surface with a flap that covers a multiple-BFD imprinted plate, or a flap on a helmet surface that covers a model BFD crater. Such a model could be made of hard plastic, or, a softer coating could be applied. While the thickness of the flap would affect the absolute readings, the relative readings between labs and operators would not be affected. This artifact would be used to confirm that any change (e.g., hardware, software, or operator) still resulted in a consistent measurement.

Evaluation Item 4: The exposure of the clay backing material to rain and other outside environmental conditions as well as the effect of high oven temperatures during storage and conditioning

ATC demonstrated to the committee that the ovens used to condition the clay prior to testing had been moved inside the testing range building to preclude exposure of clay to rain and other outside weather.

The issue of high oven temperatures is more complicated. Back in the late 1970s, Roma Plastilina #1, a modeling clay developed for artists, calibrated at room temperature when used for testing applications. Subsequently, the composition of the clay changed over time to meet artists' needs. As the composition slowly changed, experimentation has shown the only way that the calibration standard (25 mm ± 3 mm) could be achieved was by heating the clay. As a rule of thumb, experienced test operators at NIJ-certified commercial testing laboratories, who have decades of experience with clay, reported to the committee that the clay has had to be heated about an additional 1°F each year to achieve calibration standards.

Currently, there is an informal experimental effort at both ATC and the commercial labs to attempt to restrict the temperature of the clay boxes to about 105°F.[68] (Interestingly, for the Phase I body armor testing demonstrations at ATC in December 2009, the temperature goal was approximately 104°F.) However, if the clay formulation continues to change to meet artist expectations, it is likely that the temperature of the clay will have to be increased or decreased to allow the clay to properly calibrate.

Accordingly, the committee found that DOT&E and the Army should urgently develop a standard formulation for ballistics backing material that allows calibration at room temperature, as was the case when body armor testing using clay began. Additionally, as a medium-term goal, the committee found that DOT&E and the Army should research an alternative to the current Roma Plastalina #1 that not only calibrates at room temperature but also minimizes the effect of human working (thixotropy).

By standardizing a ballistics backing material formulation that minimizes the variation caused by temperature and thixotropy the committee feels that the body armor community will have eliminated a significant amount of the variation in the body armor testing process. Historically (and incorrectly) all variation in the testing process has been assumed to be due to body armor plates inconsistencies. The practical result of a standard ballistics backing material formulation as just described could be the production of lighter body armor that still provides the same level of soldier protection.

[68]Confirmed during personal communication between Larry Lehowicz, Committee chair, and Travis Humiston, ATC, October 28, 2010.

Evaluation Item 5: The use of an additional series of clay calibration drops when the first series of clay calibration drops does not pass required specifications

ATC test operators demonstrated the clay calibration process during each phase of the study. Concerning the actions to repair and recondition clay boxes, which include using additional clay calibration drops when the first series of clay calibration drops did not pass specifications, the Phase I committee agreed that test operators were following "procedures that are consistent with standard practice by artists and others for filling space without entrapping air. That is, small additions are made sequentially and each is heavily sheared by hand to express any entrapped air. This procedure represents good practice." (NRC, 2009, p. 15)

Rheologists on the committee appreciate that there are many variables and unknowns in clay calibration that need to be investigated by DOT&E and the Army. The Phase II report (NRC, 2010) contains several findings and recommendations pertaining to calibration and the drop test. Some of the committee's calibration-related findings are these:

- Until a standard ballistics backing material that calibrates at room temperature is developed, the body armor testing community should conduct a posttest calibration drop to ensure that the clay has remained in calibration.
- Evaluate measurement differences when calibrations are made on the side of the box and when they are made at the center of the box.
- Conduct calibration experiments with a gas gun to better replicate the projectile velocity and penetration depth that create a BFD.
- Ensure pre- and postshot calibration consistency.
- Investigate the effects of box size and shape on calibration results.

Action Item 1: Determine whether those practices that deviated from established testing protocols during First Article Testing will be continued during future testing and change the established testing protocols to reflect those revised practices

The committee chair was told by senior Army staff members that the Army would appoint a knowledgeable person to conduct an independent assessment of the practices mentioned in the GAO report.

Action Item 2: Evaluate and re-certify the accuracy of the laser scanner to the correct standard with all software modifications incorporated and include in this analysis a side-by-side comparison of the laser measurements of actual back-face deformations with those taken by digital caliper to determine whether laser measurements can meet the standard of the testing protocols

The committee spent considerable time with ATC and NIJ-certified commercial testing labs to better understand the advantages and disadvantages of the digital caliper and the laser scanner. See Chapter 5 and Appendix M.

REFERENCES

ASTM (American Society for Testing and Materials). 2008. ASTM E29-08 Standard Practice for Using Significant Digits in Test Data to Determine Conformance with Specifications. West Conshohocken, Pa.: ASTM International.

DOT&E (Director, Operational Test and Evaluation). 2010. Standardization of Hard Body Armor Testing. Memorandum dated April 27, 2010. Arlington, Va.: Director, Operational Test and Evaluation.

GAO (United States Government Accountability Office). 2009. GAO-10-119. Independent Expert Assessment of Army Body Armor Test Results and Procedures Needed Before Fielding. Washington, D.C.: Government Accountability Office.

Huber, J. 2009. Internal Operating Procedure No. 001: Measurement of Backface Deformation [BFD] Using Faro® Quantum Laser Scan Arm and Geomagic® Qualify® for Hard and Soft Body Armor. Aberdeen Proving Ground, Md.: Aberdeen Test Center.

NRC (National Research Council). 2009. Phase I Report on Review of the Testing of Body Armor Materials for Use by the U.S. Army: Letter Report. Washington, D.C.: National Academies Press.

NRC. 2010. Phase II Report on Review of the Testing of Body Armor Materials for Use by the U.S. Army. Washington, D.C.: National Academies Press.

Appendix G

Determining the Necessary Level of Precision for Body Armor Testing

The new body armor testing protocol assesses body armor performance via two metrics, the probability of no penetration and backface deformation (BFD). Assessing the probability of no penetration is relatively straightforward in the sense that the outcome is binary (either a plate is penetrated or it is not), and the determination of that binary outcome is fairly clear-cut. Measuring and assessing BFD is less clear-cut because, as described in Chapter 4, the test methodology, particularly the use of Roma Plastilina #1 as a recording medium, introduces variability into the BFD measurement process. In this appendix the committee focuses on BFD measurement.

INTRODUCTION

As discussed in Chapter 7, there are two types of mistakes that are possible when interpreting the results of live-fire armor testing. The first is grievous—the inadvertent acceptance of armor from a manufacturer that should have been rejected. The second is the nonacceptance (failing) of armor from a manufacturer that should have been accepted. Both errors are theoretically possible in any type of statistically based test, though as will be discussed, for all practical purposes the existing armor testing protocol prevents the first type of mistake. This is true as long as the mean is accurate independent of the level of precision in the test.

For BFD, the protocol involves subjecting a set of 60 samples from a lot to live-fire testing. If on the first shot the 90 percent upper tolerance limit calculated with 90 percent confidence is in excess of 44 mm, the manufacturer fails the first article test (FAT). In the hypothetical case of armor with no variance and a test with no variance, the armor is accepted as long as the sample mean of the BFD calculated from the 60 samples is less than 44 mm. An increase in variance from any source (the armor itself, backing material behavior, or measurement) causes a spread in the data. If these effects are normally distributed, and if we assume that the distribution is perfectly estimated from the 60 shots (i.e., there is no sampling error), then a manufacturer will pass the FAT if the 90th percentile of the distribution is less than 44 mm and will fail if it is greater than 44 mm.

Under these conditions, if we assume that the variability of the armor, measured in terms of standard deviations, is 1 mm, then body armor designs with a true mean BFD (μ_{pop}) less than 42.7 mm will pass FAT. Furthermore, under these conditions a design that just passes FAT with μ_{pop} = 42.7 mm will cause 10 percent of the individual body armor plates to experience BFDs > 44 mm. In fact, 5 percent of the plates will have BFDs > 44.3 mm and 1 percent will have BFDs > 45 mm.

This effect can be illustrated as follows. First consider a hypothetical armor system that has negligibly small variance in true performance. If the backing material exhibits a variance characterized by a standard deviation $\sigma_{bck\ matl}$, then the true population of BFDs will be

$$f_{length'x'} = \frac{1}{\sigma_{pop}\sqrt{2\pi}} \exp\left(\frac{-(x-\mu_{pop})^2}{2\sigma_{pop}^2}\right) \qquad \text{(G.1)}$$

where μ_{pop} is the mean BFD for that particular armor system (which depends on the mechanical response of the armor and the elastic recovery in the backing material) and $\sigma_{pop} = \sigma_{bck\ matl}$. For μ_{pop} = 42.7 mm and with σ_{pop} = 1 mm, Figure G-1 shows the distribution of BFDs for the hypothetical population of body armor.

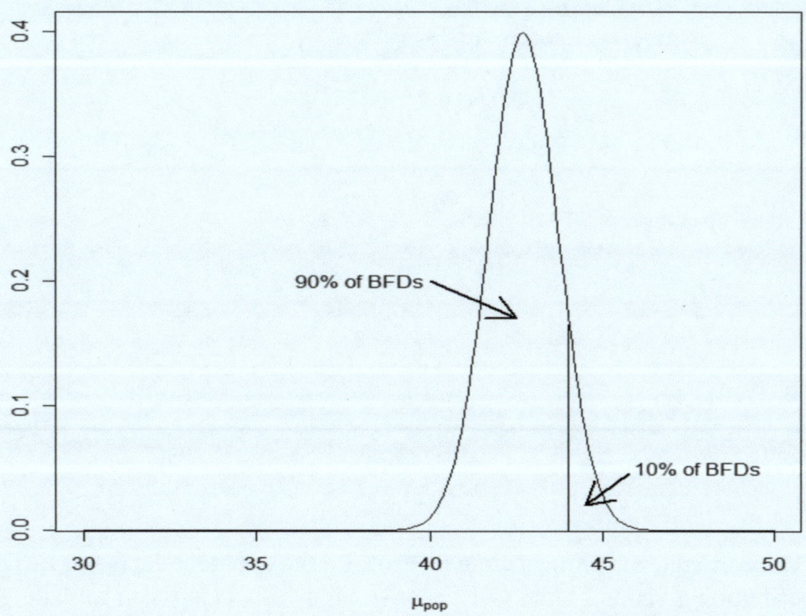

FIGURE G-1 Plot of normally distributed BFDs from a design that just meets the 90 percent upper tolerance limit requirement with μ_{pop} = 42.7 mm and σ_{pop} = 1 mm.

The impact of employing a measurement technique that adds significant variability is illustrated in Figure G-2. In this illustration, we assume the measurement technique adds 2 mm to the standard deviation, so that $\sigma_{pop} = 3$ mm. To pass the FAT, again assuming no sampling error, the manufacturer must now have a design in which $\mu_{pop} \leq 40.15$ mm. The dotted line curve represents a population of body armor that, with $\sigma_{pop} = 3$ mm, can just pass FAT with $\mu_{pop} = 40.15$ mm. Comparing the two curves, we see that to decrease the likelihood of failing FAT, the manufacturer must shift the entire distribution to the left so that the area under the curve falling to the right of the 44 mm is less than 10 percent. If the additional variance reflected in the curve is due to the performance of the backing material during testing rather than variance in the armor, then it is proper to refer to this as a "design penalty." In other words, armor that would appropriately prevent injury can be inadvertently rejected during testing due to variance associated with the test (i.e., variance in the backing material and, in addition, in the measurement methods).

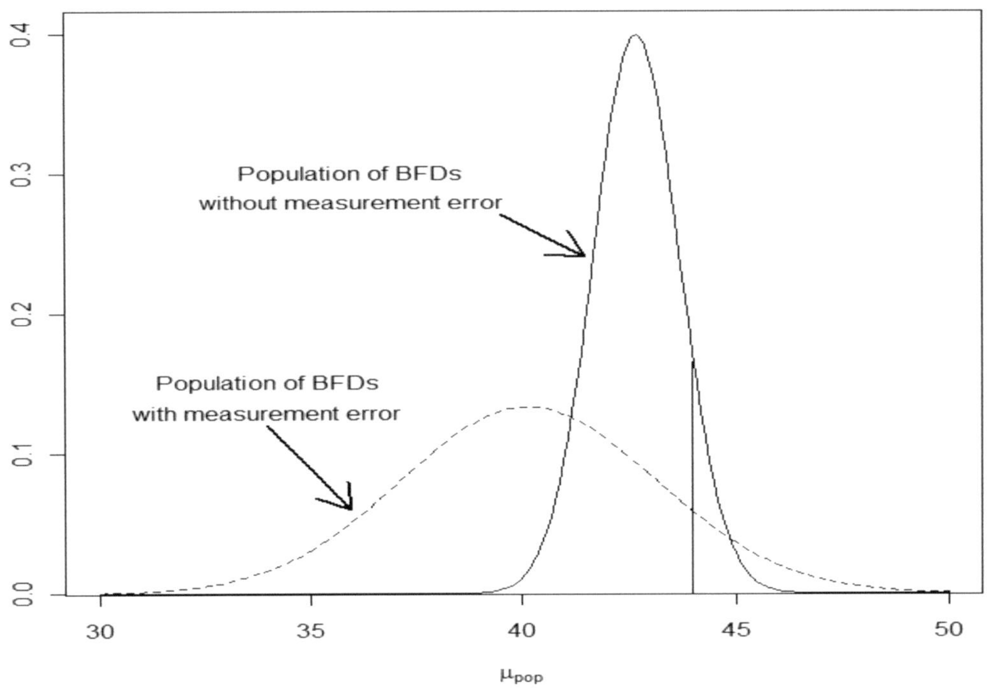

FIGURE G-2 The consequence of measurement error on the apparent depths of BFDs. The normal distribution given in Figure G-1 is convoluted with a Gaussian spreading function to represent measurement error. The result is another normal distribution with a larger standard deviation, given by the dotted line. Both populations are just able to pass the FAT (assuming no sampling error) as their 90th percentiles are just below 44 mm.

This means the test is more conservative, in the sense that, if anything, armor that might have been adequate is rejected. Measurement error will not increase the chance of armor with a given mean BFD, μ_{pop}, being accepted. Various people have referred to this as, for instance, erring on the safe side or taking the soldiers' perspective.

This inadvertent rejection of adequate armor, however, also is a problem. It decreases the availability of armor directly due to its rejection; concomitantly it increases cost and produces a degree of randomness in the testing that can diminish vendor confidence, because indistinguishable lots of armor will sometimes pass and sometimes fail.

The relative contribution of variance in the clay and finite precision in the measurement technique is therefore important and can be assessed quantitatively. In particular, it is possible to identify the relative importance of the precision of the measurement.

RELATIONSHIP BETWEEN BACKING MATERIAL VARIANCE AND MEASUREMENT PRECISION

The process of characterizing the cavities produced in the plastic backing material during live-fire ballistic testing of hard body armor is the same as measuring critical dimensions of manufactured articles. In all such problems there exists a trade-off between cost (capital equipment, operating costs, and throughput rate, and so on) and precision. Therefore the field of engineering economics has long employed rigorous methods for making decisions about the value of precision relative to its cost. Development efforts frequently have the goal of achieving high precision at low cost. In deployment the decision is one of selecting amongst available techniques to achieve maximum utility.

That is, with a few substitutions (in square brackets in following extract), an expression of the principles employed in specifying metrology tools in the context of manufacturing can be generalized to apply to measurement deformation results (Besfamil'naya, 1974, pp. 458-460):

> If a planned level [of dimensional reproducibility] is to be justified economically, it is not only necessary to select, or focus upon, the measurable variables suitable for evaluating [dimensions], but also to assign the appropriate precision, and consequently also the cost of measurements and inspection. That explains why one of the central aspects of . . . quality assurance . . . is optimized selection of monitoring and measuring equipment, as well as the characteristics set forth in technical specifications for new instruments and gages as to precision, reliability, cost, etc.

One advantage of employing a rigorous decision-making process is that it permits a meaningful choice to be made under current circumstance and creates a framework to identify how precision requirements must evolve as specifications and capabilities change.

Three ideas are developed below. The first is a general discussion of stacked errors—that is, how variability of backing material deformation and variability in the results of the measurement tools combine to give the general result. Secondly, the available data associated with the use of RP #1 is considered in the context of justifiable precision. Finally, the need for precision in the context of improving the backing material or shifting to an alternative technology is discussed.

As described in Chapter 3, the cavities produced in the backing medium during both low-velocity calibration drop tests and high-velocity tests of stingballs and soft armor are all characterized by substantial variance. Thus, the problem to be considered is how much the variance of the measurement technique adds to an already recognized noisy signal (i.e., the depth of the cavities in the backing medium).

The initial portion of the following analysis assumes a normal distribution for both the population of cavity depths under identical conditions and for the error associated with the measurement techniques (calipers, laser scanner, or the like). Subsequently, it is shown that the same result obtains without the assumption of normality.

If we have a population of lengths that follow a normal distribution, the frequency of lengths is given by

$$f_{length'x'}(x) = \frac{1}{\sigma_{pop}\sqrt{2\pi}} \exp\left(\frac{-\left(x - \mu_{pop}\right)^2}{2\sigma_{pop}^2} \right) \qquad \text{(G.2)}$$

where μ_{pop} is the mean length in the population and σ_{pop} is the standard deviation of the population. When integrated over a particular range, this function gives the probability that any given element in the population is in that range.

If the lengths are measured with a tool that yields an instrumental spreading function of the form

$$f_{instrument}(y) = \frac{1}{\sigma_{ins}\sqrt{2\pi}} \exp\left(\frac{-(y)^2}{2\sigma_{ins}^2} \right)$$

then the function for the apparent length z ($z = x + y$) that will be recorded is the convolution of these two functions. One of the properties of the Gaussian distribution function is that when convoluted with another Gaussian function the result also is a Gaussian. Assuming there is no systematic error in the measurement tool (i.e., that it is not biased) the mean remains the same and the standard deviations are added in quadrature:

$$f_{apparent\ length'z'}(z) = \frac{1}{\sqrt{\left(\sigma_{pop}^2 + \sigma_{ins}^2\right)}\sqrt{2\pi}} \exp\left(\frac{-\left(z - \mu_{pop}\right)^2}{2\left(\sigma_{pop}^2 + \sigma_{ins}^2\right)} \right)$$

This is equivalent to the result obtained when there are independent sources of error. Of course we can rewrite this as

$$f_{apparent\ length'z'}(z)=\frac{1}{\sigma_{pop}\sqrt{\left(1+\left(\frac{\sigma_{ins}}{\sigma_{pop}}\right)^2\right)}\sqrt{2\pi}}\exp\left(\frac{-(z-\mu_{pop})^2}{2\sigma_{pop}^2\left(1+\left(\frac{\sigma_{ins}}{\sigma_{pop}}\right)^2\right)}\right)$$

and defining $\sigma_{eff}=\sigma_{pop}\left(1+\left(\frac{\sigma_{ins}}{\sigma_{pop}}\right)^2\right)$ once again yields the simple Gaussian form back:

$$f_{apparent\ length'z'}(z)=\frac{1}{\sigma_{eff}\sqrt{2\pi}}\exp\left(\frac{-(z-\mu_{pop})^2}{2\sigma_{eff}^2}\right)$$

So, one way to assess the effect of high or low precision of the instrument on a signal that has noise is to compare the standard deviation of the "apparent length" distribution to that of the native population.

The fractional change is given by

$$\frac{\sigma_{eff}-\sigma_{pop}}{\sigma_{pop}}=\frac{\sigma_{pop}\sqrt{\left(1+\left(\frac{\sigma_{ins}}{\sigma_{pop}}\right)^2\right)}-\sigma_{pop}}{\sigma_{pop}}=\sqrt{\left(1+\left(\frac{\sigma_{ins}}{\sigma_{pop}}\right)^2\right)}-1 \quad \text{(G.3)}$$

These results can be illustrated graphically (Figure G-3). The purpose of this is to emphasize the contribution of the measurement error relative to variation in cavity formation, given by σ_{ins} and σ_{pop}, respectively. Therefore, the fractional increase in standard deviation, $\frac{\sigma_{eff}-\sigma_{pop}}{\sigma_{pop}}$, is plotted as a function of $\frac{\sigma_{pop}}{\sigma_{ins}}$, that is, the inverse of the ratio on the right-hand side of equation G.3. When this ratio is 1, the variance associated with the production of the cavity and that of the measurement technique are the same. If the standard deviation of the population (in our case determined by RP #1) is held constant, larger values for this ratio indicate progressively more precise measurement techniques. As Figure G-3 makes evident, there is little practical benefit to improving the measurement technique past the point of $\sigma_{ins}=0.1\times\sigma_{pop}$.

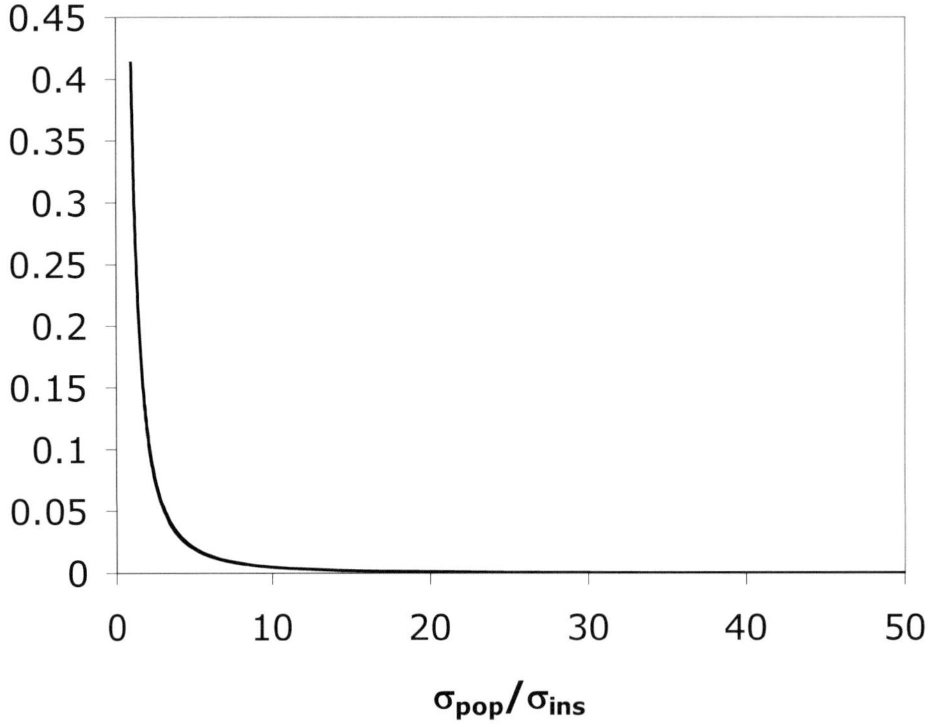

FIGURE G-3 The relationship between measurement error and the overall variance in armor testing. This figure can be helpful in guiding decisions about the useful precision of measurement devices. As long as the standard deviation is one order of magnitude smaller than that associated with armor performance and backing material combined, it is as precise as possible. Further increases in precision cannot be justified. Thus, when the system variance is high (e.g., 3 mm) even modest precision measurements (e.g., 0.3 mm) are as good as perfection. However, there is a portion of the curve where the fractional increase rises markedly. One important conclusion from this diagram is that replacing the existing backing material, RP #1, with a material having a significantly lower variance, say 1 mm, would require a corresponding increase in the measurement technique, down to 0.1 mm, to keep the relative impact of measurement error constant.

Specifically, if we assume $\sigma_{pop} = 3$ mm, all BFD measuring devices with standard deviations of less than 0.3 mm are equivalently precise from a practical perspective in terms of yielding useful data for decision making.

The above analysis assumes that the distributions of both depths and measurements are normal, largely for purposes of clarity. It is not necessary to make this assumption to reach this conclusion. However, this result is more general and holds whenever the length X is probabilistically independent of the measurement error Y. That is, equation G.3 follows from the fact that the variance of the sum of two independent random variables is equal to the sum of their variances (see Appendix M). Thus, the result described here is not dependent on the assumption of normality.

KEY POINT 1: The useful precision of a device used to measure BFD is fundamentally related to the variance in the signal due to the performance of the armor itself and the variance in the medium recording the BFD. Increased precision adds cost without adding benefit if it exceeds a well defined value, i.e., $\sigma_{pop}/\sigma_{ins} > 10$.

Based on information briefed to the committee, $3\text{ mm} \le \sigma_{pop} \le 4\text{ mm}$. Therefore there is no advantage for the current system to have a measurement tool with precision $\sigma_{ins} < 0.3$ to 0.4 mm.

Variance in the recording medium in particular, and from the entire test methodology in general, is parasitic and tends to cloud the data, preventing the variance in the armor from being manifest in the test. It also creates a condition that either leads to adequate armor being rejected, overdesign of the armor itself, or both.

As recommended in this report, the main path to improvement of armor testing is to replace RP #1 with a material(s) of lower variance. When this is done, it is crucial that the precision of the BFD be commensurate with that of the improved simulant materials. That is, when the simulant material indents are more reproducibly created, the precision of their measurement becomes more important.

ESTIMATION OF THE RELATIVE PRECISION OF MEASUREMENT TECHNIQUES

With the aggregate set of data provided to the committee, it is possible to make a quantitative assessment for the techniques employed in cavity measurement in RP #1 on the range. First, the data provided in Figure 4-7a in Chapter 4 indicate the standard deviation of drop test results is between 2 and 3 mm depending on the particular clay box.

For illustration, we can use the calculated measurement standard deviations on RP #1 from Walton (2008): 0.82 mm and 0.10 mm for the digital caliper and the Faro laser scanner, respectively.[69]

Thus, for the Faro arm the ratio of $\sigma_{ins}/\sigma_{pop}$ is 0.03, whereas for the digital caliper the same ratio is 0.27. In practical terms, the Faro results in an essentially perfect measurement that is, the variation in recorded measurements of cavity depth will be indistinguishable from the variation observed in the modeling clay.

The digital caliper yields a population of measurements that has a standard deviation that is 3.5 percent larger than that of the BFD population. Thus the variability of the BFDs is dominated by the inherent properties of the RP #1,

[69]Appendix M describes possible issues with the methods of Walton et al. (2008).

although this dominance does not mean the precision of the measuring tool is not without cost.

The significance of the limited precision of the caliper is dependent on the design strategy of the armor manufacturer. A worked example helps to illustrate this. Consider the DOT&E BFD protocol requirement: For the first shot test of 60 plates, the 90 percent upper tolerance limit must be less than 44 mm with a 90 percent confidence to pass FAT.

The one-sided upper confidence limit is

$$Y_U = \overline{Y} + k_1 S$$

where

$$\overline{Y} = \frac{1}{60}\sum_{i=1}^{60} Y_1 \text{ and } \overline{Y} = \sqrt{\frac{1}{59}\sum_{i=1}^{60}\left(Y_1 - \overline{Y}\right)^2}$$

The k_1 factor, assuming the data are normally distributed, is

$$k_1 = \frac{z_{1-p} + \sqrt{z_{1-p}^2 - ab}}{a}$$

where

$$a = 1 - \frac{z_{1-y}^2}{2(N-1)} \text{ and } b = z_{1-p}^2 - \frac{z_{1-y}^2}{N}.$$

For the case under consideration, $N = 60$ and $z_{1-p} = z_{1-\gamma} = z_{1-0.9} = 1.281552$, so $a = 0.986$ and $b = 1.615$. Thus, the upper tolerance limit is $Y_U = \overline{Y} + 1.53 \times S$, and a manufacturer will pass FAT if $Y_U < 44$ mm and fail otherwise.

The question, then, is What is the effect on the probability a manufacturer passes or fails FAT for $Y \sim N(\mu, 9)$, that is, $\sigma_{\text{eff}} = 3$ mm (perfect Faro measurement) versus $Y \sim N(\mu, 9.64)$, $\sigma_{\text{eff}} = 3 \times 1.035 = 3.105$ mm (caliper measurement)?

Because both the sample mean and sample standard deviation are stochastic, rather than derive an analytical expression to calculate the probability a manufacturer passes or fails, we simulate it. That is, we simulate sets of 60 random Y values from the appropriate distribution, from them calculate many Y_U values, and from those Y_U values estimate the probability of failing (by calculating the fraction of Y_U values greater than or equal to 44 mm) for various values of μ, $35 \leq \mu \leq 46$.

Figure G-4 shows the result, where we see the probability of failure for the hypothetical perfect measurement $\sigma = 3.00$ mm, compared to that calculated taking into account the measurement error of the digital caliper $\sigma = 3.11$ mm.

N

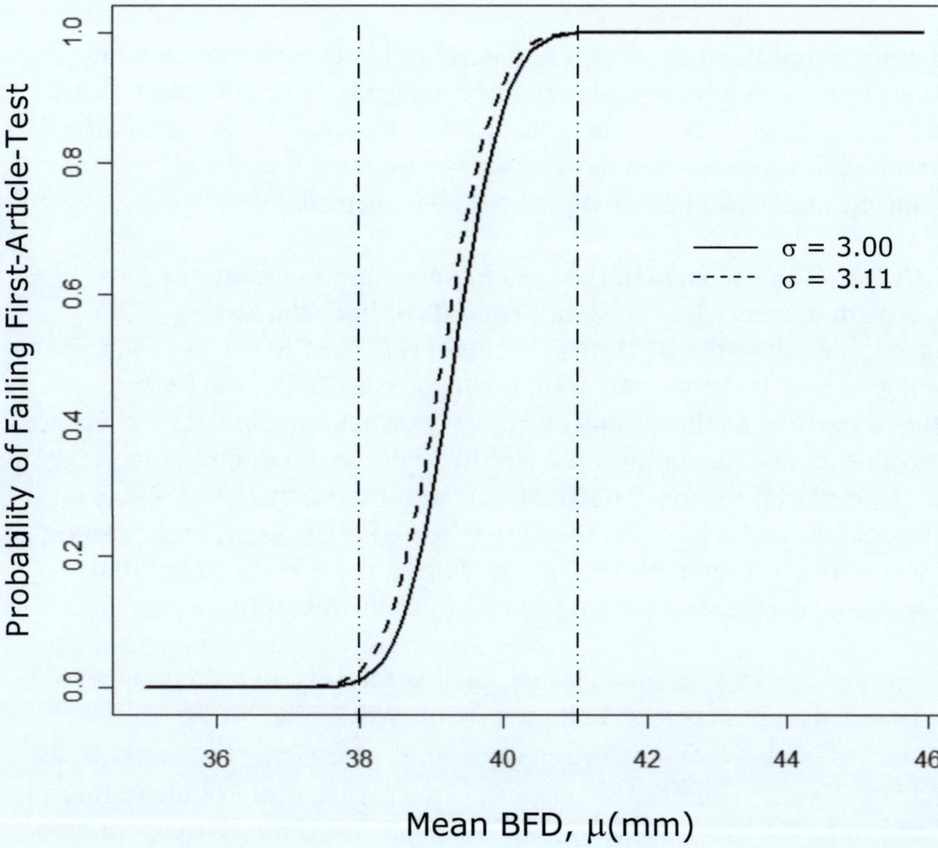

FIGURE G-4 How improving the performance of armor relates to the probability of passing FAT assuming a lot size of 60 plates. The 90 percent upper tolerance limit must be less than 44 mm with a 90 percent confidence to pass FAT. The solid curve represents how shifting the mean of the armor performance affects the probability of armor being rejected when the standard deviation associated with the population of BFDs is 3 mm (assumed to be dominated by the behavior of the RP #1, the measurement method adding no significant breadth to the distribution, as would be expected of the Faro laser scanner). The vertical lines represent 1σ and 2σ below the 90 percent upper tolerance limit cutoff criterion of 44 mm. These results show that as long as the mean performance of the armor is within 1σ of the criterion, the lot will virtually always fail. However, when the mean performance of the armor is slightly more than 2σ below the critical value, the armor will virtually always be accepted. This graph, then, demonstrates the fundamental role of testing variance in limiting armor design. If the testing method has a high variance it imposes a design penalty to pass the test. Importantly, this design penalty is not realized as value in the field. When the critical BFD is appropriately chosen based on injury probability, having to design 2σ below that represents only waste—unnecessary increase in mass or cost. The dotted curve shows how the probability of failing a FAT is increased by the addition of measurement error to the variance of the results—that is, a larger design penalty. For this particular case the standard deviation of the measurement is assumed to be 0.82 mm (the value associated with the digital caliper). The errors add in quadrature, yielding a net standard deviation of 3.11 mm.

Before contrasting the two curves it is helpful to consider the general shape of the curve. The form is essentially two plateaus. When the mean value for the BFD is less than 1σ below the critical value, that armor will essentially always be rejected, whereas when the average is a bit more than 2σ below the critical value the armor will be virtually always be accepted.

KEY POINT 2: The variance in the system determines whether or not armor is accepted, where the "system" consists of both the testing methodology, including the BFD measurement tool used in the test, and the armor being tested. If the overall system variance is dominated by the armor, this is correct, as the accept-or-reject decision appropriately prevents highly variable armor (meaning some will be underperforming) from being accepted. However, if the system variance is dominated by the variance in the tissue simulant material (as is the case when RP #1 is used) and/or some other aspect of the testing methodology, including the measurement tool, then Figure G-4 represents a design penalty in units of length.

In other words, even though an armor design gives a BFD 90 percent upper tolerance level < 44 mm in theory, to be certain of passing the test the manufacturer will seek a design with a mean BFD ≤ (44 – 2σ). To the extent that σ, the standard deviation observed in the test, is mainly the result of the testing methodology (e.g., Roma Plastilina #1), the manufacturer is forced to overdesign the armor. Conversely, to the extent that the variation induced by the testing methodology can be decreased, the manufacturer has more design space for the armor while still being able to pass FAT.

And, in fact, the available data indicate that the variance in the system is significantly influenced by the behavior of RP #1. The data presented in Figures 4-7 and 4-8 give the standard deviation for the drop tests in the clay of somewhere between 1 and 3 mm. The observed standard deviation reported by Ceradyne for an armor FAT or LAT ranges between 2.93 and 3.82 mm, with most results clustered around 3 mm.[70]

There are perhaps two complementary ways to view the difference between the two curves. Firstly, the digital caliper always has a higher chance of rejection—that is, the same armor tested with a less precise measuring device is more likely to be rejected.

In looking at Figure G-4, the difference in the probabilities of failure seem small, and, in fact, if the armor designers have as a design target armor with a mean value that is less than 2σ below the cutoff, there is no practical difference. However, if the armor manufacturers are seeking to be on the low end of the rising portion of the curve, there will be a significant difference.

Figure G-5 shows the difference between the two curves (i.e., the probability of failing using the caliper minus the probability of failing using the Faro for each value of μ). This figure shows that for a range of μ values, the increase in the

[70]David Reed, President, North American Operations, Ceradyne, Inc., "Pragmatics of Body Armor Testing Manufacturer's View," presentation to the committee, August 9, 2010.

probability of failure is significant. For $\mu \approx 39$ mm, the difference is somewhere around 11 percent; for $38.5 \leq \mu \leq 40$ or so, the difference is at least 5 percent.

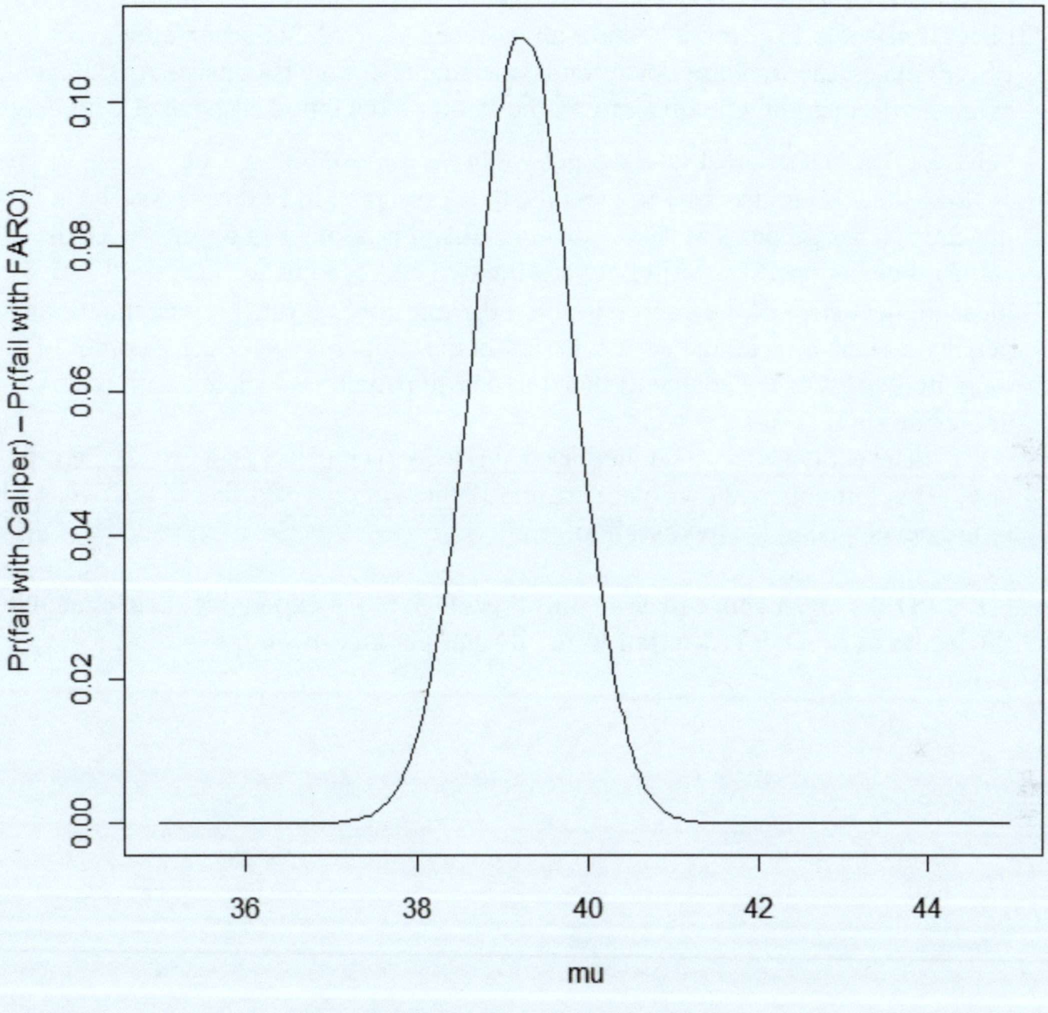

FIGURE G-5 Plot of the difference between the two FAT failure curves given in Figure G-4. This confirms that the maximum distinction occurs for $-2\sigma_{sys} < (\mu_{lot} - z_{crit}) < -1\sigma_{sys}$. In that range, the probability of rejection can be as much as 11 percent higher.

There are data available to check whether or not the order-of-magnitude effect expected by this analysis is consistent. The presentation by Ceradyne gives the average BFD and standard deviations for data sets from ESAPI and XSAPI plates measured by the Aberdeen Test Center (ATC) and H.P. White Laboratory with both caliper and laser scanner.[71]

First, the standard deviations are between 2.93 and 3.82 mm. Therefore the assumption of a baseline standard deviation of roughly 3 mm in order to estimate the significance of effects is a reasonable choice.

Second, the standard deviations of the populations obtained using laser scanning are roughly 10 percent lower than that using the caliper (for the XSAPI plates, one set is 15 percent lower with laser scanning, the other 4 percent; for the ESAPI plates, the standard deviation is unchanged). This is consistent with the estimates from adding, in quadrature, the standard deviation associated with clay behavior and that of the digital caliper—that is, $\sqrt{3^2 + 0.82^2} = 3.11$.

Now, there is another way to view the data presented in Figure G-3. The plateau is a design penalty. For typical results, it is about 4 to 6 mm (twice the standard deviation). The shift between the two curves indicates that the measurement error of the caliper is about 0.2 mm (or 200 μm). Thus, the design penalty associated with the error inherent in the caliper is well over an order of magnitude smaller (3-5 percent) than the design penalty associated with other aggregate sources of variance.

The equivalence of an increased variance to an offset is shown in Figure G-6. This formalism allows comparison to other sources of offset in the data such as are developed in the next section.

KEY POINT 3: A more precise and therefore more expensive measurement device leads to a net reduction in the design penalty of no more than 5 percent.

[71]Ibid.

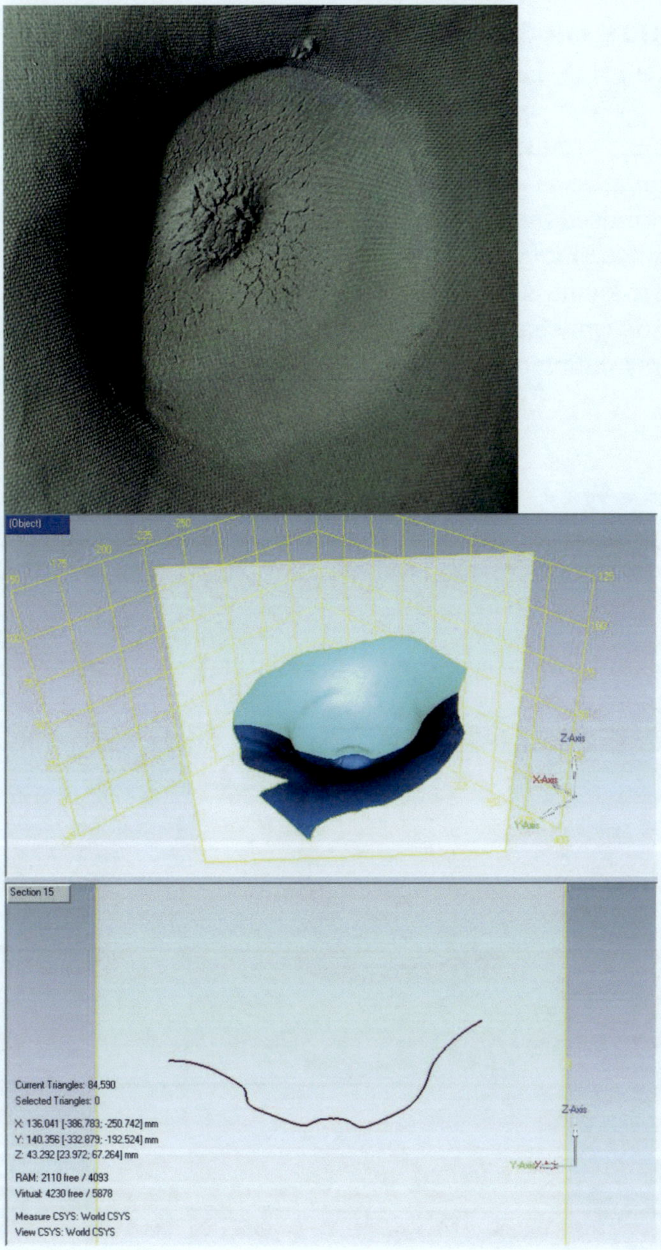

FIGURE G-6 Photograph, laser scan, and cross section of cavity in RP #1 produced by armor testing that illustrates typical roughness that characterizes the surface of the depression. SOURCE: Rice et al., 2010.

INTRODUCTION OF BIAS OR OFFSET DUE TO HIGH SPATIAL RESOLUTION LASER SCANNING

The surfaces of the craters produced during live-fire testing are rough (see Figure G-6). Much of this roughness is not going to be sensed by either the roughly 3 mm (1/8 in.) probe tip used by ATC or, certainly, the roughly 19 mm used by some commercial labs (see Figure G-7).

However, much of the roughness is imaged by the laser scanner, which not only has greater depth resolution (in what is usually called the *z*-direction), but also has much greater spatial resolution (nominally in the plane perpendicular to the *z*-direction).

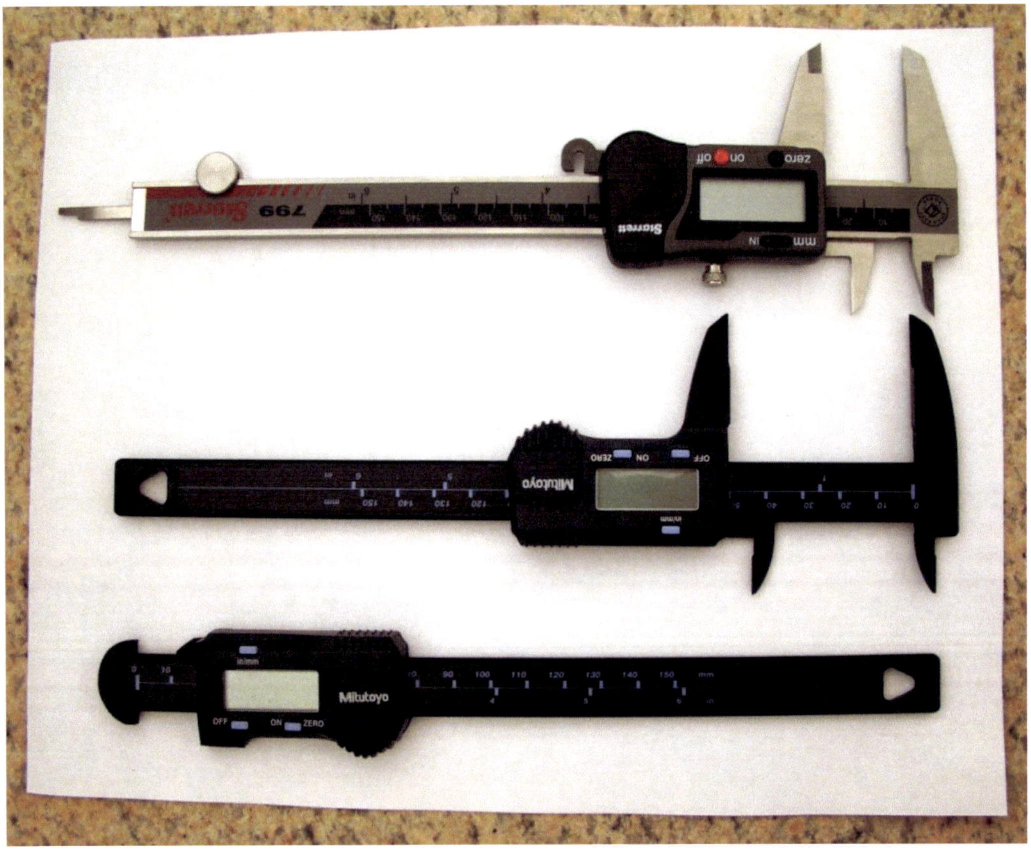

FIGURE G-7 Digital calipers used in armor testing. The ATC standard caliper with the small end (3 mm) is shown at top. Caliper used by commercial testers (H.P. White Laboratory and Chesapeake Testing) with the large 19 mm tip is shown at bottom. The dimension of the wide tip was measured by Chesapeake Testing at the request of the committee. (The center caliper is not used.) SOURCE: H.P. White Laboratory.

Figure G-8 shows two images provided by the ATC. The image on the left shows the roughness detected at the maximum resolution of the scanner raw data. The data on the right represent the same data set after digital filtering (the point set appears to have been tessellated with triangles to create a surface). The maximum depression in the first case is 42.841 mm whereas in the second it is reported as 41.819 mm. That is, there is a difference between the two in excess of 1 full millimeter. Significantly, the smoothed surface still appears to reflect significant surface roughness that would not be taken into account by the digital caliper probe tip. This roughness of the scanner data is a function of the smoothing algorithm employed.

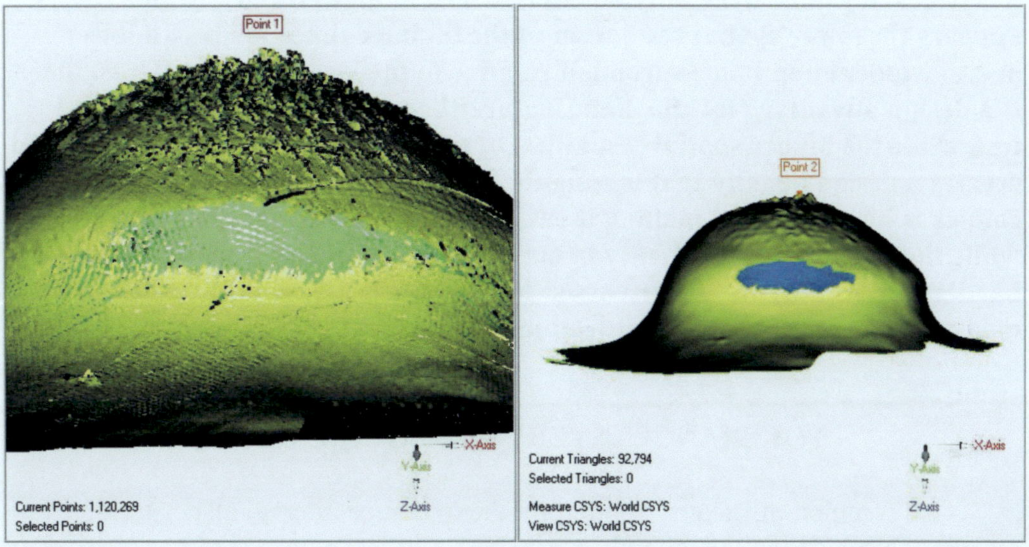

FIGURE G-8 Two images of typical BFD cavities in RP #1 produced by the Faro laser scanner. The figure on the left is generated from the point cloud raw data oriented such that the armor would have been toward the base of the figure and deformation would be upward. Significant roughness on the surface is visible. The figure on the right was smoothed and tessellated to create a surface. Much of the roughness was removed in this procedure, but much remains that would likely not be sensed by an operator using a digital caliper. SOURCE: W. Scott Walton, Shane Esola, and Barbara J. Gillich, ATC, "Laser Scanner Certification," presentation to the committee, October 6, 2010.

This estimate of a systematic shift of roughly 1 mm or so in apparent depths is consistent with the assertion in the D. Reed presentation and the data therein on the XSAPI plates.[72] Ceradyne calculated average BFDs from the H.P. White Laboratory data: 37.84 mm for the digital caliper and 38.78 mm for the laser scanner. From the ATC data the respective averages were 38.16 mm and 39.41 mm. That is, a net offset is seen between the digital caliper and the laser scanner of approximately 1 mm in tests carried out by a commercial lab and by the Army.

Thus, the apparent offset associated with the way the laser scanner is currently operated appears to be an effective design penalty on the order of 1 mm, or five times larger than that produced by the lower measurement precision of the digital caliper.

KEY POINT 4 When one measurement technique is substituted for another it is necessary to analyze fully how the two techniques differ. In this case, it appears the lower spatial resolution of the digital caliper leads to a lower average indent depth measurement relative to the laser scanner. Thus, there is a design advantage for the manufacturer because it is easier to pass the test, while the higher spatial resolution of the laser as it is currently operated creates a design penalty that is roughly five times larger. In short, the digital caliper is biased toward making it easier for a manufacturer to pass the test while the laser, as it is currently being operated, is biased toward making it harder to pass the test. As discussed in Chapter 5, a best-utility measurement tool should not introduce bias into the measurement.

TOLERANCE INTERVAL REQUIREMENTS

A common mistake is to interpret the tolerance interval test criterion as meaning that each individual body armor BFD (either in the tested sample or the larger population) must be less than 44 mm. Rather, the requirement is, roughly speaking, that one will be very sure that the 90th percentile of all BFDs in the population is less than 44 mm.

Consider, for example, a manufacturer that passes the first article test and whose armor population has a mean BFD performance of 40 mm with a standard deviation of 3 mm. Then the probability that any individual plate produced by this manufacturer will have a BFD greater than 50 mm is 0.000429. While seemingly small, such a probability means that one plate in about 2,300 will have a BFD greater than 50 mm. Given all the plates the Department of Defense procures, that translates into a not insignificant number of plates that will perform in this region. Thus under the current DOT&E protocol it is possible to see BFDs at or above 50 mm.

Figure G-9 compares the probability a manufacturer will pass the first article, first shot BFD test (90 percent upper tolerance limit less than 44 mm with 90 percent confidence) for various true BFD levels, μ, against the probability that

[72]Ibid.

a plate from the same population will have a BFD greater than 50 mm. The plot shows that the probability a manufacturer can pass FAT recedes virtually to zero before the probability a plate has a BFD > 50 mm becomes large. This is appropriate assuming that 50 mm is the threshold at which serious injury may start to occur. That said, note that at $\mu = 40$ mm there is still some chance that a manufacturer would pass FAT (roughly one chance in 10 or so). If such a manufacturer were to pass FAT then, as previously described, the probability that any one plate could have a BFD > 50 mm is 0.000429.

KEY POINT 5: The 44 mm tolerance interval requirement does not mean that all plates will have BFDs less than 44 mm. Rather, the 44 mm tolerance interval requirement in the DOT&E protocol is designed to ensure that the vast majority of plates will have BFDs less than 50 mm, but some small number may exceed 50 mm.

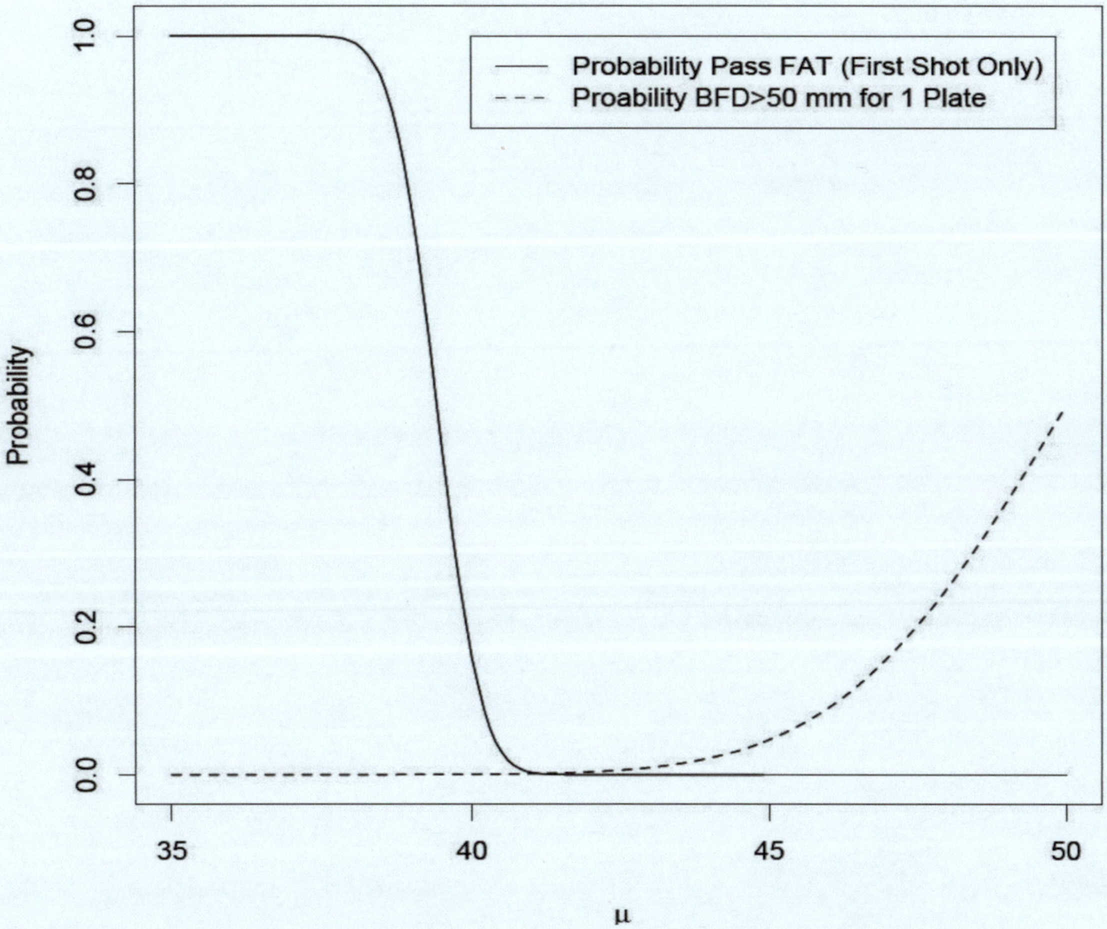

FIGURE G-9 The probability a manufacturer will pass the first article, first shot BFD test (solid line) for various population mean BFD levels (μ) versus the probability that a plate will have a BFD greater than 50 mm from the same population (dotted line).

REFERENCES

Besfamil'naya, L.V. 1974. How precision of monitoring and measuring instruments and gages affects cost utilization factors. Measurement Techniques 17(3):458-460.

Rice, K., M. Riley, A. Forster, S.D. Phillips, C.M. Shakarji, D. Sawyer, C. Blackburn, B. Borchardt, and J. Filliben. 2010. Dimensional metrology issues of Army body armor testing. National Institute of Standards and Technology, Final report: February 17.

Walton, S., A. Fournier, B. Gillich, J. Hosto, W. Boughers, C. Andres, C. Miser, J. Huber, and M. Swearingen. 2008. Summary Report of Laser Scanning Method Certification Study for Body Armor Backface Deformation Measurements. Aberdeen Proving Ground, MD: U.S. Army Aberdeen Test Center.

Appendix H

Statistical Tolerance Bounds

Suppose that a manufacturer must demonstrate that a high percentage of units (say, 100p percent) meet a particular specification with some level of confidence—say, $100(1 - \alpha)$ percent. Since every unit cannot be tested, any statement must be based on a sample of units (say, *n*).

The statistical tool to make this demonstration is called a "tolerance bound." An easy way to think about a tolerance bound is as follows. (The result is stated for upper confidence bounds, as those are the most relevant for the assessment of body armor.)

> An upper $100(1 - \alpha)$ percent tolerance bound for 100p percent of a population is the same as an upper $100(1 - \alpha)$ percent confidence bound for the 100p percentile of the population distribution.

More formally, the interpretation of an upper tolerance bound is as follows: "If we calculated an upper tolerance bound from many independent groups of random samples, $100(1 - \alpha)$ % of the bounds would, in the long run, correctly include 100p percent of the population."

The procedure and/or formula for calculating the one-sided tolerance bound varies depending on the underlying population distribution. This distribution is unknown, and it must be estimated from historical data. If the underlying distribution is normal, then the tolerance bound is calculated as

$$T_p = \bar{x} + (s \times g'_{(1-\alpha;p,n)})$$

where $\bar{x}$ is the sample mean, s is the sample standard deviation of n data points, and g' is a factor to adjust the width of the interval. It is available in many software packages, in Odeh and Owen (1980) or in less extensive tabulations in Hahn and Meeker (1991). Notice that the value of g' depends on the desired confidence, the desired percentage of units, and the sample size.

If the data are not normal, the formulas for tolerance bound calculations are worked out for a number of distributions: see, for example, Krishnamoorthy and Mathew (2009) and Young (2010).

Another option for constructing the tolerance bound is to use a "nonparametric tolerance interval." The nonparametric tolerance intervals do not make assumptions about the underlying population distribution. However, the cost of not making distributional assumptions is that the nonparametric methods require larger sample sizes to achieve the same length of bound. In other words, for the same sample size, an upper nonparametric tolerance bound will tend to be higher.

The procedure for calculating a nonparametric tolerance bound is as follows:

- Order the sample data $x_1, \ldots, x_n$ from smallest to largest. Denote the ordered set of data as $x_{(1)}, \ldots, x_{(n)}$. One of the sample data points will be chosen as the nonparametric bound.
- Find the smallest integer *k* so that $P(X \leq k - 1|n, p) \geq 1 - \alpha$, $0 < k \leq n + 1$, where *X* is a binomial(*n*,*p*) random variable. If $k = n + 1$, then use the $x_{(n)}$ as the order statistic. Otherwise, $x_{(k)}$ is the upper tolerance bound.
- The actual confidence level is $P(X \leq k - 1|n, p)$. Note that it may not be possible to find a tolerance bound with the values of *p* and α that are desired. In particular, the smallest sample size needed to have 100(1 − α) percent confidence that the largest observation in the sample will exceed at least 100p percent of the population is $n = \log(\alpha)/\log(p)$.

In Figure H-1, the bell curve represents the population distribution and the solid vertical line is the 95th quantile of the population distribution. In practice, both of these are unknown. The dotted vertical line is the specification. Since the population distribution is unknown, we test a sample from the population. The small circles are 50-sample observations.

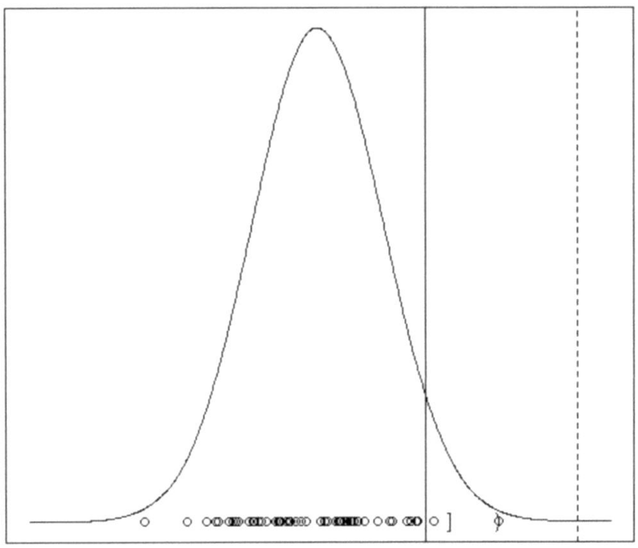

FIGURE H-1 Ninety-fifth quintile distribution.

The right bracket (]) shown in the figure is the calculated 90 percent normal tolerance bound for 95 percent of the population. To calculate this bound, we have to assume that the sample data are from a normal distribution. We make this assumption based on the 50 samples and any previous data that we have collected.

The right parenthesis) in the figure is the calculated 90 percent one-sided nonparametric tolerance bound for 95 percent of the population. Notice that the nonparametric tolerance bound is equal to the maximum observation.

In practice, one would calculate only one tolerance bound. If the tolerance bound is lower than the specification (as it is in this case), the test is "passed." More formally, we are 90 percent confident that 95 percent of the population is below the specification.

As another example, suppose that we have 15 observations: 1.57, −0.57, −1.19, 0.08, 0.83, −1.55, 1.14, 0.63, −0.11, 1.64, 0.79, −0.44, 0.27, 1.18, −0.47. We want to calculate a 90 percent confidence bound for 95 percent of the population. We have $\bar{x} = 0.2533333$ and s = 0.9725935.

Using the equation for the normal one-sided tolerance bound, we have $g' = 2.068$ and T_p= 0.2533333 + 2.068(0.9725935) = 2.264657.

Using the nonparametric tolerance interval, we order the observations from smallest to largest: −1.55, −1.19, −0.57, −0.47, −0.44, −0.11, 0.08, 0.27, 0.63, 0.79, 0.83, 1.14, 1.18, 1.57, 1.64.

We find that $k = 16$, so our one-sided tolerance bound is 1.64. However, our confidence level is $P(X \leq 14|15,0.95) = 0.5397$, or 54 percent. We require at least $n = \log(0.1)/\log(0.95) = 45$ samples to achieve a 90 percent confidence level using the largest sample value as our upper tolerance bound.

REFERENCES

Hahn, G.J., and W.Q. Meeker. 1991. Statistical Intervals: A Guide for Practitioners. New York, N.Y.: John Wiley & Sons.

Krishnamoorthy, K., and T. Mahew. 2009. Statistical Tolerance Regions: Theory, Applications, and Computation. New York, N.Y.: John Wiley & Sons.

Odeh, R.E., and D.B. Owen. 1980. Tables for Normal Tolerance Limits, Sampling Plans, and Screening. New York, N.Y.: Marcel Dekker, Inc.

Young, D. 2010. tolerance: An R package for estimating tolerance intervals. Journal of Statistical Software 36(5).

Appendix I

Analytical Approaches for Comparing Test Protocols

Existing armor has proven effective on the battlefield. That is, while statistical rigor was lacking in the Army's original FAT and LAT protocols, there is no known evidence that substandard body armor has been sent to the field. Thus, it is not unreasonable to assume that the manufacturers and their lots of body armor that passed the Army's original first article testing (FAT) and lot acceptance testing (LAT) produced body armor that met the required (or at least necessary) performance standards.

Of course, it is possible that, given the lack of statistical rigor, substandard body armor passed both FAT and LAT or, conversely, fully acceptable body armor failed either FAT or LAT. It is also possible that body armor has failed in the field and the evidence of such failure has been lost, or that substandard body armor that inadvertently passed FAT and/or LAT simply has not been put to the ultimate test. These outcomes are all impossible to determine.

ASSESSING MANUFACTURER RISK

In spite of these unknowns, one way to assess the impact of the new Office of the Director, Operational Test and Evaluation (DOT&E) protocols for body armor on manufacturers is to use historical test data for body armor that passed earlier FAT and LAT tests. The idea is to draw on actual historical test data to gain some insight into how manufacturers would fare under the new DOT&E protocol. Such an analysis is based on the following assumption:

> All body armor that successfully passed the Army's original FAT and LAT protocols was, in fact, fit for use in the field.

Analytical Approach

Given that the foregoing assumption is correct, the effect of the DOT&E protocol on manufacturers can be assessed as follows:

- Simulate a test under the new DOT&E protocol by randomly drawing (with replacement) from an appropriate pool of historical test data (at a minimum, by manufacturer and type of plate) using only data from passed tests.

- Given the pseudo data, determine whether the manufacturer (lot) would have passed or failed the FAT (LAT).
- Repeat steps 1 and 2 as necessary and appropriate to estimate rate of pass/fail for a given manufacturer/type of plate combination.
- Compare and contrast the estimated rates to the historical passing rates to evaluate whether the new DOT&E protocol is likely to result in higher failure rates and, if so, under what conditions and by what magnitude.

This type of simulation cannot determine whether the new DOT&E protocols will decrease the government's risk of buying substandard body armor. What it can do is to assess, from the manufacturers' perspective, whether the new DOT&E protocol will result in higher FAT/LAT failure rates for body armor that probably would have passed under the previous Army protocol.

Discussion

From a manufacturers' economic risk perspective, this is a very reasonable comparison since, from that perspective, the issue is not how the armor performs in the field but whether the new DOT&E protocol increases test failure rates (and thus costs) for existing products and processes.

Note that this is essentially a nonparametric approach to evaluating manufacturer risk. That is, by using actual historical data drawn from passed tests, one does not have to make any parametric assumptions about the distribution of backface deformation (BFD), nor estimate the probability of penetration, nor try to model whether and how the two measures are jointly distributed.

However, this approach does depend on having sufficient historical data from which to resample. If insufficient data are available, then a parametric approach may be taken in which distributions are fit to the BFD and penetration data, and those distributions are used to simulate future data.

COMPARING PROTOCOLS

Another way to assess the impact of the new DOT&E protocols also uses historical test data for body armor. This approach can be used to compare two protocols. Instead of considering only manufacturer and design combinations that passed the historic protocols, consider a representative range of manufacturer and design results. For illustration, the committee compares the new DOT&E protocol and a historic Army FAT protocol.

Analytical Approach

- Using historical data, simulate enough test data to satisfy the protocol with the largest sample size. This can be done either nonparametrically (by randomly drawing with replacement from an appropriate pool of historical test data) or parametrically (by using summary statistics from previous tests).
- Note that the simulations may require several summary statistics (or appropriate data to resample), including probability of first and second shot complete and partial penetrations by shot order and first and second shot BFD by shot order. Depending on the criteria for each protocol, data may be simulated (for example, data on partial penetrations) that will be used to assess only one protocol.
- If the two protocols require a different sample size, randomly select the smaller sample size from the simulated test data.
- Compute whether or not the test would have been passed or failed under each protocol.
- Repeat the simulation as necessary and appropriate to estimate the four pairs of probabilities: (1) pass under protocol 1/pass under protocol 2, (2) pass under protocol 1/fail under protocol 2, (3) fail under protocol 1/pass under protocol 2, or (4) fail under protocol 1/fail under protocol 2.
- Repeat the simulation for a range of representative manufacturer and design results.

Discussion

This type of analysis may point to specific design characteristics that are advantaged or disadvantaged by particular protocols.

Appendix J

Contemporary Methods for Assessing Behind-Armor Blunt Trauma in Live Animals

This appendix provides an overview of behind-armor blunt trauma (BABT) assessment methodologies.[73] Several research groups from at least four nations have adopted a pig model for live-animal testing, and a fairly standard instrumentation package has evolved. This was agreed upon by a meeting in Koblenz in 1998 involving the principal countries involved in this work (Mayorga et al., 2010). The protocols for these studies require animals to be anesthetized and to have approval from animal care and welfare review boards.

Pigs weighing up to 60 kg are used because of their availability, thorax size that mimics human anatomy, and ease of instrumentation. Each pig is intubated and properly infused with supportive electrolytes. Respiration, blood pressure, electrocardiogram, blood oxygen saturation, and temperature are monitored. Most experiments used the North Atlantic Treaty Organization (NATO) 7.62 mm projectile fired at full charge from 10 meters at a velocity of approximately 820 m/sec. The pigs are shot over the eighth rib. The rib cage is instrumented with accelerometers and pressure sensors close to the impact point. (This is a major protocol defect because the placements of pressure transducers are affected by motion sensor saturation.) Each animal is examined by autopsy using a standard procedure including photography, with specific attention directed to the thoracic organs and the presence of trauma to the abdominal viscera.

This is the protocol followed by the NATO group and is not a protocol that will allow observations of pressure waves or measurement of pressure transmission and pathophysiology of organs such as brain, heart, and intestines. The protocol for most of the NATO experiments does not allow observing effects beyond the acute stage of 30 min. Thus, to answer the question of remote damage from blunt trauma a more extensive protocol is needed. Essential elements of such a protocol are shown in Figure J-1.

---

[73]The committee is grateful to Miriam D. Budinger and Robert Smith of Lawrence Berkley National Laboratory, who provided information for the preparation of this appendix.

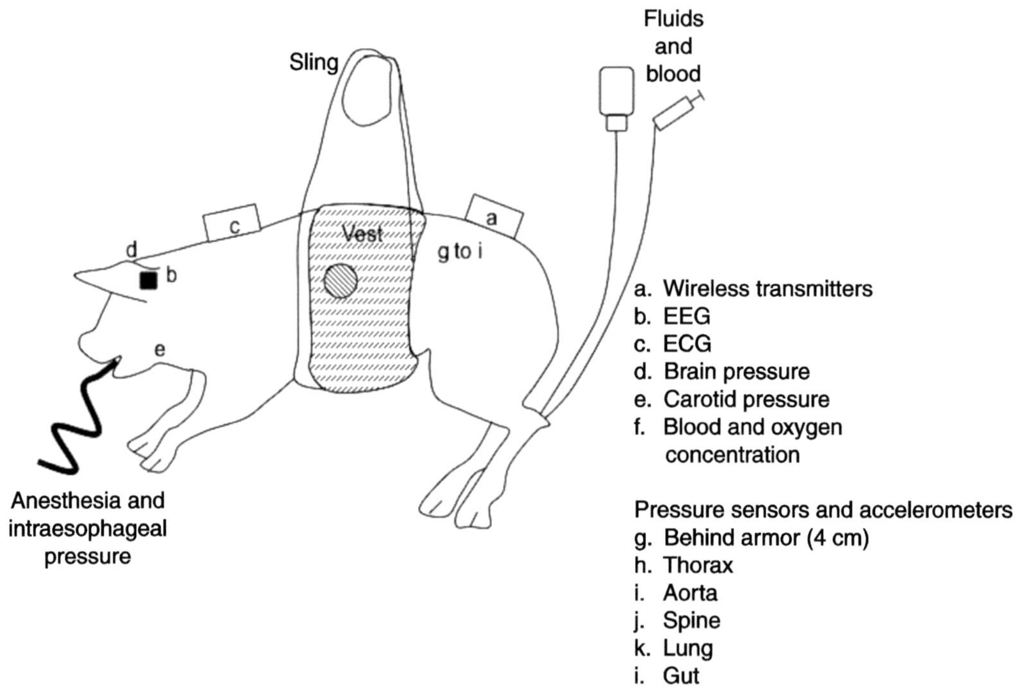

FIGURE J-1 Comprehensive protocol for live-animal live-fire tests.

COMPREHENSIVE PROTOCOL FOR ASSESSING DAMAGE TO TISSUES REMOTE FROM THE SITE OF BLUNT TRAUMA

It is important to maintain surveillance of animals for as long as feasible after the live-fire exposure. Previous studies under NATO protocols lasted for a little as 30 min. Other studies have observed animals up to 8 hr before termination. The humane guidance has been to terminate the animals before they recover from the anesthesia, and this is recommended if the animals have received substantial trauma. Under conditions that produce minimal trauma—for example, EKG, EEG, and respirations—and no hemoptisis, permission to observe the animals longer term should be pursued, as the resulting medical information would be crucial for development of personnel protective armor.

Theory and model studies cannot predict long-term consequence for blunt trauma to live organisms.

Pathology

Conventional gross and microscopic histopathology studies should be routine at the termination of animal studies. Measurements of blood-brain barrier or small vessel permeability changes are extremely important according to original studies by Suneson et al. (1987), wherein Evans blue dye injected before live-fire tests showed small vessel leakage at autopsy. Important additions to the study of the brain are the search for T-tau hyperphosphorylated protein as well as measurements of c-Fos and c-Myc expression and deposition of β-APP (Blennow et al., 1995; Blennow et al., 2010; Säljö et al., 2002). Other important assays for nerve damage include glial fibrillar acidic protein and fibrillar light protein. The timing for these measurements after trauma is important as previous studies might have waited too long (e.g., 7 days) to see some chemical manifestation of nerve damage from spinal fluid samples.

Detection of Brain Pathology from Transmitted Shock Pressures in Animals

Perhaps of greatest importance is the need for a method that can detect one or more of the following subtle and often microscopic changes in vivo through noninvasive imaging for large animals.

- Early epidural and subdural hematomas less than 5 mm wide at the cortical-skull boundary;
- Early signs of edema, such as flattening of the sulci, changes in MR T1, changes in acoustic reflection (impedance), microwave reflective power (dielectric coefficient), or electrical activity (impedance, potential difference dynamics);
- Axonal damage in the brain stem and corpus callosum with local edema and water diffusion changes;
- Brain surface contusion before frank edema occurs;
- Brain blood flow changes;
- Local brain blood volume changes due to local vascular dilatation or vascular tears at the cortical-skull boundary (epidural and subdural hematomas less than 5 mm wide).

Quantification of Pressure Wave Dispersion

Two general categories for measurement are pressure wave dispersion imaging and pressure transducer implantation. The early imaging studies included the spark gap optical methods of Harvey and McMillen (1947) and cineradiography applied to ballistic trauma to the head (Butler et al., 1945) and to nerve and bone (Puckett et al., 1946). Modern instrumentation for cineradiography, while expensive to deploy in live-fire tests in live large animals, is an important approach. Quantification of pressure distribution does require more invasive instrumentation. While straightforward for low frequency measurements, instrumentation for very high frequency response is needed for these applications, and miniaturization is essential for minimizing trauma during implantation. In addition the pressure sensor must be insensitive to accelerations and temperature changes. Successful recordings of pressures in the brain have been reported in small animals by Chavko et al. (2007), who used a miniature fiberoptic transducer implanted in the brain and invented by Pinet et al. (2005). The probe is an order of magnitude smaller than conventional piezoelectric sensors and is able to withstand harsh environments (Pinet et al., 2005).

Electrical Pathophysiology

Electroencephalography and electrical impedance tomography are two techniques that might be used to assess central nervous system integrity through measurement of electrical properties both during the acute phase of ballistic trauma and during posttrauma intervals up to months. Both approaches require sensitive instruments and are plagued by electrode coupling noise. However, in previously successful large- and small-animal experiments, EEG measurements and impedance measurements (Drobin et al., 2007; Cooper, 1996; Olsson et al., 2006; Klein and Krop-Van Gastel, 1993) have shown the kinetics of brain physiologic response to blunt trauma to the chest.

MRI Imaging

Of the four methods that have known efficacy in the examination of the brain in vivo (EEG, X-ray CT, PET, and MRI), MRI is the one that can provide noninvasive information specific to most of the relevant pathologies.

MRI can provide a wealth of information regarding organ changes associated with ballistic trauma to the body, as has already been shown in studies of blast-injured veterans (Van Boven et al., 2009). Specific capabilities for noninvasive measurements are as follows.

- *Brain contusion.* Edema is an expected early sign of contusion. It will appear as a bright signal on T2-weighted or fluid attenuation inversion recovery MRI. T1-weighted protocols might give as sensitive a diagnosis as other protocols.
- *Brain edema.* Edema resulting from vascular compromise (i.e., air emboli from lung damage), pressure impulse transmitted from the periphery to the brain, or ischemic damage from other causes can be detected by MRI diffusion weighted imaging sequences by fluid attenuation inversion recovery, and possibly by T1-weighted protocols.
- *Hemorrhage.* Early signs of hemorrhage usually occur due to tears in the tributary surface veins that bridge the brain surface to the dural venous sinus. T2-weighted MRI can show the accumulation of blood as a bright signal initially, with an evolution to a dark signal in 2 to 3 days and back again to a bright signal within the first 2 weeks (Taber et al., 2003).
- *Neuronal disruption.* Neural axon injury might be the most subtle yet the most important pathology that requires early imaging for diagnosis

(Mayorga, 1997). Experience has shown that this pathology might be seen in the corpus callosum and brain stem. Diffusion weighted imaging and T1-weighted protocols might be of extreme importance in this diagnosis (Huisman, 2010). But the choice of MR protocol is important here as it has been shown that susceptibility-weighted MRI depicts significantly more small hemorrhagic lesions than does conventional gradient echo MRI and therefore has the potential to improve the diagnosis of diffuse axonal injury (Tong et al., 2003).

The MRI system should be able to perform the above imaging studies in addition to standard structural sequences, including gradient echo as well as spin echo approaches to achieve the desired physiologic contrast signals.

Instrumentation availability and costs vary widely from a permanent magnet system for small animals at less than \$0.5 million to elaborate systems that combine magnetic resonance with PET at over \$2 million. Most studies can be enabled through collaboration with medical studies.

PET and SPECT Imaging

Metabolic and quantitative flow imaging using positron emission tomography or single photon tomography can provide sensitive metrics of pathological changes in most of the body organs of medium to large animals. The methods are noninvasive and can be repeated over the course of hours or days. Whereas PET and SPECT are readily available in medical centers, not all experimentalists will have these instruments and the required radioisotopes available, particularly for small animal studies. The spatial resolution in instruments designed for animal studies can be 2 mm or less. Normally the spatial resolution for large animals and human subjects is 5 to 6 mm.

The tracers available allow studies of blood flow, glucose uptake (commonly interpreted as cerebral metabolism), dopamine transporters and receptors, muscarinic system activity, and blood-brain permeability. Recent human studies in boxers showed patterns of hypometabolism using as a marker the accumulation of F-18 deoxyglucose, but one must be careful not to interpret hypometabolism when the reason for less apparent tracer uptake is tissue atrophy rather than a decrease in the metabolic uptake mechanism (Provenzano et al., 2010). Thus, metabolic and neurochemical studies should be accompanied by MR anatomical studies and, in some cases, by flow studies since compromised flow will lead to an apparent decrease in uptake, particularly when studying the neurochemical systems.

PET and SPECT instrumentation for small animal studies is available from a number of vendors. Large animal studies can be accomplished through collaborators at medical institutions where the requisite approvals for use of radionuclides are already in place.

Ultrasound Brain Blood Flow Measurements

Measurements of blood flow in the brain basal arteries and the carotids by transcranial Doppler are surrogates for estimating cerebral vascular resistance and are effective methods for detection of vasospasm associated with abnormally high velocities (Jaffres et al., 2005; Visocchi et al., 2002). These measurements rely on some skill of the operator. Vascular spasm can occur late after brain injury and this will result in a change in the flow characteristics with eventual change in electrical impedance (Armonda et al., 2006; Kochanowicz et al., 2006; Oertel et al., 2005; Fritz et al., 2005; Harting et al., 2010).

Ultrasound instrumentation is generally more available than the other radiological imaging systems for human studies. Specialized small animal systems are now available to the researcher.

Short- and Long-Term Cardiac Responses

During the first hour after blunt trauma to the chest, temporary cardiac arrhythmias have been observed in previous live-fire tests on animals protected by vests. The longer term as well as short term changes in heart contractions are unknown but will be important to determine for current and future protective vest designs. Thus in some experiments direct and continuous measurements of intrathoracic cardiac and aortic pressures and dimensions are recommended using radiotelemetry. These techniques are well known and can be reliably implemented in unrestrained animals.

GLOSSARY

Acoustic impedance. A material property that relates to its resistance to the propagation of sound pressure. It is the square root of the product of the tissue modulus of elasticity and the tissue density. The equivalent definition is impedance equals the product of tissue density and the speed of sound in that tissue (e.g., 1,480 m/sec for solid tissue and water, 5,900 m/sec in steel, 9,900 m/sec in alumina).

Atmospheric pressure. The pressure exerted at sea level from atmospheric gases is measured as 14.7 pounds per square inch or, in SI units, as about 100 kilopascals (kPa).

Backface deformation (BFD). The extent to which the back material of the body armor is displaced by low- or high-velocity ballistic impacts.

Behind-armor blunt trauma (BABT). When body armor is impacted by a high-velocity bullet but not perforated, some of the energy of the bullet will enter the body. The interaction of this energy with the thoracic region of the human body may or may not cause an injury. If injury is caused, it is referred to as behind-armor blunt trauma (BABT). In the past, it was considered as trauma to the ribs and lungs but now includes trauma anywhere in the body.

Cineradiography. Method for acquiring x-rays at very rapid rate so that the dynamics of a process can be recorded when tissue opaqueness does not allow rapid light photography.

Decibel. A measure of the amount of some physical parameter relative to a reference base. For blast pressure it is force per area (newtons/m^2), and the ratio of the measured pressure to the base value for human perception of sound is so great that the decibel is reported as 20 times the logarithm of the ratio. Sound intensity is power per area (watts/m^2) and the decibel is 10 times the logarithm of the ratio. Conversion of decibel sound pressure level, dB(SPL), to pascal, ρ(Pa) units is

$$\rho(\mathrm{Pa}) = 2 \times 10^{-5} \times 10^{\mathrm{dB}/20}$$

where the factor 2×10^{-5} is the minimum pressure for human sound detection in newtons/m^2, or pascals. Thus the pressure for normal conversation at 60 dB is 0.02 Pa and for a passing truck at 100 dB is 2 Pa. Pressures from behind armor are in the range of 500 kPa, or 208 dB.

Kinetic energy. The energy associated with the velocity and mass of a body (projectile). It is ½ mass × (velocity)^2. The unit is the joule (J). Projectiles deliver 033 to 13 kJ depending on the bullet used.

Magnetic resonance imaging (MRI). An imaging method that shows tissue anatomy based on water content and local environment characteristics. Diffusion-weighted imaging (DWI) and diffusion tensor imaging (DTI) are forms of MRI that allow definition of structural properties of tissue based on water diffusion directional preferences.

National Institute of Justice (NIJ) Standard. The NIJ 0101.04 standard stipulates the maximum deformation a soft armor vest can undergo without penetration is 44 mm as measured in a clay substrate after a live fire test of the armor.

Overpressure. The blast pressure from a bomb, artillery discharge, bazooka, or other explosion. Overpressure is defined as the pressure from the blast over atmospheric pressure; it is usually followed by an underpressure.

Pascal. Unit for pressure equivalent to 1 newton/m^2 (1 pascal is 0.0001 atmospheric pressure). A gigapascal (GPa) is a unit of pressure equal to a billion pascals. A kilopascal (kPa) is a unit of pressure equal to a 1,000 pascals (100 kPa is 1 atmosphere of pressure).

Positron emission tomography (PET). An imaging method that provides quantative information on metabolism, flow, and neurochemical receptors using radionuclides usually obtained from a cyclotron. The method is useful for imaging metabolism and function in the brain, lungs, and other organs.

Power. Energy per time. Unit is the watt, which is 1 Joule/sec.

Pressure. Force per area (newton /m^2 = pascal).

Single photon emission tomography (SPECT). This is an imaging method similar to PET; however, it uses radionuclides generally obtainable without the need for a cyclotron. The method is useful for imaging metabolism and function in brain, lungs, and other organs.

Strain. The relative change in dimension ΔL/L in response to a stress, where L= is a length measure.
Stress. The force per area applied to a material. Units are newton/m^2 and are usually reported as pascals.
Stress waves. Compression waves in a material due to an impulse or sudden load change.
Underpressure. The negative pressure relative to atmospheric pressure experienced by personnel following the blast pressure from an explosion.
Young's modulus. A measure of the stiffness of elastic material, it is defined as the ratio of the uniaxial stress or force per area over the strain or the fractional length change in the direction of the stress. The dimension is given as pascals or pounds per square inch (psi). For example, steel has a Young's modulus of 200 GPa, Kevlar of about 100 GPa, and polyethylene of 3 GPa.
7.62 mm x 51 mm bullet. A rifle bullet similar to the .30-06 bullet in dimensions and performance. Another model is the 7.62 mm × 61 mm bullet.

REFERENCES

Armonda, R., R. Bell, A. Vo, G. Ling, T. DeGraba, B. Crandall, J. Ecklund, and W. Campbell. 2006. Wartime traumatic cerebral vasospasm: Recent review of combat casualties. Neurosurgery 59(6):1215-1225.

Blennow, K., M. Johnson, N. Andreasen, L. Rosengren, A. Wallin, P. Hellström, and H. Zetterrberg. 2010. No neurochemical evidence for brain injury after blast overpressure by repeated explosions or firing heavy weapons. Acta Neurologica Scandinavica. Available online <http://dx.doi.org/10.1111/j.1600-0404.2010.01408.x>. Last accessed January 24, 2011.

Blennow, K., K. Walllin, H. Agren, C. Spenger, J. Siegfied, and E. Vanmechelen. 1995. Tau protein in cerebrospinal fluid: A biochemical marker for axonal degeneration in Alzheimer's disease? Molecular and Chemical Neuropathology 26(3):231-245.

Butler, E., W. Puckett, E. Harvey, and J. McMillen. 1945. Experiments on head wounding by high velocity missiles. Journal of Neurosurgery 2(4):358-363.

Chavko, M., W. Koller, W. Prusaczyk, and R. McCarron. 2007. Measurement of blast wave by a miniature fiber optic pressure transducer in the rat brain. Journal of Neuroscience Methods 159(2):277-281.

Cooper, G. 1996. Protection of the lung from blast overpressure by thoracic stress wave decouplers. The Journal of Trauma: Injury, Infection, and Critical Care 40(3S):105S-110S.

Drobin, D., D. Gryth, J. Persson, D. Rocksen, U. Arborelius, L. Olsson, J. Bursell, and B. Kjellstrom. 2007. Electroencephalogram, Circulation, and Lung Function After High-Velocity Behind Armor Blunt Trauma. The Journal of Trauma 63(2):405-413.

Fritz, H., B. Walter, M. Holzmayr, M. Brodhun, S. Patt, and R. Bauer. 2005. A pig model with secondary increase of intracranial pressure after severe traumatic brain injury and temporary blood loss. Journal of Neurotrauma 22(7):807-821.

Harting, M., C. Smith, R. Radhakrishnan, K. Aroom, P. Dash, B. Gill, C. Cox Jr. 2010. Regional differences in cerebral edema after traumatic brain injury identified by impedance analysis. Journal of Surgical Research 159(1):557-564.

Harvey, E., and H. McMillen. 1947. An experimental study of shock waves resulting from the impact of high velocity missiles on animal tissues. Journal of Experimental Medicine 85(3):321-328.

Huisman, T. 2010. Diffusion-weighted and diffusion tensor imaging of the brain, made easy. Cancer Imaging 10(1A):S163-S171.

Jaffres, P., J. Brun, P. Declety, J. Bosson, B. Fauvage, A. Schleiermacher, A. Kaddour, D. Anglade, C. Jacquot, and J. Payen. 2005. Transcranial Doppler to detect on admission patients at risk for neurological deterioration following mild and moderate brain trauma. Intensive Care Medicine 31(6):785-790.

Klein, H.C., W. Krop-Van Gastel, K.G. Go, and J. Korf. 1993. Prediction of specific damage or infarction from the measurement of tissue impedance following repetitive brain ischaemia in the rat. Neuropathology and Applied Neurobiology 19(1):57-65.

Kochanowicz, J., J. Krejza, Z. Mariak, M. Bilello, T. Lyson, and J. Lewko. 2006. Detection and monitoring of cerebral hemodynamic disturbances with transcranial color-coded duplex sonography in patients after head injury. Neuroradiology 48(1):31-36.

Mayorga, M. 1997. The pathology of primary blast overpressure injury. Toxicology 121(1):17-28.

Mayorga, M., I. Anderson, J. van Bree, P. Gotts, J-C. Sarron, and P. Knudsen. 2010. Thoracic response to undefeated body armor. BP 25, F-9220.1 Neuilly-sur-Seine Cedex, France: North Atlantic Treaty Organization (NATO), Research and Technology Organisation.

Oertel, M., W. Boscardin, W. Obrist, T. Glenn, D. McArthur, T. Gravori, J. Lee, and N. Martin. 2005. Posttraumatic vasospasm: The epidemiology, severity, and time course of an underestimated phenomenon: a prospective study performed in 299 patients. Journal of Neurosurgery 103(5):812-824.

Olsson, T., M. Broberg, K. Pope, A. Wallace, L. Mackenzie, F. Blomstrand, M. Nilsson, and J. Willoughby. 2006. Cell swelling, seizures and spreading depression: An impedance study. Neuroscience 140(2):505–515.

Pinet, E., A. Pham, and S. Rioux. 2005. Miniature fiberoptic pressure sensor for medical applications: An opportunity for intra-aortic balloon pumping (IABP) therapy. Pp. 234-237 in Proceedings - SPIE 5855. Bruges, Belgium: SPIE.

Provenzano, F.A., B. Jordan, R.S. Tikofsky, C. Saxena, R.L. Van Heertum, and M. Ichise. 2010. F-18 FDG PET imaging of chronic traumatic brain injury in boxers: A statistical parametric analysis. Nuclear Medicine Communications 31(11):952-957.

Puckett, W., H. Grundfest, W. Mcelroy, and J. McMillen. 1946. Damage to peripheral nerves by high velocity missiles without a direct hit. Journal of Neurosurgery 3(4):294-305.

Säljö, A., F. Bao, J. Shi, A. Hamberger, H. Hansson, and K. Haglid. 2002. Expression of c-Fos and c-Myc and deposition of b-APP in neurons in the adult rat brain as a result of exposure to short-lasting impulse noise. Journal of Neurotrauma 19(3):379-385.

Suneson, A., H. Hansson, T. Seeman. 1987. Peripheral high-energy missile hits cause pressure changes and damage to the nervous system: Experimental studies on pigs. The Journal of Trauma 27(7):782-789.

Taber, K., S. Rauch, R. Lanius, and R. Hurley. 2003. Functional magnetic resonance imaging: Application to posttraumatic stress disorder. The Journal of Neuropsychiatry and Clinical Neurosciences 15(2):125-129.

Tong, K., S. Ashwal. B. Holshouser, L. Shutter, G. Herigault, E. Haacke, and D. Kido. 2003. Hemorrhagic shearing lesions in children and adolescents with posttraumatic diffuse axonal injury: Improved detection and initial results. Radiology 227(2):332-339.

Van Boven, R., G. Harrington, D. Hackney, A. Ebel, G. Gauger, J. Bremner, M. D'Esposito, J. Detre, E. Haacke, C. Jack Jr., W. Jagust, D. Le Bihan, C. Mathis, S. Mueller, P. Mukherjee, N. Schuff, A. Chen, and M. Weiner. 2009. Advances in neuroimaging of traumatic brain injury and posttraumatic stress disorder. Journal of Rehabilitation Research & Development 46(6):717-757.

Visocchi, M., A. Chiaretti, D. Cabezza, and M. Meglio. 2002. Hypoflow and hyperflow in diffuse axonal injury. Prognostic and therapeutic implications of transcranial Doppler sonography evaluation. Journal of Neurosurgical Sciences 46(1):10-17.

Appendix K

Phase I Findings

This appendix contains the Phase I study findings. Phase I resulted in four findings that were submitted to the Office of the Director, Operational Test and Evaluation in the Phase I letter report (NRC, 2009):

Finding 1. The procedure documented in "Internal Operating Procedure No. 001: Measurement of Backface Deformation [BFD] Using Faro® Quantum Laser Scan Arm and Geomagic® Qualify® for Hard and Soft Body Armor" (Aberdeen Proving Ground, Md., Aberdeen Test Center, September 1, 2009) adequately describes the appropriate use of the laser scanning system.

Finding 2. Surface profilometry by a laser scanning system (including the testing protocols, facilities, and instrumentation) as currently implemented by the Army (or similar equipment), if used in accordance with the Army's procedures, is a valid approach for determining the contours of an indent in a nontransparent clay material at a level of precision adequate for the Army's current ballistic testing of body armor.

Finding 3. The digital caliper is adequate for measurements of displacements created in clay by the column-drop performance test: there is a well-defined reference plane, and one can visually see the surface of the clay, given that the depression is relatively shallow (approximately 22 to 28 mm) and fairly smooth.

Finding 4. The column-drop performance test (including the testing protocols, facilities, and instrumentation) is a valid method for assessing the part-to-part consistency of clay boxes used in body armor testing.

REFERENCE

NRC (National Research Council). 2009. Phase I Report on Review of the Testing of Body Armor Materials for Use by the U.S. Army: Letter Report. Washington, D.C.: National Academies Press.

Appendix L

Phase II Recommendations

This appendix contains the Phase II study recommendations. Phase II resulted in nineteen findings and one overarching finding that were submitted to the Office of the Director, Operational Test and Evaluation in the Phase II letter report (NRC, 2010). The recommendations are summarized in Box L-1 and listed here in full:

Recommendation 1: The Army's medical and testing communities should be adequately funded to expedite the research necessary both to quantify the medical results of blunt force trauma on tissue and to use those results as the updated mathematical underpinnings of the back face deformation (BFD) body armor testing methodology.

Recommendation 2: The Army should develop ballistic testing performance specifications and properties that will lead to a short-term, standard replacement for the current Roma Plastilina #1 oil-based modeling clay.

Recommendation 3: Rheological and thermogravimetric measurements should be carried out to better understand the properties and behaviors of clay as it is being prepared and worked.

Recommendation 4: If it is demonstrated to achieve improved part-to-part consistency of the clay compared to hand preparation procedures, a mechanical compounding machine for clay preparation should be acquired, experimented with, and used by the Aberdeen Test Center.

Recommendation 5: In-box mechanical conditioning might obviate the need for precise temperature control and reduce the need for hand working of the clay. Mechanical working methods should be tested.

Recommendation 6: Since oil-based modeling clay is time and temperature sensitive, a post-drop calibration test is needed to validate that the clay remains within specification at the end of a body armor test. The Army should add this requirement for a post-drop calibration test of the clay to its Test Operating Procedure (TOP 10-2-210).

Recommendation 7: The spatial variation of modeling clay is significant and three-dimensional. The response of the clay appears to depend on temperature, shear history, and proximity to the edge. Given the confounding effect of box geometry, the Aberdeen Test Center should perform a systematic set of column-drop performance tests as experiments to assess the consequence of variation due to the shape and size of the frame that defines the clay box. These tests should determine if a circular box of approximately the same area as the current box reduces the spatial variation that affects ballistic testing, or if a larger box area eliminates the clay edge effects that affect ballistic testing.

Recommendation 8: As an alternative to the current column-drop performance test the Army should quickly develop and experiment with a gas gun calibrator, or equivalent device, that delivers impactors to the surface of clay boxes and that determines local variation within a clay box at speeds and depths corresponding to those involved in the generation of the backface deformation. These experiments should be used to estimate as accurately as possible the variation of backface deformation measurements both within a given box and between boxes, under realistic testing conditions using existing test protocols.

Recommendation 9: While the committee applauds the Aberdeen Test Center efforts to understand and attempt to measure the dynamics associated with the creation of a backface deformation, the signal-to-noise ratio of the flash x-ray cineradiography approach should be thoroughly analyzed to determine if the desired spatial and temporal resolution can be achieved.

Recommendation 10: To better understand and measure the forces that create the backface deformation the Army should experiment with inserting microscopic temperature and displacement sensors into the clay near the site of the backface deformation.

Recommendation 11: The Army should consider experimenting with high-speed photographic analysis of backface deformation in ballistic gelatin as an alternative for providing needed information on the forces that shape the backface deformation.

Recommendation 12: The Army should conduct rheology and other studies on ballistic gelatin as a mid-term alternative to modeling clay due to its properties, which include the ability to directly record BFD using high-speed photography and the elimination of the effects of shear history, time, and temperature on the response of the backing material. However, correlation studies and tests are needed to better understand the differences in the extent of deformation and dynamics among gelatin and alternative clay formulations.

Recommendation 13: The Army should perform rheology and other evaluations on microcrystalline wax mixtures as a possible long-term replacement for Roma Plastilina #1 as a backing material for ballistic testing. Studies are needed to optimize the composition of the mixtures to produce the desired properties. In addition, correlation studies are needed to compare the response of the microcrystalline wax mixtures to the current material and/or ballistic gelatin.

Recommendation 14: The ad hoc clay working group should be empowered and adequately resourced to gather information, influence research, and develop working-level consensus across body armor testing organizations for the uniform application of National Institute of Justice standards across participating test organizations.

Recommendation 15: The Department of Defense Director of Operational Test & Evaluation (DOT&E) and the National Institute of Justice (NIJ), in collaboration with the military services, unified commands, other governmental organizations, NIJ-certified laboratories, and appropriate nongovernmental and commercial organizations should convene a nationally recognized group to review all appropriate considerations and develop recommendations that could lead to a single national body armor testing standard to achieve more uniform testing results.

Recommendation 16: Before adopting the proposed statistically based protocol, the Department of Defense Director, Operational Test & Evaluation (DOT&E), should explicitly compare the risks of the proposed protocol and those of the existing Army and U.S. Special Operations Command (USSOCOM) protocols, in order to establish which test plan increases soldier safety while balancing the manufacturer's risk and incentives to overdesign. The committee notes that the USSOCOM first article test protocol may not be intended as a comprehensive technical test, and clarifying this issue would also help in the comparison of the protocols.

Recommendation 17: The committee recommends that testers and statisticians continue to work together as a team (1) to quantify in a statistically rigorous manner the amount of variation in backface deformation attributable to the testing process and that attributable to the plates, and (2) to ensure these results are appropriately reflected in an updated protocol. In particular, the statisticians involved with developing and implementing the statistically based protocol should be involved with the experimentation recommended in Recommendations 2-8. It would be helpful for statisticians to be part of the process of understanding and quantifying test system variation.

Recommendation 18: The Department of Defense should develop standard statistically based body armor Lot Acceptance Testing (LAT) protocols that incorporate aspects of MIL-STD-1916, particularly those related to quality control and improvement and switching procedures. Adopting and incorporating modern statistical process control methods into the manufacturing processes is specifically recommended so that plate quality can be managed and assessed prior to lot acceptance testing. This could potentially reduce testing effort and costs. Note that while MIL-STD-1916 states that the "sampling plans and procedures of this standard are not intended for use with destructive tests," these aspects of the military standard are relevant to body armor LAT testing.

Recommendation 19: The Department of Defense (DoD) Director, Operational Test & Evaluation (DOT&E) should provide briefings to and receive feedback from all stakeholders in DoD (military service Program Executive Officers, testers, users) and non-DoD organizations (National Institute of Justice, National Institute of Standards and Technology, certified private testing laboratories, vendors) concerning the statistically based protocol. This feedback, as well as the results of the experiments and analyses proposed in this report, should be used as due diligence to carefully and completely assess the effects, large and small, of the proposed statistically based protocol before it is formally adopted across the body armor testing community. DOT&E should act on feedback from the community to improve the proposed protocol as necessary, to ensure that testing terms and concepts make sense to a nontechnical audience, and it should promote the use of statistically based protocols in future national standards for body armor testing, as appropriate.

Overarching Recommendation: The committee applauds DOT&E for assuming a national-level leadership role in bringing the body armor test community together. The committee recommends that the DOT&E (1) work with Congress, DoD, the military services, and other organizations to find the resources necessary to implement the recommendations described in this report and summarized in [Box L-1] and (2) oversee, review, track, and assist designated action organizations with implementing these recommendations. This approach should result in more consistent test results that will provide equally survivable but lighter-weight body armor to our military service members and civilian police forces.

Box L-1 Phase II Recommendations to Improve Body Armor Testing

Achieving Greater Part-to-Part Consistency in Clay

1. Quantify the Medical Results of Blunt Force Trauma on Tissue and Incorporate Results into the BFD Methodology
2. Determine Short-Term Standard Clay Specification
3. Conduct Rheological and Thermogravimetric Measurements
4. Procure and Experiment with a Clay Compounding Machine
5. Examine Technologies for "In Box" Mechanical Clay Working
6. Modify TOP 10-2-210 Procedures to Add a Post-calibration Drop (ATC, 2008)
7. Experiment with Various Clay Box Sizes and Shapes
8. Develop and Experiment with a Gas Gun Calibrator or Equivalent Device

Analyzing Backface Deformation Dynamics

9. Analyze the Signal-to-Noise of Flash X-Ray Cineradiography
10. Experiment with Microscopic Temperature and Displacement Sensors in Clay
11. Experiment with the High-Speed Photographic Analysis of BFD Creation in Ballistic Gelatin

Determining Possible Replacements for Modeling Clay

12. Study Ballistic Gelatin as a Mid-Term Alternative to Modeling Clay
13. Study Microcrystalline Waxes as a Long-Term Alternative to Modeling Clay or Ballistic Gelatin.

Achieving a Single National Clay Standard for Body Armor Testing

14. Empower and Resource the Ad Hoc Clay Working Group
15. Convene a Nationally Recognized Group to Establish a Single National Standard for Handling and Validating Clay

Implementing Statistically Based Protocols

16. Compare the Proposed Statistically Based Protocol with the Existing USSOCOM Protocol
17. Quantify the Variation in the Body Armor Test Process and Incorporate in the Protocol
18. Develop a Statistically-Based LAT Protocol
19. Conduct Due Diligence Before Implementing and Formally Adopting a Set of Statistically Based Protocols

SOURCE: NRC, 2010

REFERENCE

NRC (National Research Council). 2010. Phase II Report on Review of the Testing of Body Armor Materials for Use by the U.S. Army: Letter Report. Washington, D.C.: National Academies Press.

Appendix M

Estimating the Accuracy and Precision of the Digital Caliper and Faro Laser

This appendix presents the datasets available to the Committee for assessing the accuracy and precision of the digital caliper and Faro laser as used in measuring backface deformation (BFD) during body armor testing. As discussed in Chapter 5 and Appendix G, both accuracy and precision are important characteristics in determining the suitability of a measurement system for use in a testing process.

During Phase III, two new data sets were presented to the committee: the side-by-side comparisons of BFD measurements made by the Aberdeen Test Center (ATC) (Table M-1) and the side-by-side comparisons of BFD measurements made by Chesapeake Testing (Table M-2).[74] The committee also had access to Walton et al. (2008), which is a summary report of the ATC experimental data from the 228-page ATC experimental data report (Hosto and Miser, 2008). The committee evaluated and reanalyzed data from all of these sources.

SIDE-BY-SIDE COMPARISONS

Tables M-1 and M-2 are datasets that were collected by ATC and Chesapeake Testing. Each measures BFDs created during a test of hard body armor. The ATC data (Table M-1, N = 91) were collected in early 2008 as part of a Program Executive Officer Soldier (PEO Soldier) product data management test. The Chesapeake Testing data (Table M-2, N = 83) were collected in February 2011 during routine PEO-funded R&D testing on a developmental design prototype (different from that used for the ATC data). Chesapeake Testing is a National Institute of Justice (NIJ)-certified ballistics laboratory and is also certified by ATC in the use of the Faro laser. Both data sets were collected using standard test operating procedures. Plots of the two data sets appear in Figures M-1 and M-2.

---

[74]The new data contained in Tables M-1 and M-2 were provided via personal communication between U.S. Army PEO Soldier and Larry G. Lehowicz, committee chair, September 7, 2011.

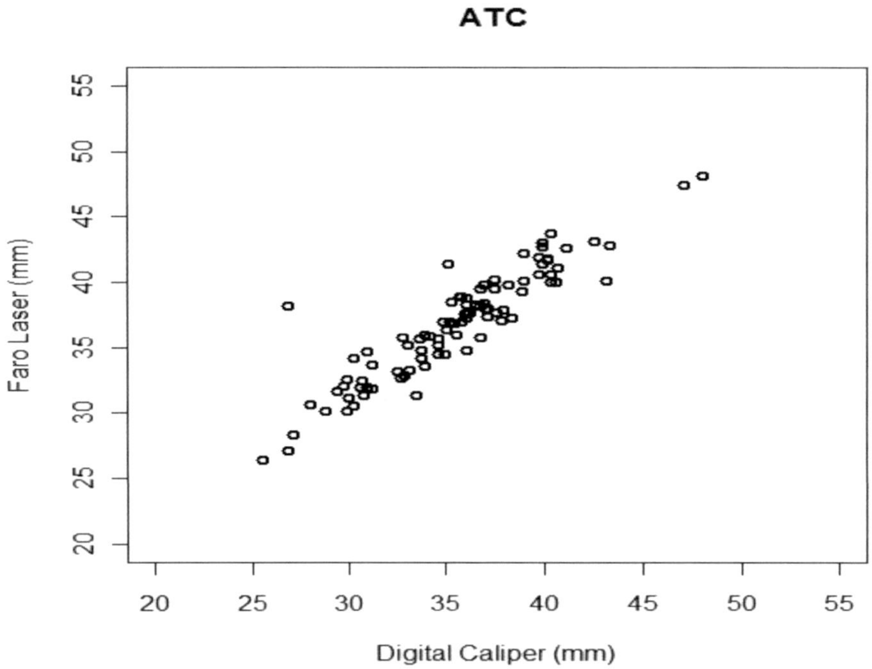

FIGURE M-1 Plot of the paired BFD measurements made by ATC.

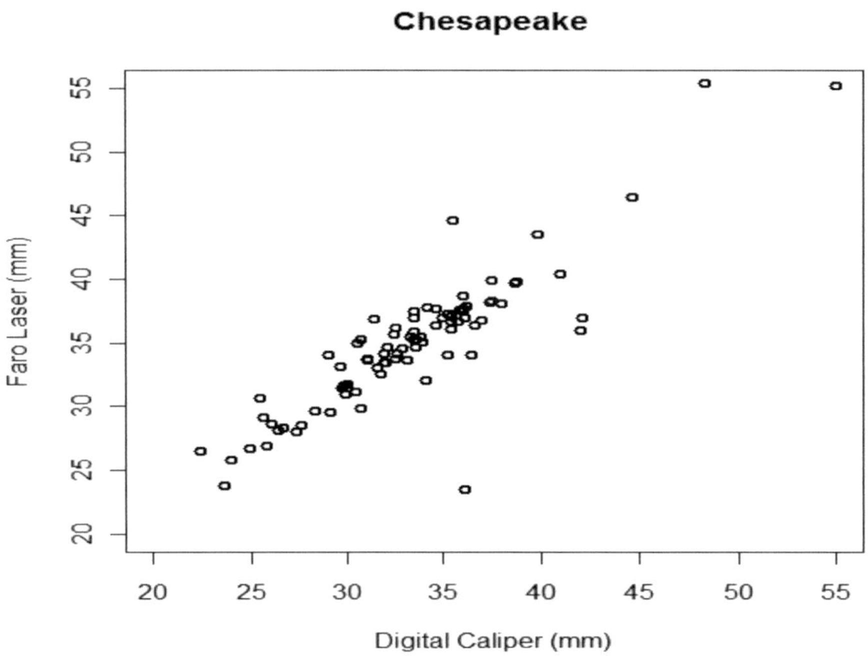

FIGURE M-2 Plot of the paired BFD measurements made by Chesapeake Testing.

Consider first the question of relative accuracy. For the ATC data, the average difference between the laser and caliper measurements is 1.36 mm. Using a paired t-test, this difference is statistically significantly different from zero ($p < .0001$). There is an outlier in the data, however, with a difference of 11.647 mm. Removing this data point, the average difference between the laser and caliper measurements is 1.25 mm, with a 95 percent confidence interval of (0.95, 1.54) mm (significantly different from zero, with $p < .0001$). For the Chesapeake Testing data, the average difference between the laser and caliper is 1.56 mm, with a 95 percent confidence interval of (0.98, 2.13) mm (significantly different from zero, with $p < .0001$). These data strongly suggest that the digital caliper and Faro laser may have systematic differences in their measurements of between 1.25 and 1.5 mm, with the laser producing a "deeper" measurement, on average.

We can also use these data to estimate the precision of the caliper and laser and to test whether the precisions of the two systems are different. A methodology for estimating precision was provided to the committee that depends on making a few assumptions.[75] The primary assumption is that the overall variance in each measurement is the sum of the variances of two independent components: that of the underlying "true value," assumed common to the two measurements, and the method-specific "measurement error." A second assumption is that the collections of measurements are roughly normal and free from outliers. It is only with roughly normally distributed observations that simple variance calculations can be relied on. Further, if outliers are present, they can distort calculations of variance and lead to incorrect conclusions. This second assumption is reasonable for the ATC data and questionable for the Chesapeake Testing data.

For the ATC data, one can calculate the variance of the laser measurements (18.0), of the caliper measurements (18.9), of the laser measurement less the caliper measurement (3.07), and of the laser plus the caliper measurements (70.7). Assume that

$$L = T + e$$
$$C = T + f$$

where L and C are the observed laser and caliper measurements, T is the true but unknown measurement value, and e and f are the laser and caliper measurement errors, respectively. Assume that T, e, and f are mutually independent and identically distributed with true variances $\mathrm{Var}(T)$, $\mathrm{Var}(e)$ and $\mathrm{Var}(f)$ respectively. We observe a small systematic difference in the two measurements, which, as long as it is constant, can be absorbed into the mean of the errors. That is, we assume the errors have constant, not necessarily zero means. It is an easy consequence of these equations and assumptions that the following hold:

$$\mathrm{Var}(L) = \mathrm{Var}(T) + \mathrm{Var}(e),$$
$$\mathrm{Var}(C) = \mathrm{Var}(T) + \mathrm{Var}(f),$$
$$\mathrm{Var}(L - C) = \mathrm{Var}(e) + \mathrm{Var}(f),$$
$$\mathrm{Var}(L + C) = 4\mathrm{Var}(T) + \mathrm{Var}(e) + \mathrm{Var}(f)$$

[75]The methodology was suggested by Terry Speed, University of California, Berkeley, to member Thomas Budinger in a personal communication, December 1, 2011.

Now we calculate the observed variances Var*(L)*, Var *(C)*, Var*(L − C)*, and Var*(L + C)* of these four quantities and use them and the above equations to obtain unbiased estimates of Var*(e)* and Var*(f)*. We take the difference between Var*(L + C)* and Var*(L − C)* and divide by 4: this estimates Var*(T)* . Then we subtract this quantity from Var*(L)* and Var*(C)* to give estimates of Var*(e)* and Var*(f)* .

Using this methodology, the estimate of the variance of the caliper is 1.99; the precision (standard deviation) is 1.41 mm, with a bootstrapped approximate 95 percent confidence interval of (0.38, 2.11) mm. The estimate of the variance of the laser is 1.09; the precision (standard deviation) is 1.04 mm, with a bootstrapped approximate 95 percent confidence interval of (0, 1.52) mm.

Because of the possible presence of outliers, the results from the Chesapeake Testing data are less reliable. However, using the same methodology, one can calculate the variance of the laser + part (29.0), the caliper + part (27.8), laser – caliper (6.9), laser + caliper (106.6). The estimate of the variance of the caliper is 2.83; the precision is 1.68 mm, with a bootstrapped approximate 95 percent confidence interval of (0, 2.47) mm. The estimate of the variance of the laser is 4.09; the precision (standard deviation) is 2.02 mm, with a bootstrapped approximate 95 percent confidence interval of (0, 3.11) mm.

Testing formally for equality of variance between the variances of the two columns (digital caliper and laser arm) in each dataset using the Pitman-Morgan test on the ATC data and nonparametric test of Sandvik and Olsson (1982) on the Chesapeake Testing data, one does not reject the null hypothesis of equal variances.

However, the probability that the data can support a conclusion that there is no significant difference between the variances of the two measurement systems is very low; that is, the statistical power for the design of the side-by-side tests is low. Power is the probability a test will reject the null hypothesis for a specific effect size, and it depends on both the effect size and the sample size. With N = 91, the power to detect the difference in precision (square root of variance) of the laser and the caliper of the size estimated by Walton et al. (2008) is only 12 percent. Thus the currently available data cannot be construed as evidence that the variances of the two measurement systems are similar.

The side-by-side ballistic tests do provide important information about the bias or absolute accuracy of the test instruments. The tests reported here reveal significant differences in accuracy. While they reveal differences in accuracy, side-by-side tests such as those reported here cannot be definitive as to which (if either) system provides desirable accuracy. The consequences of having an inaccurate test instrument on body armor testing are discussed in Chapter 5 and Appendix G. The accuracy issue is separate from the issue of relative precision. Side-by-side procedures can also provide some information about precision of the measurement procedures, although larger and more carefully designed studies are needed to provide definitive results about precision. A formal gauge repeatability and reproducibility study for the laser, caliper, and other potential measuring instruments is needed to provide reliable information about both accuracy and precision (see Recommendation 5-3).

ASSESSMENT OF OTHER TESTING RESULTS

In this section the committee assesses the results from Walton et al. (2008) and Hosto and Miser (2008) and estimates confidence limits for specific quantities. These data were collected using a different experimental design than the side-by-side data. Four BFDs were created in a mounting box using a mold. Quoting from Walton et al. (2008), "These clay molds, made from actual indentations in clay during body armor testing, had very rough surfaces, which showed the individual thread impressions from the Kevlar 'Soft Body Armor' backing. The molds also had remnants of small 'fissures' that typically form in the clay during the rapid deformation of ballistic testing." These mold-created BFDs were then repeatedly measured by various operators and instruments.

Faro Laser Precision

The original data for these estimates come from Hosto and Miser (2008). The Faro data, from Tables B-20a (depth, mm, deepest point column) in the report, are shown in Table M-3 of this appendix.

Table 2 of the NRC Phase I letter report (NRC, 2009), using the results of Walton et al. (2008), estimated the precision of the Faro laser as 0.0970 mm:

$$\begin{aligned}\hat{\sigma}_{\text{laser combined standard uncertainty}} &= \sqrt{\hat{\sigma}^2_{\text{laser operator}} + \hat{\sigma}^2_{\text{laser error}} + \hat{\sigma}^2_{\text{laser instrument spec}}} \\ &= \sqrt{0.0410^2 + 0.0817^2 + 0.0325^2} \\ &= 0.0970\end{aligned}$$

The data contain information only about the variation in the operator and the error. Here the statistical uncertainty of the laser is defined as:

$$\begin{aligned}\hat{\sigma}_{\text{laser statistical uncertainty}} &= \sqrt{\hat{\sigma}^2_{\text{laser operator}} + \hat{\sigma}^2_{\text{laser error}}} \\ &= \sqrt{0.0410^2 + 0.0817^2} \\ &= 0.0914.\end{aligned}$$

Using a parametric bootstrap, the committee estimates a 95 percent confidence interval for the laser statistical uncertainty as (0.042, 0.141) mm. Taking the upper end of the confidence interval as a worst case estimate, the actual laser precision is highly likely to be less than

$$\hat{\sigma}_{\text{laser } \textbf{worst case} \text{ combined standard uncertainty}} = \sqrt{0.141^2 + 0.0325^2} = 0.145 \text{ mm.}$$

Digital Caliper Precision

The digital caliper data, reproduced from Table B-21a (deepest point, mm, corrected depth column) in Hosto and Miser (2008), are shown here as Table M-4.

Table 2 of the NRC (2009), again using Walton et al. (2008), estimated the precision of the caliper as 0.823 mm:

$$\begin{aligned}\hat{\sigma}_{\text{caliper combined standard uncertainty}} &= \sqrt{\hat{\sigma}^2_{\text{caliper operator}} + \hat{\sigma}^2_{\text{caliper error}} + \hat{\sigma}^2_{\text{caliper instrument spec}} + \hat{\sigma}^2_{\text{correction factor}}} \\ &= \sqrt{0.4715^2 + 0.36^2 + 0.0073^2 + 0.57^2} \\ &= 0.823,\end{aligned}$$

Because the data contain information only about the variation in the operator and the error, the "statistical uncertainty" corresponds to the first two terms. The correction factor term, which is unique to the caliper, accounts for the uncertainty in the correction methodology when the deepest point is different from the aim point. (This difference is called an "offset.")

The caliper statistical uncertainty is

$$\begin{aligned}\hat{\sigma}_{\text{caliper statistical uncertainty}} &= \sqrt{\hat{\sigma}^2_{\text{caliper operator}} + \hat{\sigma}^2_{\text{caliper error}}} \\ &= \sqrt{0.4715^2 + 0.36^2} \\ &= 0.593.\end{aligned}$$

We can estimate a 95 percent confidence interval for the caliper statistical uncertainty of (0.367, 0.825) mm. Taking the lower end point of the interval as the caliper best case for statistical uncertainty, we estimate

$$\hat{\sigma}_{\text{caliper } \mathbf{best\ case} \text{ combined standard uncertainty}} = \sqrt{0.367^2 + 0.0073^2 + 0.57^2} = 0.678 \text{ mm.}$$

Turning to the correction factor term, consider the 0.57 mm uncertainty associated with the postmeasurement correction made to adjust the caliper measurements ("how a caliper measurement of a deepest point needs to be corrected to find the actual depth to the local pristine surface"). Walton et al. (2008) documents its derivation in that report's Appendix B. The correction is geometrically derived, and its uncertainty is estimated using the delta method, a standard statistical methodology for estimating the variances from complex functions. Walton et al. (2008) says that "in practice, using aim-points to reference depth measurements introduces multiple uncertainties (see Appendix B for quantification), which are found in the assumed and measured values of slope, offset, shot location on the plate and slope of the impacted surface (not quantified in this analysis)."

The data used by Walton et al. to assess this variation are from the Phase I testing of plates in 2008, not from the data in Hosto and Miser (2008).[76] The committee finds the calculation to have been done in a reasonable and correct manner.

Issues with Walton et al.

Resolution of the following issues with Walton et al. (2008) would be accomplished as part of the gauge repeatability and reproducibility studies of measuring instruments mentioned earlier and is embodied in Recommendation 5-3.

Caliper measurements were replicated while laser measurements were not.

Obviously, this is not ideal when trying to assess measurement precision. However, from the data we do have, we can estimate several components of variation. These results are taken from Table 12 of Walton et al. (2008) (with calculations replicated by the committee) or were additionally calculated by the committee.

The variation attributable to the different indentations has standard deviation 5.28 mm, the variation attributable to different operators ($\sigma^2_{\text{laser operator}}$) has standard deviation 0.040 mm, and the variation attributable to a lack of repeatability measurement-to-measurement ($\sigma^2_{\text{laser error}}$) has standard deviation 0.082 mm. If we calculate a 95 percent confidence interval for the measurement-to-measurement repeatability using these data, it is (0.045, 0.114) mm.

The measurement-to-measurement repeatability of the data is estimated using the measurements that operators make on different indentations. Without replicates, we cannot assess whether the repeatability of the operators is the same when they are measuring the same indentation multiple times as when they are measuring different indentations. However, if we make the assumption that these two variances are the same, then adding replicates does not change our variance estimate.

Sample sizes are small.

The sample sizes in the side-by-side data and the Walton et al. (2008) study are not directly comparable due to differences in study design. However, calculating confidence intervals for precision and accuracy takes into account both sample size and design differences.

---

[76]Rick Sayre, Deputy Director, OSD DOT&E Live Fire Test and Evaluation, and Tracy Sheppard, Executive Officer & Staff Specialist, OSD DOT&E, "DoD In Brief to the National Research Council Study Team," presentation to the committee, November 30, 2009.

Offsets used by Walton are excessively large.

The offsets reported in the data tables in Hosto and Miser (2008) have different statistical features from those used to calculate precision as reported in Appendix B of Walton et al. (2008). In the latter, it is reported that the 95 percent quantile of N = 654 offsets from an operationally realistic data set made up of XSAPI of all sizes first-shot data is 0.5512 in. (14 mm). The absolute value of the offsets from Realistic Clay III as reported in Hosto and Miser (2008) are shown in Figure M-3. There are two clusters of data: those below 0.5 in. are from measuring Indent 2, and those above 0.5 in. are from measuring the remaining three indents.

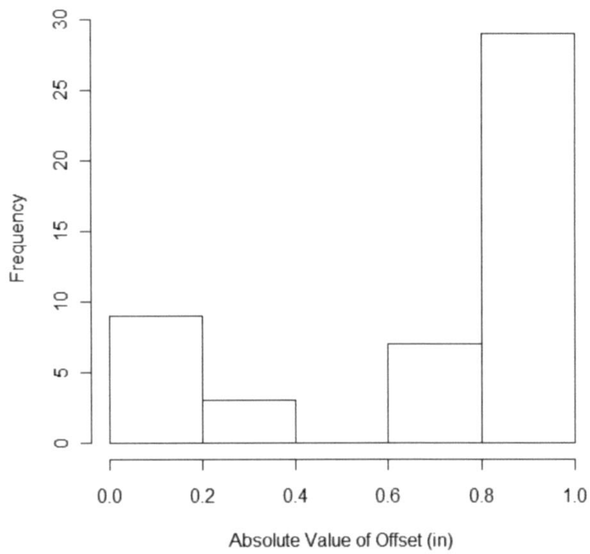

FIGURE M-3 Absolute value of offsets for caliper measurements from Realistic Clay III.

It is difficult to assess the impact that these differences could have on the accuracy and precision estimated for the caliper, although we can use the results of Walton et al. (2008) to explore the effect of some excursions. Appendix B of Walton et al. (2008) derived the 0.57 mm uncertainty associated with the postmeasurement correction using the delta method, a standard statistical approach for estimating the variance of complicated statistics—in this case, the variability for the correction factor.

One way to gain some insight into how other operationally realistic data would have affected the uncertainty estimate is to replace the Walton offset mean and variance used in the Appendix B delta method calculations with the equivalents from the ATC side-by-side data. During live-fire tests in 2008, ATC listed 41 offset measurements with a mean of 3.7 mm and a standard deviation of 3.6 mm.

Recomputing the correction factor uncertainty using these 41 edge-shot data points in place of the quantities used in Appendix B (offset 14 mm and standard deviation 0.81 mm) actually increases the correction factor uncertainty from 0.57 mm to 0.896 mm. This is because, while the mean offset is larger in Appendix B than for the ATC data, the standard deviation is substantially smaller. The latter drives the magnitude of the estimated correction factor uncertainty more than the former.

The previous calculations included 17 shots with a zero offset. One might suggest that for those shots there is no uncertainty due to the correction factor and, furthermore, that their inclusion artificially inflates the standard deviation for the nonzero offsets. Removing the data for these 17 shots results in a mean offset of 6.3 mm with a standard deviation of 2.4 mm. Recalculating using these values results in a correction factor uncertainty of 0.723, which is still larger than the Walton et al. (2008) value of 0.57 mm.

So, while intuition would suggest that smaller magnitude offsets result in improved caliper precision, using the offset mean and standard deviations from the ATC data, which has a smaller mean offset but a larger standard deviation, results in a larger uncertainty estimate.

Measurements in Walton et al. (2008) were on clay indents produced from molds of clay impressions that were made from ballistic experiments, not on actual ballistically induced clay impressions.

An advantage of this procedure is that the mold becomes a more-or-less permanent artifact that allows replicate measurements by laser, caliper, or other devices after proper validation. In its Recommendation 5-2, the committee suggests that a standard BFD artifact should be developed to assist in the assessment of measurement systems.

The Walton et al. (2008) data were not measured at 100°F.

As discussed in the report, the temperature of the clay can have an impact on the depth of the BFD created during operational testing. However, the temperature of the clay should not have an impact on the measurement precision of either the laser or the caliper because the shape and surface characteristics of the clay impression are determined by the characteristics of the mold. Those characteristics did depend on the temperature of the clay when the indents were made, and they should have been made under operational conditions. But the temperature of the clay should not have an impact on the measurement precision of either the laser or the caliper. This can be empirically verified as required.

The average caliper BFD measurement was greater than the average laser BFD measurement.

This is the reverse of what was observed in the side-by-side analysis and what has been reported to the committee as a generally observed phenomenon. The smoothing algorithm used to generate the Walton (2008) data was not specified:

	Indent 1	Indent 2	Indent 3	Indent 4
Laser	29.8	39.8	34.5	41.4
Caliper	33.5	40.2	36.4	41.3

This demonstrates reversal of the direction. Note that in calculations of precision, the sample means are subtracted from the data as variation is calculated around the sample mean.

Deriving the "Factor of 10" Heuristic

Let Z be the observed BFD, which is the sum of the true (but unobservable) BFD, Y, and the instrument measurement error X: $Z = Y + X$. Assume that Y does not affect X and vice versa. Then the variance of the observed BFD (Z) is the sum of the variances of the true BFD and measurement error—that is,

$$\sigma_Z^2 = \sigma_Y^2 + \sigma_X^2 \text{, so } \sigma_Z = \sqrt{\sigma_Y^2 + \sigma_X^2}$$

Now, we want instrument precision to have a negligible effect on the variation of the observed BFD. That is, we want $\sigma_Z \approx \sigma_Y$. This is achieved when $\sigma_X \le 0.1\sigma_Y$ (equivalently, $10\sigma_X \le \sigma_Y$), as follows. Given we want $\sqrt{\sigma_Y^2 + \sigma_X^2} \approx \sigma_Y^2$, divide both sides by σ_Y and substitute $\sigma_X = 0.1\sigma_Y$ to get $\sqrt{1.01\sigma_Y^2}/\sigma_Y \approx 1$, or $\sqrt{1.01} = 1.005 \approx 1$.

So, as long as the precision of the measuring instrument is less than one-tenth of the variation in the actual BFDs, the measurement instrument only negligibly increases the variation in the observed BFD, where "negligible" is defined as ≤ 0.005. For the current clay process with an observed BFD standard deviation of 3.5-4.5 mm or so, this means the precision of the measuring instrument, in terms of its standard deviation, should be no greater than 0.3 to 0.4 mm.

In Chapter 5, the precision requirement was relaxed to 0.5 mm. The committee estimated something on the order of a 1 percent increase in BFD variation attributable to the measurement instrument ($\sqrt{3^2 + 0.5^2}/3 = 1.014$ and $\sqrt{4^2 + 0.5^2}/4 = 1.008$). While that sounds quite small, Appendix G went on to examine the effect of relaxing instrument precision further on the likelihood of making decision errors under the Office of the Director, Operational Test and Evaluation test protocol. That part of the analysis found that relaxing the precision any further than 0.5 mm unacceptably increased the probability of accepting bad body armor and rejecting good armor.

The committee wishes to emphasize that the above derivation of the heuristic is dependent only on assuming the actual BFDs are independent of the instrument measurement error. It is not dependent on the assumption of normality of the BFDs, nor is it dependent on any information from the Walton (2008) study and its supporting data.

REFERENCES

Hosto, J., and C. Miser. 2008. Quantum FARO Arm Laser Scanning Body Armor Back Face Deformation. Report No. 08-MS-25. Aberdeen Proving Ground, Md.: Aberdeen Test Center Warfighter Directorate, Applied Science Test Division, Materials and Standards Testing Team.

NRC. (National Research Council). 2009. Phase I Report on Review of the Testing of Body Armor Materials for Use by the U.S. Army: Letter Report. Washington, D.C.: National Academies Press.

Sandvik, L. and B. Olsson. 1982. A nearly distribution free test for comparing dispersion in paired samples. Biometrika 69(2):484-485.

Walton, S., A. Fournier, B. Gillich, J. Hosto, W. Boughers, C. Andres, C. Miser, J. Huber, and M. Swearingen. 2008. Summary Report of Laser Scanning Method Certification Study for Body Armor Backface Deformation Measurements. Aberdeen Proving Ground, Md.: Aberdeen Test Center.

TABLE M-1 Side-by-Side Comparison of BFD Measurements by ATC

Data Number	Digital Caliper (mm)	Laser Arm, Smoothed (mm)
	39.92	41.398
	35.79	38.931
	35.74	38.92
	31.19	33.656
	29.94	32.526
	34.61	35.169
	30.68	32.412
	36.76	38.224
	32.68	32.623
0	36.78	35.804
1	26.87	38.187
2	34.63	35.714
3	37.46	39.456
4	28.03	30.61
5	37.58	37.71
6	35.34	38.503
7	30.99	34.638
8	37.51	40.232
9	40.33	43.751
0	34.86	37.006
1	40.38	40.047
2	30.27	34.191
3	35.22	37.017
4	29.99	31.114
5	33.61	35.71
6	33.09	33.268
7	36.57	38.336
8	39.88	43.044
9	38.96	42.177
0	37.14	37.344
1	31.17	31.874
2	28.81	30.112
3	30.99	31.905
4	38.97	40.094
5	43.29	42.804
6	38.18	39.775
7	36.08	37.664
8	39.9	42.756
9	35.02	36.424
0	36.97	39.788
1	39.74	40.598
2	35.8	37.023
3	42.54	43.165
4	27.12	28.307
5	40.18	41.751

6	34.58	34.469
7	29.94	30.144
8	48.04	48.131
9	36.11	38.75
0	33.48	31.311
1	30.58	31.988
2	30.78	31.344
3	36.93	38.403
4	29.38	31.617
5	37.1	38.029
6	36.12	37.304
7	34.95	34.496
8	34.16	35.909
9	32.87	32.859
0	36.06	34.779
1	26.84	27.09
2	33.9	33.583
3	33.77	34.147
4	30.26	30.552
5	29.75	32.086
6	33.88	35.969
7	47.1	47.497
8	36.02	37.586
9	40.37	40.598
0	35.59	35.941
1	38.36	37.334
2	32.53	33.185
3	25.53	26.371
4	33.73	34.724
5	37.05	37.961
6	40.73	41.137
7	33.02	35.199
8	36.78	39.462
9	40.58	39.999
0	40.16	41.819
1	37.93	37.881
2	37.79	37.087
3	39.73	41.94
3	43.17	40.157
5	35.52	36.928
6	35.17	41.413
7	36.12	38.33
8	32.8	35.736
9	36.24	37.67
0	38.85	39.256
1	41.1	42.662

TABLE M-2 Side-by-Side Comparison of BFD Measurements by Chesapeake Testing

BFD Data Number	Digital Caliper (mm)	Faro Laser Arm (mm)
1	32.9	34.6
2	35.7	36.7
3	30.5	35
4	38.8	39.8
5	33.5	37
6	35.4	36.7
7	33.9	35.1
8	35	37
9	31	33.7
10	31.4	36.9
11	29.1	29.5
12	32.5	33.8
13	26.4	28.1
14	33.5	35.3
15	33.3	35.5
16	25.6	29.1
17	32	33.5
18	26.1	28.6
19	22.4	26.5
20	29	34.1
21	31	33.8
22	26.7	28.3
23	34.1	32.1
24	30.7	29.8
25	36.4	34.1
26	37	36.8
27	31.6	33.1
28	31.9	34.2
29	24.9	26.7
30	42	36
31	32.4	35.7

32	48.3	55.4
33	37.5	39.9
34	34.6	36.4
35	30.7	35.3
36	35.8	37.6
37	37.4	38.2
38	27.4	28
39	29.7	31.4
40	42.1	37
41	29.9	30.9
42	27.6	28.5
43	36.6	36.4
44	35.2	34.1
45	39.8	43.5
46	36	38.7
47	25.5	30.6
48	31.9	33.5
49	32.6	34.2
50	30.4	31.1
51	36.1	23.5
52	25.8	26.9
53	28.3	29.6
54	24	25.8
55	29.8	31.6
56	37.5	38.3
57	38	38.1
58	29.6	33.2
59	30	31.7
60	38.7	39.7
61	35.2	37.3
62	23.6	23.8
63	32.1	34.7
64	31.7	32.6
65	33.5	35.9
66	44.6	46.5
67	36	37.6
68	34.6	37.7
69	32.5	36.2

70	33.8	35.5
71	35.4	37.1
72	41	40.4
73	33.6	34.7
74	36.2	37.9
75	33.6	35.3
76	35.5	44.6
77	55	55.2
78	34.2	37.8
79	33.1	33.7
80	36	37.7
81	35.4	36.1
82	33.5	37.5
83	36.1	37

TABLE M-3 Faro Data

	Impression (*j*)			
Operator (i)	**1**	**2**	**3**	**4**
1	29.8130440	39.753433	34.436170	41.442090
2	29.8898140	39.724725	34.581093	41.414619
3	29.8399260	39.951892	34.724191	41.442222
4	29.8008041	39.828403	34.407924	41.457630

SOURCE: Hosto and Miser, 2008.

TABLE M-4 Caliper Data

Operator (i)	Operator (i)	Impression (*j*) 1	2	3	4
	a	33.12	40.87	36.65	42.05
1	b	32.97	40.74	35.77	41.49
	c	33.50	41.32	36.44	40.57
	a	33.56	39.48	37.05	41.12
2	b	33.57	39.56	36.95	40.92
	c	33.84	39.53	37.28	41.22
	a	33.15	40.72	35.45	41.43
3	b	33.17	40.58	35.62	41.68
	c	33.31	40.60	35.7	41.67
	a	33.69	39.77	36.76	40.85
4	b	34.11	38.85	36.81	40.91
	c	33.81	40.36	36.85	41.39

SOURCE: Hosto and Miser, 2008.